D0143609

INTRODUCTION TO PROBABILITY THEORY
With Contemporary Applications

Lester L. Helms
University of Illinois at Urbana-Champaign

DOVER PUBLICATIONS, INC.
Mineola, New York

Copyright

Bibliographical Note

This Dover edition, first published in 2010, in an unabridged republication of the work originally published in 1997 by W. H. Freeman and Company, New York. The author has provided a new errata list for this edition.

Library of Congress Cataloging-in-Publication Data

Helms, L. L. (Lester La Verne), 1927–
Introduction to probability theory : with contemporary applications / Lester L. Helms.
Dover ed.
p. cm.
Originally published: New York : W. H. Freeman, 1997. With new errata list.
Includes bibliographical references and index.
ISBN-13: 978-0-486-47418-2
ISBN-10: 0-486-47418-6
1. Probabilities. I. Title.

QA273.H52 2010
519.2—dc22

2009034243

Manufactured in the United States by Courier Corporation
47418601
www.doverpublications.com

In memory of

David Michael Helms
1955–1990

CONTENTS

PREFACE

In addition to exposing a student to diverse applications of probability theory through numerous examples, a probability textbook should convince the student that there is a coherent set of rules for dealing with probabilities and that there are powerful methodologies for solving probability problems. Aside from routine differentiation and integration methods, as far as possible I have based this book on the following three topics from the calculus: (1) the principle of mathematical induction, (2) the existence of limits of monotone sequences, and (3) power series. With three or four exceptions, complete proofs of theorems are included for the benefit of the highly motivated student.

The transition from calculus to probability theory is not easy for the typical student. New concepts, which are not amenable to "plug and chug" methods, are introduced at each turn of the page. At the risk of verbosity, I have endeavored to err on the side of readability to make this transition easier. Even so, the student will need to have pen and scratch pad ready for writing out some of the details. Again for the benefit of the student, solutions to all the exercises are included at the end of the text. Students need immediate reassurance that they have worked a problem correctly so that they can get on with the learning process and should not be made to wait until the next class meeting. Some of the exercises are tagged with a caution symbol in the form of a hand; these should not be attempted without Mathematica or Maple V software. Some of these exercises stipulate that answers should be calculated to n decimal places to ensure that mathematical software is actually used rather than a hand calculator and a crude approximation.

At most, two or three classroom periods should be allotted to Chapter 1. A standard one-semester course might consist of Chapters 1–4, one section from

Chapter 5, and Chapters 6–7. On the other hand, an instructor who believes in the inevitability of a digitized science might offer a one-semester course based only on Chapters 1–5. There is more than enough material for a two-semester course.

The manuscript was classroom tested during the fall semester of 1994 and the spring semester of 1996. Many examples and exercises have been added to the original manuscript at the suggestion of the students. This book was written for students, and I would welcome any suggestions from them on how it might be improved via e-mail at l-helms@math.uiuc.edu.

I would like to express my appreciation to W. H. Freeman reviewers S. James Taylor of the University of Virginia and Cathleen M. Zucco of LeMoyne College for their many suggestions on how to improve the book. I also thank Mary Louise Byrd, Project Editor at W. H. Freeman, for maintaining a reasonable production schedule. I especially thank Holly Hodder, Senior Editor at W. H. Freeman, for her interest in publishing this book.

July 1996

ERRATA

1. In line 15 of page 44, change 3/16 to 5/32.

2. In line 12 of page 48, interchange .116 and .5.

3. In line 16 of page 48, change sentence beginning "Since ..." to

 That someone must be able to pass on a B allele and therefore must be of genotype OB, BB, or AB; in the first and third cases there is a 50-50 chance that the B allele will be passed to the child. The computation of $P(E|F^c)$ is similar to that of $P(E|F)$ except for an additional term $P(E \cap F^c|F_{AB})P(F_{AB})$ in the numerator. Thus,

$$P(E|F^c) = \frac{(.5)(.116) + .007 + (.5)(.038)}{.877} = .096.$$

4. Line 11^- of page 48 should read

$$I = \frac{P(E|F)}{P(E|F^c)} = \frac{.528}{.096} \simeq 5.5$$

5. In line 2 of page 49, change .89 to .85.

6. The following sentence should be added to line 2 on page 106:

 Before trying to verify (4.2) below, the reader should do Exercise 4.2.8 first.

7. The first sentence of line 14 on page 106 should read:

 A similar argument applies to the fourth starting pattern, but care must be taken with the second and third starting patterns.

8. Equation (4.2) on page 106 should be changed to read:

$$g_T(n) = g_T(n-1) - \frac{1}{8}g_T(n-3) + \frac{1}{16}g_T(n-4), \qquad n \geq 4.$$

9. The display equation following Equation (4.3) on page 106 should read:

$$\begin{aligned}&\hat{g}_T(t) - 1 - t - t^2 - t^3\\ &= t(\hat{g}_T(t) - 1 - t - t^2) - \frac{t^3}{8}(\hat{g}_T(t) - 1) + \frac{t^4}{16}\hat{g}_T(t).\end{aligned}$$

10. The last display equation following Equation (4.3) on page 106 should read:

$$\hat{g}_T(t) = \frac{16 + 2t^3}{16 - 16t + 2t^3 - t^4}.$$

11. The 30 on lines 3 and 4 of page 107 should be changed to 18.

12. Exercise 4.2.8 on page 107 should be replaced by

Consider a sequence of Bernoulli trials with probability of success $p = 1/2$, define a waiting time T by putting $T = n$ if the word 101 appears for the first time at the end of the nth trial, and let $g_T(n) = P(T > n)$. Show that

$$g_T(n) = \frac{1}{2}g_T(n-1) + \left(\frac{1}{2}g_T(n-1) - \frac{1}{4}g_T(n-2)\right) + \frac{1}{8}g_T(n-3)$$

for $n \geq 3$. Using the fact that $g_T(0) = g_T(1) = g_T(2) = 1$, find the generating function $\hat{g}_T$ and $E[T]$.

13. The following sentences should be added to Definition 5.2 on page 154:

(3) The state j is periodic of period $d(j)$ if $d(j)$ is the greatest common divisor of the set $\{n \in N : p_{jj}^{(n)} > 0\}$ and is called *aperiodic* if $d(j) = 1$; (4) the chain $\{x_n\}_{n=0}^{\infty}$ is *aperiodic* if each state is aperiodic.

14. In line 6^- on page 154, insert "an aperiodic" before the word irreducible.

15. In line 8^- of page 168, add right parenthesis.

16. In line 6 of page 172, replace "with" by "by".

17. In line 9^- of page 179, replace "and" by "<".

18. In line 5^- of page 179, replace "and" by "<".

19. The D in line 3 on page 238 (after Figure 7.1) should have an exponent 2 as in D^2.

20. Solution 2.5.3 on page 308 should be 17/24.

21. Solution 4.2.8 on page 319 should read:

Let $T = n$ if the word 101 occurs on the nth trial for the first time. Then

$$\hat{g}_T(t) = \frac{8 + 2t^2}{8 - 8t + 2t^2 - t^3}$$

and $E[T] = \hat{g}_T(1) = 10$.

22. Solution 4.2.9 on page 319 should read:

$$P(T \leq 11) = 1 - P(T > 11) = 1 - g_T(11) = \frac{435}{1024} = .4248.$$

INTRODUCTION TO PROBABILITY THEORY
With Contemporary Applications

CHAPTER 1

CLASSICAL PROBABILITY

1.1 BEGINNINGS

As far back as 3500 B.C., devices were used in conjunction with board games to inject an element of uncertainty into the game. A heel bone or knucklebone of a hooved animal was commonly used. Dice made from clay were in existence even before the Greek and Roman empires. Just how the outcomes of these devices were measured or weighted, if at all, is unknown. It may be that the outcomes were ascribed to fate, the gods, or whatever, with no attempt being made to associate numbers with outcomes.

At the end of the fifteenth century and beginning of the sixteenth century, numbers began to be associated with the outcomes of gaming devices, and by that time empirical odds had been established for some devices by inveterate gamblers. In the first half of the sixteenth century, the Italian physician and mathematician Girolamo Cardano (1501–1576) made the abstraction from empiricism to theoretical concept in his book *Liber de Ludo Alea,* "The Book of Games of Chance," which was published posthumously in 1663. An English translation of this book by Sydney Gould can be found in *Cardano, The Gambling Scholar* by Oystein Ore (see the Supplemental Reading List at the end of this chapter). Among other things, Cardano calculated the odds of getting various scores with two dice and with three dice.

During the period 1550–1650, several mathematicians were involved in calculating the chances of winning at gambling. Sometime between 1613 and 1623, Galileo (1564–1642) wrote a paper on dice without alluding to any prior work, as though the calculation of probabilities had become commonplace by then. Some historians mark 1654 as the birth of the theory of probability. It was in this year that a gambler, the Chevalier de Méré, proposed several problems to Blaise Pascal (1623–1662), who in turn communicated the problems to Pierre de Fermat (1601–1665). Thus began a correspondence between Pascal and Fermat about probabilities that some authors claim to be the beginning of the theory of probability. On the heels of this correspondence, in 1667, another seminal work appeared: *De Ratiociniis in Ludo Alea* by Christianus Huygens (1629–1695), in which the concept of expectation was introduced for the first time.

In the following sections, a theory of probability will be developed (as opposed to *the* theory of probability, since there are several approaches to probability) as expeditiously as possible using contemporary notation and terminology.

1.2 BASIC RULES

Game playing and gambling have been common forms of recreation among all classes of people for hundreds of years. In fact, the desire to win at gambling was a primary driving force in the development of probability theory during the sixteenth century. As a result, much of the early work dealt with dice and with answering questions about perceived discrepancies in empirical odds. Some of the gamblers were quite astute at recognizing discrepancies on the order of 1/100. The point is that the gamblers of the sixteenth century were aware of some kind of empirical law according to which there was predictability about the frequency of occurrence of a specified outcome of a game, even though there was no way of predicting the outcome of a particular play of the game.

Consider an experiment or game in which the outcome is uncertain and consider some attribute of an outcome. Let A be the collection of outcomes having that attribute. Suppose the experiment or game is repeated N times and $N(A)$ is the number of repetitions for which the outcome is in A. The ratio $N(A)/N$ is called the *relative frequency* of A. The fact that the ratio $N(A)/N$ seems to stabilize near some real number p when N is large, written

$$\frac{N(A)}{N} \to p,$$

is an *empirical law*. This law can no more be proved than Newton's law of cooling can be proved. Of course, the number p depends upon A, and it is customary to denote it by $P(A)$, so that the empirical law is usually written

$$\frac{N(A)}{N} \to P(A).$$

The number $P(A)$ is called the probability of A. This tendency of relative frequencies to stabilize near a real number is illustrated in Figure 1.1. The

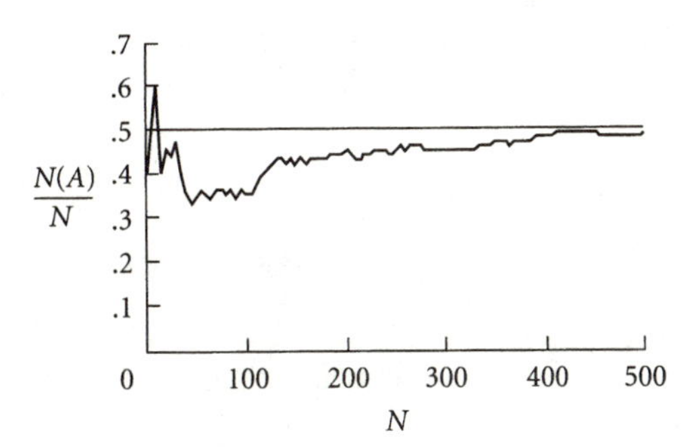

FIGURE 1.1 Relative frequencies.

graph depicts the relative frequency of getting a head in flipping a coin N times for N up to 500, calculated at multiples of 5 and rounded off. The number 500 was chosen in advance of any coin flips.

Just what was Cardano's contribution? It was the observation that for most simple games of chance, the probability of a particular outcome is simply the reciprocal of the total number of outcomes for the game, an observation that seemed to agree with empirical odds established by the gamblers. For example, if the game consists of rolling a fair die (i.e., a nearly perfect cubical die), then the outcomes 1, 2, 3, 4, 5, 6 represent the number of pips on the top surface of the die after coming to rest, and so each outcome has an associated probability of 1/6.

Cardano also considered the roll of two dice; for purposes of argument, a red die and a white die. There are six outcomes for the red die. Each of the outcomes of the red die can be paired with one of six outcomes for the white die, and so there are a total of 36 possible outcomes for the roll of the two dice. Thus, Cardano assigned each outcome a probability of 1/36.

Cardano's assignment of probabilities is universally accepted for most simple games of chance: rolling a die, rolling two dice, . . . , rolling n dice; flipping a coin, flipping a coin two times in succession, . . . , flipping a coin n times in succession; flipping n coins simultaneously; and dealing a hand of n cards from a well-shuffled deck of playing cards.

Consider two collections of outcomes A and B having no outcomes in common; i.e., A and B are *mutually exclusive*. If $A \cup B$ denotes the collection of outcomes in A or in B and $N(A \cup B)$ denotes the number of times the outcome is in $A \cup B$ in N repetitions of the game, then $N(A \cup B) = N(A) + N(B)$. Since

$$\frac{N(A \cup B)}{N} = \frac{N(A)}{N} + \frac{N(B)}{N},$$

it follows from the empirical law that

$$P(A \cup B) = P(A) + P(B) \tag{1.1}$$

whenever A and B are mutually exclusive. Note also that $0 \leq N(A)/N \leq 1$, and it follows from the empirical law that

$$0 \leq P(A) \leq 1. \tag{1.2}$$

In particular, if Ω is the collection of all outcomes of the game, then $N(\Omega) = N$ and

$$P(\Omega) = 1. \tag{1.3}$$

The properties of probabilities expressed by Equations 1.1, 1.2, and 1.3 embody the basic rules for more general probability models.

Returning to Cardano's assignment of probabilities, if A consists of the outcomes $\omega_1, \omega_2, \ldots, \omega_k$ and $N(\omega_i)$ is the number of times the outcome is ω_i in N repetitions of a game, then $N(A) = N(\omega_1) + \cdots + N(\omega_k)$ and $P(A) = P(\omega_1) + \cdots + P(\omega_k)$ by the empirical law. Letting $|A|$ denote the number of outcomes in A,

$$P(A) = \frac{|A|}{|\Omega|}. \tag{1.4}$$

We now have the basic rules for calculating probabilities associated with simple games of chance. Such calculations are reduced to counting outcomes. The reader should develop a systematic procedure for identifying and labeling outcomes, as in the following example.

EXAMPLE 1.1 Consider an experiment in which a coin is flipped three times in succession. The outcomes can be labeled using three-letter words made up from an alphabet of H and T (or 1 and 0). The label TTH stands for an outcome for which the first two flips resulted in tails and the third in heads. All possible outcomes can be listed: HHH, THH, HTH, HHT, HTT, THT, TTH, TTT. Consider the attribute "the number of heads in the outcome is 2." If A is the collection of outcomes having this attribute, then $|A| = 3$, and so $P(A) = 3/8$. ■

EXAMPLE 1.2 Consider an experiment in which a bowl contains five chips numbered 1, 2, 3, 4, 5. The chips are thoroughly mixed and one of them is selected blindly, the remaining chips are thoroughly mixed again, and then one of the remaining chips is selected blindly. An outcome of this experiment can be labeled using a two-letter word made up from an alphabet consisting of the digits 1, 2, 3, 4, 5, with the proviso that no digit can be repeated. All possible outcomes can be listed: 12, 13, 14, 15, 21, 23, 24, 25, 31, 32, 34, 35, 41, 42, 43, 45, 51, 52, 53, 54. Consider the attribute "the first digit is less than the second." If A is the collection of outcomes with this attribute, then $|A| = 10$, and so $P(A) = 10/20 = 1/2$. ■

EXERCISES 1.2

The last problem requires the principle of mathematical induction, which states: If $P(n)$ is a statement concerning the positive integer n that satisfies (i) $P(1)$ is true and (ii) $P(n+1)$ is true whenever $P(n)$ is true, then $P(n)$ is true for all integers $n \geq 1$.

1. Consider an experiment in which a coin is flipped four times in succession. If A is the collection of outcomes having two heads, determine $P(A)$.
2. If a coin is flipped n times in succession, what is the relationship between $|\Omega|$ and n?
3. Consider four distinguishable coins (e.g., a penny, a nickel, a dime, and a quarter). If the four coins are tossed simultaneously and A consists of all outcomes having two heads, determine $P(A)$.
4. If four coins of like kind are tossed simultaneously and A consists of all outcomes having three heads, determine $P(A)$.
5. A simultaneous toss of two indistinguishable dice results in a configuration (if the positions of the two dice are interchanged, a new configuration is not obtained). What is the total number of configurations?
6. Use the principle of mathematical induction to prove that

$$1+2+\cdots+n = \frac{n(n+1)}{2}$$

and

$$1^2+2^2+3^2+\cdots+n^2 = \frac{n(n+1)(2n+1)}{6}$$

for every integer $n \geq 1$.

1.3 COUNTING

In the previous section, outcomes were given labels that were words made up using a specified alphabet. This procedure is just a special case of more general schemes.

If $a_1, \ldots, a_n$ are n distinct objects, A will denote the collection consisting of these objects, written $A = \{a_1, \ldots, a_n\}$. If $B = \{b_1, \ldots, b_m\}$ is a second collection, we can form a new collection, denoted by $A \times B$, which consists of all *ordered pairs* (a_i, b_j) with $i = 1, \ldots, n$ and $j = 1, \ldots, m$. Since we can form a rectangular array with n rows and m columns in which the element of the ith row and jth column is (a_i, b_j), the total number of ordered pairs in the array is $n \times m$. Therefore,

$$|A \times B| = n \times m. \tag{1.5}$$

More generally, let $A_1, \dots, A_r$ be r collections having $n_1, \dots, n_r$ members, respectively. We can then form the collection $A_1 \times \cdots \times A_r$ of *ordered r-tuples* $(a_1, \dots, a_r)$ where each a_i belongs to A_i, $i = 1, \dots, r$. In this case,

$$|A_1 \times \cdots \times A_r| = n_1 \times n_2 \times \cdots \times n_r. \tag{1.6}$$

The proof of this result requires a mathematical induction argument. The essential step in the argument is as follows if we agree that the ordered r-tuple $(a_1, \dots, a_r)$ is the same as the ordered pair $((a_1, \dots, a_{r-1}), a_r)$. By the induction argument, the number of ordered $(r-1)$-tuples $(a_1, \dots, a_{r-1})$ is $n_1 \times \cdots \times n_{r-1}$, and it follows from Equation 1.5 that the number of ordered r-tuples is $(n_1 \times \cdots \times n_{r-1}) \times n_r$.

In the particular case that all of the $A_1, \dots, A_r$ are the same collection A and $|A| = n$, then the number of ordered r-tuples $(a_1, \dots, a_r)$ where each a_i belongs to A is given by

$$|\underbrace{A \times \cdots \times A}_{r \text{ times}}| = n^r. \tag{1.7}$$

Ordered r-tuples of the type just described have another name in probability theory. The collection A is called a *population* and the ordered r-tuple $(a_1, \dots, a_r)$ is called an *ordered sample of size r with replacement* from a population A of size n. Such an ordered sample can be thought of as being formed by successively selecting r elements from A with each element being returned to A before the next element is chosen.

Theorem 1.3.1 *The number of ordered samples of size r with replacement from a population of size n is n^r.*

EXAMPLE 1.3 Suppose a die is rolled three times in succession. The outcome can be regarded as an ordered sample of size 3 with replacement from the population $A = \{1, 2, 3, 4, 5, 6\}$. In this case $n = 6$ and $r = 3$, and so the total number of outcomes is $6^3 = 216$. ■

EXAMPLE 1.4 Two dice are thrown 24 times in succession. The outcome can be described as an ordered sample of size 24 from a population of size 36 with replacement. The total number of outcomes is 36^{24}. ■

If in forming an ordered sample from a population A we choose not to return an element to A, then we obtain an *ordered sample of size r without replacement* from a population A of size n.

Theorem 1.3.2 *The number of ordered samples of size r without replacement from a population of size n is $n(n-1) \times \cdots \times (n-r+1)$.*

Again a mathematical induction argument is needed to prove this result. Simply put, there are n choices for the first member of the sample, $n-1$ choices for the second, and upon making the rth choice there are only $n-(r-1) = n-r+1$ choices for the rth member of the sample, and so the total number of choices is $n(n-1) \times \cdots \times (n-r+1)$. Because the latter product arises quite frequently, it is convenient to introduce a symbol for it; namely,

$$(n)_r = n(n-1) \times \cdots \times (n-r+1). \tag{1.8}$$

Note that

$$(n)_n = n(n-1) \times \cdots \times 2 \times 1 = n!$$

EXAMPLE 1.5 The game of solitaire is played with a deck of 52 cards. The game commences with 28 cards placed on a table in a prescribed order as drawn from the deck and constitutes an ordered sample of size 28 without replacement from a population of size 52. The number of such samples is $(52)_{28} = 52 \times 51 \times \cdots \times 25$. ■

In some cases, order is irrelevant. For example, it is not necessary to hold the cards of a poker hand in the order in which they are dealt from a deck. An *unordered sample* of size r from a population A of size n is just a subpopulation of A having r members. $C(n, r)$ will denote the number of such unordered samples. Such a sample is also called a *combination of n things taken r at a time.*

Theorem 1.3.3 *The number of unordered samples of size r from a population of size n is*

$$C(n,r) = \frac{(n)_r}{r!} = \frac{n(n-1) \times \cdots \times (n-r+1)}{r!}.$$

A convincing argument can be made as follows. The number of ordered samples of size r without replacement from the population of size n is $(n)_r$. Each such ordered sample can be obtained by first selecting an unordered sample of size r from the population, which can be done in $C(n, r)$ ways, and then taking an ordered sample of size r without replacement from the subpopulation of size r. Since the latter can be done in $r!$ ways, $(n)_r = C(n, r) \times r!$

EXAMPLE 1.6 Suppose a poker hand of 5 cards is dealt from a well-shuffled deck of 52 cards. Since a poker hand can be regarded as unordered, the total number of poker hands is $C(52, 5) = 2,598,960$. ■

Note that

$$C(n,r) = \frac{n(n-1) \times \cdots \times (n-r+1)}{r!} = \frac{n!}{r!(n-r)!}$$

since $n! = (n(n-1) \times \cdots \times (n-r+1)) \times (n-r)!$. As a matter of notational convenience, 0! is equal to 1 by definition and $C(n,0) = 1$ if calculated formally using the last displayed equation. Another commonly used notation for $C(n,r)$ is $\binom{n}{r}$. The two will be used interchangeably. It is implicit in the above definition of $C(n,r)$ that $0 \le r \le n$. Again as a matter of notational convenience, we will put $C(n,r) = \binom{n}{r} = 0$ if $r < 0$ or if $r > n$. The $\binom{n}{r}$ are called binomial coefficients because of their association with the binomial theorem:

$$(a+b)^n = \sum_{k=0}^{n} \binom{n}{k} a^k b^{n-k}. \tag{1.9}$$

This theorem can be used to derive useful relationships connecting the coefficients. For example, putting $a = b = 1$,

$$2^n = \sum_{k=0}^{n} \binom{n}{k} = \binom{n}{0} + \binom{n}{1} + \cdots + \binom{n}{n}. \tag{1.10}$$

EXAMPLE 1.7 If n is a positive integer with $n \ge 2$, then

$$1 - \binom{n}{1} + \binom{n}{2} - \cdots \pm \binom{n}{n} = 0$$

where the coefficient of $\binom{n}{n}$ is + or − as n is even or odd, respectively. This can be seen as follows. Taking $b = 1$ in Equation 1.9,

$$(a+1)^n = \sum_{k=0}^{n} \binom{n}{k} a^k.$$

Taking $a = -1$,

$$\begin{aligned} 0 = (-1+1)^n &= \sum_{k=0}^{n} (-1)^k \binom{n}{k} \\ &= 1 - \binom{n}{1} + \binom{n}{2} - \cdots + (-1)^n \binom{n}{n}. \ \blacksquare \end{aligned}$$

EXAMPLE 1.8 If m and n are positive integers and t is any real number, then

$$(1+t)^{m+n} = (1+t)^m(1+t)^n.$$

Applying the binomial theorem three times,

$$\sum_{k=0}^{m+n}\binom{m+n}{k}t^k = \left(\sum_{i=0}^{m}\binom{m}{i}t^i\right)\left(\sum_{j=0}^{n}\binom{n}{j}t^j\right)$$
$$= \sum_{i=0}^{m}\sum_{j=0}^{n}\binom{m}{i}\binom{n}{j}t^{i+j}.$$

Collecting terms with a factor of t^k,

$$\sum_{k=0}^{m+n}\binom{m+n}{k}t^k = \sum_{k=0}^{m+n}\left(\sum_{i=0}^{k}\binom{m}{i}\binom{n}{k-i}\right)t^k.$$

Equating corresponding coefficients of t^k, it follows that

$$\binom{m+n}{k} = \sum_{i=0}^{k}\binom{m}{i}\binom{n}{k-i}. \quad \blacksquare \qquad (1.11)$$

Returning to Equation 1.8, note that the right side of the equation makes sense if n is any real number. For any real number x and any positive integer r, we define

$$(x)_r = x(x-1)\times\cdots\times(x-r+1)$$

and

$$C(x,r) = \binom{x}{r} = \frac{(x)_r}{r!} = \frac{x(x-1)\times\cdots\times(x-r+1)}{r!}.$$

We also define $(x)_0 = 1$ and $C(x,0) = \binom{x}{0} = 1$; for any negative integer r, we define $\binom{x}{r} = 0$.

EXAMPLE 1.9 If r is a nonnegative integer, then $\binom{-1}{r} = (-1)^r$ since

$$\binom{-1}{r} = \frac{(-1)_r}{r!} = \frac{(-1)(-2)\times\cdots\times(-1-r+1)}{r!}$$
$$= \frac{(-1)^r r!}{r!} = (-1)^r. \blacksquare$$

There are many more equations relating to binomial coefficients. The reader interested in pursuing the subject further should consult the books by Feller and Tucker (see the Supplementary Reading List at the end of the chapter).

It was tacitly assumed in the preceding discussions that the elements of a sample are distinguishable. But there are probability models in physics in which some elementary particles behave as though they are indistinguishable. A general scheme for dealing with such particles can be described as follows. Consider r indistinguishable balls and n distinguishable boxes numbered 1, $2, \ldots, n$. If the balls are distributed among the boxes in some way, the result is called a *configuration*. If a ball in Box 1 is interchanged with a ball in Box 2, the configuration does not change. The total number of configurations can be calculated using the following device. The label $* \mid * \mid *** \mid ** \mid *$ signifies that there are a total of five boxes with one ball in Box 1, one ball in Box 2, three balls in Box 3, two balls in Box 4, and one ball in Box 5, which is to the right of the last vertical bar. In general, there are $(n-1)+r = n+r-1$ symbols in the label because the number of vertical bars is one less than the number n of boxes. A label is completely specified if r of the $n+r-1$ positions are selected to be filled by asterisks; i.e., by selecting a subpopulation of size r. The total number of ways of selecting such subpopulations is $C(n+r-1, r)$.

Theorem 1.3.4 *The total number of ways of distributing r indistinguishable balls into n boxes is* $\binom{n+r-1}{r}$.

EXAMPLE 1.10 Suppose two dice, indistinguishable to the naked eye, are tossed. To determine the total number of possible configurations, consider boxes numbered $1, 2, \ldots, 6$ and consider the dice as balls that are placed into the boxes. In this case, $n = 6$ and $r = 2$, so that the total number of configurations is $\binom{6+2-1}{2} = 21$. $\blacksquare$

Dice do not behave as though they are indistinguishable; in fact, two dice behave as though there are 36 outcomes with each having the same probability.

EXERCISES 1.3

The reader should review Maclaurin and Taylor series expansions before starting on these problems.

1. If m and n are positive integers with $0 \le n \le m$, show that $\binom{m}{n} = \binom{m}{m-n}$ by (a) expressing both sides in terms of factorials and (b) interpreting each side as the number of ways of selecting a subpopulation.
2. If n and r are positive integers with $1 \le r \le n$, show that $C(n,r) = C(n-1,r) + C(n-1,r-1)$. (This equation validates the triangular array

$$\begin{array}{ccccccccc} & & & & 1 & & & & \\ & & & 1 & & 1 & & & \\ & & 1 & & 2 & & 1 & & \\ & 1 & & 3 & & 3 & & 1 & \\ 1 & & 4 & & 6 & & 4 & & 1 \\ \vdots & & \vdots & & \vdots & & \vdots & & \vdots \end{array}$$

commonly called "Pascal's Triangle" in Western cultures.)
3. If n is a positive integer, show that the Maclaurin series expansion of the function $f(t) = (1+t)^n$ is

$$f(t) = \sum_{k=0}^{\infty} \binom{n}{k} t^k.$$

How does this result relate to the binomial theorem?
4. If α is any real number, show that the Maclaurin series expansion of the function $f(t) = (1+t)^\alpha$ is

$$(1+t)^\alpha = \sum_{k=0}^{\infty} \binom{\alpha}{k} t^k$$

(which is valid for $|t| < 1$).
5. If n is a positive integer, use the binomial expansion of $(1+t)^n$ to show that

$$n2^{n-1} = \sum_{k=1}^{n} k\binom{n}{k} = \binom{n}{1} + 2\binom{n}{2} + \cdots + n\binom{n}{n}.$$

6. If n is a positive integer with $n \ge 2$, show that

$$n(n-1)2^{n-2} = 2 \cdot 1\binom{n}{2} + 3 \cdot 2\binom{n}{3} + \cdots + n(n-1)\binom{n}{n}.$$

7. A die is tossed n times in succession. What is the probability that a 1 will not appear?
8. A coin is flipped $2n$ times in succession. What is the probability that the number of heads and tails will be equal?
9. A rectangular box in 3-space is subdivided into $2n$ congruent rectangular boxes numbered $1, 2, \ldots, 2n$. (a) If n indistinguishable particles are distributed in the $2n$ boxes, what is the total number of configurations? (b) If all configurations have the same probability of occurrence, what is the probability that boxes numbered $1, 2, \ldots, n$ will be empty?
10. If $x > 0$ and k is a nonnegative integer, show that

$$\binom{-x}{k} = (-1)^k \binom{x+k-1}{k}.$$

The remaining problems are too tedious to do manually. Mathematical software such as Mathematica or Maple V is appropriate.

11. A coin is flipped 20 times in succession. Find the probability, accurate to six decimal places, that the number of heads and tails will be equal.
12. Suppose 20 dice, indistinguishable to the naked eye, are tossed. What is the total number of possible configurations?

1.4 EQUALLY LIKELY CASE

This section will address only probability models of the type described in the previous section. For such models, calculating probabilities consists of two steps: counting the total number of outcomes and counting the number of outcomes in a given collection. In doing so it is important either to make a complete list of all outcomes or to give a precise mathematical description of all such outcomes.

Consider an experiment in which two dice, one red and one white, are tossed simultaneously. The outcome of the experiment is a complicated picture that can be recorded only partially by a camera. It is not necessary to go that far, however, since we are interested only in the number of pips showing on the two dice, and that information can be summarized by creating a name or label for it; e.g., the ordered pair (i, j), $1 \leq i, j \leq 6$ can be used as a label for the outcome in which the red die shows i and the white die shows j. Each such outcome is an ordered sample of size 2 with replacement from a population of size 6. The total number of such outcomes is $6^2 = 36$. The collection of all 36 labels is shown in Figure 1.2.

In tossing two dice, we are not usually interested in the number of pips on each die but rather in the sum of the two numbers; i.e., the score. For example, consider the score of 4. We can identify the outcomes with a score of 4 as those in the third diagonal from the upper left corner in Figure 1.2;

(1,1)	(1,2)	(1,3)	(1,4)	(1,5)	(1,6)
(2,1)	(2,2)	(2,3)	(2,4)	(2,5)	(2,6)
(3,1)	(3,2)	(3,3)	(3,4)	(3,5)	(3,6)
(4,1)	(4,2)	(4,3)	(4,4)	(4,5)	(4,6)
(5,1)	(5,2)	(5,3)	(5,4)	(5,5)	(5,6)
(6,1)	(6,2)	(6,3)	(6,4)	(6,5)	(6,6)

FIGURE 1.2 Outcomes for two dice.

namely, $(1, 3)$, $(2, 2)$, and $(3, 1)$. If A is the collection of these outcomes, then $P(A) = 3/36 = 1/12$. In general, if x is one of the scores $2, 3, \ldots, 12$ and $p(x)$ denotes the probability of the collection of outcomes having the score x, then $p(x)$ can be calculated in the same way. The results are shown in Figure 1.3:

x	2	3	4	5	6	7	8	9	10	11	12
$p(x)$	$\frac{1}{36}$	$\frac{2}{36}$	$\frac{3}{36}$	$\frac{4}{36}$	$\frac{5}{36}$	$\frac{6}{36}$	$\frac{5}{36}$	$\frac{4}{36}$	$\frac{3}{36}$	$\frac{2}{36}$	$\frac{1}{36}$

FIGURE 1.3 Scores for two dice.

Coin flipping is an experiment that most people have performed. Consider an experiment in which a coin is flipped n times in succession. An outcome of this experiment can be labeled by an n-letter word using an alphabet made up of T and H (or 0 and 1). For example, TTHTHH is the label for an outcome of an experiment of flipping a coin six times in succession with tails occurring on the first, second, and fourth flips and heads appearing on the remaining flips. If n is large it is impractical to make a list of all outcomes, but we can count the total number of outcomes because an outcome is an ordered sample of size n with replacement from a population {T,H} of size 2. Thus, the total number of outcomes is 2^n by Theorem 1.3.1.

EXAMPLE 1.11 Suppose a coin is flipped 10 times in succession. The total number of outcomes is $2^{10} = 1024$. What are the chances that there will be three heads in the outcome? To answer this question, let A be the collection of outcomes having three heads. Since each outcome has the same probability assigned to it, we need only count the number of outcomes having three heads. A label for an outcome consists of 10 letter positions that are filled by H's or T's. There are $C(10, 3)$ ways of selecting three positions to be filled with H's and the remaining seven positions with T's. Thus,

$$P(A) = \frac{\binom{10}{3}}{2^{10}} = \frac{15}{128}. \blacksquare$$

Notice in the wording of the question posed in this example that "three heads" is used rather than the "*exactly* three heads" that is commonly used in elementary algebra books. Three means exactly three; the prefix "exactly" is redundant.

In both the two-dice and the coin-flipping experiments, the outcomes were regarded as ordered samples with replacement. The following example illustrates counting ordered samples without replacement.

EXAMPLE 1.12 (The Birthday Problem) Consider a class of 30 students. Each student has a birthday that can be any one of the days numbered $1, 2, \ldots, 365$. Assume that the 30 birthdays of the students constitute an ordered sample of size 30 with replacement from a population of size 365 and that all outcomes have the same chance of occurring. What are the chances that no two of them will have the same birthday? Let A be the collection of outcomes for which there are no repetitions of birthdays; i.e., A is an ordered sample of size 30 without replacement from a population of size 365. Thus, $|A| = (365)_{30}$ and

$$P(A) = \frac{(365)_{30}}{365^{30}}$$

which is equal to .29 rounded to two decimal places. Thus, it is unlikely that no two will have the same birthday. ■

A poker hand can be considered an ordered sample without replacement or an unordered sample as far as calculating probabilities is concerned, provided there is total adherence to whichever of the two is adopted. It is customary to consider poker hands as unordered samples. In counting outcomes, it is important to not introduce order.

EXAMPLE 1.13 Suppose a poker hand of 5 cards is dealt from a well-shuffled deck of 52 playing cards. What is the probability of getting a royal flush; i.e., 10,J,Q,K,A of the same suit? Regarding a poker hand as an unordered sample of size 5 from a population of size 52, the total number of outcomes is $\binom{52}{5}$. Let A be the collection of outcomes that are royal flushes. We can form a royal flush in the following way. We first select a suit from among the four suits, which can be done in four ways; having selected the suit, the royal flush is then completely determined. Thus, $P(A) = 4/\binom{52}{5} = .0000015$. ■

A common mistake is to introduce order into the following example where there should be none.

EXAMPLE 1.14 Consider a poker hand as described in the previous example. What is the probability of getting two pairs; i.e., a hand of the type

$\{x, x, y, y, z\}$ where x, y, and z are distinct face values? There are 13 face values. We first choose a subpopulation of size 2, which can be done in $\binom{13}{2}$ ways, to specify the face value for each of the pairs. We then choose the face value for the singleton card from among the remaining 11 face values, which can be done in 11 ways. All face values have now been selected. Since there are four cards having the face value of the singleton, there are four choices for the singleton. We now go to the lower of the face values of the two selected for the pairs. Since there are four cards having that face value, we select a subpopulation of size 2, which can be done in $\binom{4}{2}$ ways. Having done this we now select a subpopulation of size 2 from the four cards having the other face value for a pair, which also can be done in $\binom{4}{2}$ ways. If A is the collection of outcomes that have two pairs, then

$$P(A) = \frac{\binom{13}{2} \times 11 \times 4 \times \binom{4}{2} \times \binom{4}{2}}{\binom{52}{5}},$$

which is approximately 1/20. ■

It might appear that order was introduced into this calculation when we chose to look first at the pair with the lower face value, but the order was already there once the two face values were chosen. In choosing the face values for the two pairs, it would have been incorrect to say that this could be done in 13×12 ways, because this would regard "a pair of jacks and a pair of kings" as different from "a pair of kings and a pair of jacks".

An unordered sample of size r from a population of size n is called a *random sample* if each sample has the same probability $1/\binom{n}{r}$ of occurring. A poker hand is a random sample of size 5 from a population of size 52.

We will conclude this section on counting by looking at a commonly used sampling model. Consider a population consisting of n_1 Type 1 individuals and n_2 Type 2 individuals. The population size is then $n = n_1 + n_2$. Suppose a random sample of size r is selected from the population. Since the population contains individuals of both types, we can ask for the probability that the random sample will contain k Type 1 individuals where $0 \leq k \leq r$. Of course, k cannot exceed the number of Type 1 individuals, so we must also have $k \leq n_1$; i.e., $0 \leq k \leq \min\{r, n_1\}$. Let A be the collection of samples having

k individuals of Type 1. Then

$$P(A) = \frac{\binom{n_1}{k}\binom{n - n_1}{r - k}}{\binom{n}{r}}$$

for $0 \le k \le \min\{r, n_1\}$.

EXAMPLE 1.15 On a given day, a machine produces 100 items. Assuming that 10 of the items are defective, what is the probability that a random sample of size 5 from the output will contain 3 defective items? Let A be the collection of samples having 3 defective items. Then

$$P(A) = \frac{\binom{10}{3}\binom{90}{2}}{\binom{100}{5}}. \quad \blacksquare$$

More generally, suppose a population of size n contains n_1 individuals of Type 1, n_2 individuals of Type 2, ..., n_k individuals of Type k. If a random sample of size r is taken from the population, what is the probability that the random sample will contain r_1 Type 1 individuals, r_2 Type 2 individuals, ..., r_k Type k individuals? Let A be the collection of such samples. Then

$$P(A) = \frac{\binom{n_1}{r_1}\binom{n_2}{r_2} \times \cdots \times \binom{n_k}{r_k}}{\binom{n}{r}}$$

where $r = r_1 + r_2 + \cdots + r_k$.

EXAMPLE 1.16 If a bridge hand of 13 cards is dealt from a well-shuffled deck of 52 playing cards, what is the probability that the hand will contain three hearts, five diamonds, two spades, and three clubs? Since the hand is a random sample of size 13 from a population of size 52, the total number of outcomes is $\binom{52}{13}$. Let A be the collection of samples as described. Then

$$P(A) = \frac{\binom{13}{3}\binom{13}{5}\binom{13}{2}\binom{13}{3}}{\binom{52}{13}}. \quad \blacksquare$$

EXERCISES 1.4

In sampling problems, the student should first decide whether the sample is unordered or ordered and, in the latter case, whether with replacement or without replacement.

1. Instead of the usual dice, consider two (regular) tetrahedral dice with faces bearing 1,2,3,4 pips. If the two tetrahedral dice are rolled simultaneously, find the probability $p(x)$ that the total score will be x where x can be one of the integers $2, \ldots, 8$.
2. If three tetrahedral dice are rolled simultaneously, find the probability $p(x)$ that the score will be one of the integers $3, \ldots, 12$.
3. If three cubical dice are rolled simultaneously, find the probability $p(x)$ that the total score will be x where x can be one of the integers $3, 4, \ldots, 18$.
4. If you purchase a single ticket for a lottery in which a random sample of size 6 is selected from the population $\{1, 2, \ldots, 54\}$, what is the probability that you hold the winning ticket?
5. In some state lotteries, a winning ticket must have six numbers between 1 and 48 listed in the same order as the numbers were successively drawn at random without replacement. What is the probability that the purchaser of a single ticket will hold the winning ticket?
6. If 1000 raffle tickets are sold, of which 50 are winning tickets and you purchase 10 tickets, what is the probability that you will have 2 winning tickets?
7. In a group of four people, what is the probability that no two will have the same birth month?
8. If a poker hand of 5 cards is dealt from a well-shuffled deck of 52 playing cards, what is the probability of getting a full house (i.e., 3 cards with the same face value and 2 cards with the same face value)?
9. If a poker hand of 5 cards is dealt from a well-shuffled deck of 52 playing cards, what is the probability of getting a straight flush (i.e., 5 cards in sequence in the same suit with the ace counting as a 1 or as the highest card)?
10. In a fish-tagging survey, 100 bass are netted, tagged, and released. After waiting long enough for the tagged fish to disperse, a second sample of 100 bass is taken, of which 5 are observed to be tagged. If the number of bass in the lake is n, what is the probability that a random sample of size 100 will contain 5 tagged fish? If you were asked to estimate the number of bass in the lake, what would you estimate?

1.5 OTHER MODELS

In the early stages of probability theory, a controversy arose between M. de Roberval and Blaise Pascal over the assignment of equal probabilities to outcomes. The basic issue of the controversy can be described as follows. Suppose a coin is flipped until a head appears with a maximum of two flips. It was argued by M. de Roberval that the outcomes H,TH,TT are equally likely and each should be assigned probability 1/3; Pascal, however, *reasoned* that they should be assigned the probabilities 1/2, 1/4, and 1/4, respectively, on the grounds that the coin could be flipped twice and the result of the second flip simply ignored after getting a head on the first flip; thus, the two outcomes HH and HT would have probabilities adding to 1/2.

Whether or not outcomes should be assigned equal probabilities in the case of simple games of chance depends on what one calls an outcome. Consider, for example, rolling two dice. If we declare the score obtained an outcome, then the possible outcomes are $2, 3, \ldots, 12$, and we have previously seen that these outcomes should not be assigned equal probabilities but rather those given in Figure 1.3. This suggests that we should have available a more general model.

Consider an experiment with a finite number of outcomes $\omega_1, \omega_2, \ldots, \omega_n$ and let $\Omega = \{\omega_1, \omega_2, \ldots, \omega_n\}$. For each $i = 1, 2, \ldots, n$, let $p(\omega_i)$ be a weight associated with ω_i satisfying

$$(i) \quad 0 \leq p(\omega_i) \leq 1.$$

$$(ii) \quad \textstyle\sum_{i=1}^{n} p(\omega_i) = 1.$$

The weight $p(\omega_i)$ will be called the probability of ω_i. If $A = \{\omega_{i_1}, \ldots, \omega_{i_k}\}$ is a collection of outcomes, we define

$$P(A) = \sum_{j=1}^{k} p(\omega_{i_j});$$

i.e., $P(A)$ is the sum of the weights of the outcomes in A. It can be seen that Equation 1.1 is satisfied as follows. If $A = \{\omega_{i_1}, \ldots, \omega_{i_k}\}$ and $B = \{\omega_{j_1}, \ldots, \omega_{j_\ell}\}$ have no outcomes in common, then $A \cup B$ consists of the outcomes $\omega_{i_1}, \ldots, \omega_{i_k}, \omega_{j_1}, \ldots, \omega_{j_\ell}$ and $P(A \cup B)$ is the sum of the weights associated with the latter outcomes. Thus,

$$\begin{aligned} P(A \cup B) &= p(\omega_{i_1}) + \cdots + p(\omega_{i_k}) + p(\omega_{j_1}) + \cdots + p(\omega_{j_\ell}) \\ &= [p(\omega_{i_1}) + \cdots + p(\omega_{i_k})] + [p(\omega_{j_1}) + \cdots + p(\omega_{j_\ell})] \\ &= P(A) + P(B). \end{aligned}$$

Similarly, Equations 1.2 and 1.3 are satisfied.

EXAMPLE 1.17 Consider an experiment for which the outcome can be described by a four-letter word using the alphabet 0, 1. If ω is such an outcome, a weight $p(\omega)$ can be associated with ω by forming a product in which each

1 in ω is replaced by 1/3 and each 0 by 2/3. For example, if $\omega = 1110$, then $p(\omega) = 1/3 \cdot 1/3 \cdot 1/3 \cdot 2/3 = (1/3)^3(2/3)^1$. Note that the exponent of 1/3 is just the sum of the digits in ω and the exponent of 2/3 is 4 minus the sum of the digits in ω. There are 16 outcomes and 16 associated weights. It is tedious to do so, but the 16 outcomes and weights can be listed and the sum of the weights shown to be 1. ■

EXAMPLE 1.18 (*n* Bernoulli Trials) Fix $0 < p < 1$. Let $q = 1 - p$ and let n be any positive integer. Let Ω be the collection of all words of length n using the alphabet $\{0, 1\}$. We can think of 0 and 1 as an encoding of failure and success or tail and head, respectively, in n repetitions of a basic experiment in which the probability of success is p and the probability of failure is q. If $\omega = \{x_j\}_{j=1}^n$ is an element of Ω, we associate with ω the weight

$$p(\omega) = p^{\sum_{j=1}^n x_j} q^{(n - \sum_{j=1}^n x_j)}.$$

Clearly, $p(\omega) \geq 0$ for each $\omega \in \Omega$. We need only verify that the sum of all the weights is 1. Each outcome $\omega = \{x_j\}_{j=1}^n$ such that $\sum_{j=1}^n x_j = k$ has associated weight $p^k q^{n-k}$. The number of such outcomes is equal to the number of ways of selecting k of the n letter positions to be filled with 1's, which is $\binom{n}{k}$. Thus, the sum of the weights of outcomes with $\sum_{j=1}^n x_j = k$ is $\binom{n}{k} p^k q^{n-k}$. If we now add the sums of these weights for $k = 0, 1, \ldots, n$, we obtain by the binomial theorem

$$\sum_{k=0}^n \binom{n}{k} p^k q^{n-k} = (p + q)^n = 1.$$

This model goes by the name *n Bernoulli trials.* We have just seen that if we let A_k be the collection of outcomes having k successes, then

$$P(A_k) = \binom{n}{k} p^k q^{n-k}. \quad ■$$

EXAMPLE 1.19 A distributor plans to use an optical character recognition scanner to transfer the contents of a catalog of parts to a computer. Each part has a 12-digit part number and a 13-digit stock number. The probability that the scanner will misread a digit depends upon the digit being scanned; e.g., it is more likely that an 8 will be misread as a 3 than as a 1. Assuming that the maximum probability that a digit will be misread is .01, what is the probability that the part number and stock number will be recorded without error? We can view this experiment as a succession of 25 trials in which success is interpreted to mean that a digit is read correctly and failure is interpreted to mean that a digit is misread. Assuming that probabilities are assigned in accordance with the Bernoulli model with $p = .99$, let A be the collection of outcomes having no misreads. A consists of a single outcome with probability

$$P(A) = (.99)^{25} = .78. \blacksquare$$

EXAMPLE 1.20 Consider an experiment in which a coin is flipped until a head appears for the first time with a maximum of five flips. The outcomes can be labeled H, TH, TTH, TTTH, TTTTH, TTTTT with weights $P(H) = 1/2, P(TH) = 1/4, P(TTH) = 1/8, P(TTTH) = 1/16, P(TTTTH) = 1/32$, and $P(TTTTT) = 1/32$. ■

The weights attached to the outcomes in this example were constructed using Pascal's line of reasoning.

The previous example suggests a coin-flipping experiment in which a coin is flipped until head appears for the first time, at which time the experiment terminates. The outcomes of this experiment can be described as an infinite sequence of labels H, TH, TTH, TTTH, ... Does this list describe all outcomes? What about the possibility that a head never appears? We can include or not include an unending label TTT. . . for this possibility as we choose. We will not include such a label on the grounds that in any instance in which this experiment has been performed, the experiment terminates in a finite number of steps. By analogy with the previous example, we can assign weights as follows: $p(H) = 1/2, p(TH) = 1/4, p(TTH) = 1/8, \ldots$. Note that even if we included the outcome with label TTT. . . , there would be no weight left for it because

$$p(H) + p(TH) + \cdots = \sum_{n=1}^{\infty} \frac{1}{2^n}$$

and the sum of this geometric series is 1.

This model suggests an even more general model. Consider an experiment for which there is an infinite sequence of outcomes $\omega_1, \omega_2, \ldots$. Let $\Omega = \{\omega_1, \omega_2, \ldots\}$, and for each $i \geq 1$ let $p(\omega_i)$ be a weight associated with each ω_i satisfying

$$(i) \quad 0 \leq p(\omega_i) \leq 1.$$

$$(ii) \quad \textstyle\sum_{i=1}^{\infty} p(\omega_i) = 1.$$

If $A = \{\omega_{i_1}, \ldots, \omega_{i_m}\}$ is any finite subcollection of outcomes, we define

$$P(A) = \sum_{k=1}^{m} p(\omega_{i_k});$$

if $A = \{\omega_{i_1}, \omega_{i_2}, \ldots\}$ is an infinite sequence of outcomes, we define

$$P(A) = \sum_{k=1}^{\infty} p(\omega_{i_k}).$$

Rather than making the distinction between finite sums and infinite sums as in the last two equations, we usually just write $P(A) = \sum p(\omega_{i_k})$, the range of k being clear from the description of A. Again, Equations 1.1, 1.2, and 1.3 are satisfied.

EXAMPLE 1.21 A pair of dice are rolled until a score of 6 appears for the first time, at which time the experiment is terminated. A typical outcome can be labeled by the word $* * \cdots * 6$ where $*$ represents a score other than 6; if there are n asterisks preceding the 6 with $n \geq 0$, the weight or probability associated with the outcome is

$$p(\underbrace{* * \cdots *}_{n \text{ times}} 6) = \left(\frac{31}{36}\right)^n \left(\frac{5}{36}\right)$$

Note that the weights are nonnegative, and since the weights constitute the terms of a convergent geometric series,

$$\sum_{n=0}^{\infty} \left(\frac{31}{36}\right)^n \left(\frac{5}{36}\right) = 1. \blacksquare$$

The last model we will describe involves the concept of *conditional probability*. Consider two collections of outcomes A and B associated with an experiment. Before performing the experiment, we have some notion of what $P(A)$ should be. Instead of performing the experiment and observing the outcome, an impartial observer views the outcome and relates only partial information to us; namely, that the outcome is in B. Quite often in a situation like this, we would adjust our estimate of the chance that the outcome is in A. For example, suppose the experiment consists of selecting a person at random from a given population consisting of men and women in equal numbers. Before performing the experiment, the probability that the selected person is a man is 1/2. But if the experiment is performed and an impartial observer tells us only that the selected person is color-blind, then we would adjust our estimate of the probability that the person is a man to be much higher, because color blindness is much more prevalent in men than in women.

To see how probabilities should be changed in the light of partial information, we go back to the empirical law. Suppose the experiment in question is repeated N times. Since the impartial observer conveys information to us when the outcome is in B, we can ignore all repetitions for which the outcome is not in B. Let $A \cap B$ denote the collection of outcomes that are in both A and B. The number of outcomes for which the outcome is in B is $N(B)$, and among these $N(A \cap B)$ are also in A. The relative frequency of occurrence of outcomes in A among those in B is $N(A \cap B)/N(B)$. This ratio should stabilize near the new probability when N is large. Thus,

$$\frac{N(A \cap B)}{N(B)} = \frac{N(A \cap B)/N}{N(B)/N} \to \frac{P(A \cap B)}{P(B)}.$$

Of course, $P(B)$ must be positive for the quotient to be defined. This new probability is called *the conditional probability of A given B* and is denoted by $P(A \mid B)$. We therefore define

$$P(A \mid B) = \frac{P(A \cap B)}{P(B)}. \tag{1.12}$$

Note that

$$P(A \cap B) = P(A \mid B)P(B). \tag{1.13}$$

EXAMPLE 1.22 Two dice are rolled and we are informed that the score is 6. What is the probability that there is a 3 on each die? Let A be the collection of outcomes for which there is a 3 on each die and let B be the collection of outcomes for which the score is 6. Then $P(A \mid B) = P(A \cap B)/P(B) = (1/36)/(5/36) = 1/5$. ■

Note that the conditional probability can be viewed in the following way. As soon as we are told that the score is 6, we are dealing with the population $\{(5,1), (4,2), \ldots, (1,5)\}$. Since there are only five outcomes in this new population, the probability of the outcome $(3,3)$ is $1/5$.

There are probability models for which the probability mechanism is not specified by giving the probability of each outcome but rather by a mixture of such probabilities and conditional probabilities.

EXAMPLE 1.23 Suppose a bowl contains 10 red chips and 5 white chips. An experiment consists of selecting a chip at random from the bowl. If the drawn chip is red, it and 5 other red chips are returned to the bowl; if the chip is white, it is discarded. A second chip is then selected at random from the bowl. What is the probability that both chips will be red? This model is not described in such a way that the probability of each outcome is known; it is described in terms of probabilities of some outcomes and conditional probabilities. Let B be the collection of outcomes for which the first chip selected is red and let A be the collection for which the second chip is red. Then $P(A \mid B) = 3/4$ and $P(B) = 2/3$ so that

$$P(B \cap A) = P(A \mid B)P(B) = \frac{1}{2}$$

by Equation 1.13. ■

EXERCISES 1.5

The reader should review infinite series, sums of infinite series, and infinite geometric series before doing the following exercises.

1. Determine the sum of the series $\sum_{n=0}^{\infty}(1/2)^{3n}$.
2. Determine the sum of the series $\sum_{n=4}^{\infty}(1/4)^{2n}$.
3. Suppose a pair of dice are rolled until a score of 7 appears for the first time, whereupon the experiment ends. An outcome with n scores different from 7 followed by a 7 is assigned probability $(5/6)^n(1/6), n \geq 0$. What is the probability that the experiment will terminate on an odd number of rolls of the dice?
4. If a pair of dice are rolled, what is the probability that the score will be greater than or equal to 8?
5. Suppose a coin is flipped 10 times in succession. For $i = 1, 2, \ldots, 10$ let A_i be the collection of outcomes for which there is a head on the ith flip. Calculate $P(A_1), P(A_2), P(A_1 \cap A_2)$, and $P(A_2 \mid A_1)$. How are the first three probabilities related? How are the second and fourth related? How do these numbers change if the 10 is replaced by 20?
6. In the notation of Problem 5, calculate $P(A_j \mid A_i)$ for $1 \leq i < j \leq n$.
7. Bowl 1 contains 10 red chips and 5 white chips. Bowl 2 contains 10 red chips and 10 white chips. A chip is selected at random from Bowl 1, transferred to Bowl 2, and then a second chip is selected at random from Bowl 2. What is the probability that both chips will be red?
8. A pair of dice, one red and one white, are rolled. Let A be the collection of outcomes for which the number of pips on the red die is less than or equal to 2 and let B be the collection of outcomes for which the number of pips on the white die is greater than or equal to 4. Calculate $P(A \mid B)$. What does this say about the partial information "the number of pips on the white die is greater than or equal to 4"?
9. A man has n keys of which one will open his lock and the others will not. If he tries the keys randomly one at a time, what is the probability that the lock will be opened on the $r$ths try where $1 \leq r \leq n$?
10. Consider an experiment in which the outcomes are the positive integers $1, 2, \ldots$. For each $k \geq 1$, let

$$p(k) = \frac{1}{k(k+1)} = \frac{1}{k} - \frac{1}{k+1}.$$

Can the $p(k)$ serve as weights for a probability model?

SUPPLEMENTAL READING LIST

1. F. N. David (1962). *Games, Gods, and Gambling.* New York: Hafner Publishing Co.
2. W. Feller (1957). *An Introduction to Probability Theory and Its Applications,* 2nd ed. New York: Wiley.
3. Oystein Ore (1953). *Cardano, The Gambling Scholar.* Princeton, N.J.: Princeton University Press.
4. M. A. Todhunter (1965). *A History of the Mathematical Theory of Probability.* New York: Chelsea.
5. A. Tucker (1984). *Applied Combinatorics.* New York: Wiley.

CHAPTER 2

AXIOMS OF PROBABILITY

2.1 INTRODUCTION

Rules for calculating probabilities associated with simple games of chance were developed in the works of P. R. de Montmort (1678–1719) and A. de Moivre (1667–1754). These rules also began to be applied in mortality tables and life insurance calculations as early as the late seventeenth century. Most of the effort during this period was concentrated on specific problems dealing with combinations. But eventually problems required more than combinatorial methods, and powerful tools had to be developed for their solutions.

Terms such as "gain" and "duration of play" were commonly used during this period and evolved into an abstract concept known as a "chance variable," much like "momentum" in mechanics. In any particular application, a chance variable was defined in some natural way, not as a mathematical entity but rather by its properties.

The publication of *Foundations of the Theory of Probability* by A. N. Kolmogorov in 1933 marked the beginning of a rapid development of probability theory and its application to diverse fields, particularly during and immediately after World War II.

The reader interested in alternatives to the axiomatic probability model discussed in this chapter should read the book by Hamming listed in the Supplemental Reading List.

The content of this chapter is rather abstract. A real appreciation of probability theory cannot be gained without some firsthand experience with a random device. Experiment with flipping a coin many times—you will be surprised by some of the facets of randomness.

2.2 SET THEORY

A typical exercise in probability theory involving two dice will start out "Let A be the event 'the score is 11.'" For our purposes, this layman's description of A is an abbreviated form of "Let A be the collection of outcomes ω with score 11," and consequently A is a subcollection of all possible outcomes. Later, we will define an event to be a subcollection of the collection of all possible outcomes.

We saw in Chapter 1 that probability theory pertains to collections of outcomes. Such collections are called *sets.* One starting point for developing mathematics is the set N of natural numbers 1, 2, . . . , which is denoted by $N = \{1, 2, 3, \ldots\}$ and eventually leads to the set R of real numbers. Algebraic and order properties of real numbers will be taken for granted.

A primitive notion of set theory is that of *membership.* We write $x \in X$ if x is a member or element of the set X. If x is not a member of X, we write $x \notin X$. If X and Y are two sets, we write $X \subset Y$ if $x \in X$ implies $x \in Y$ and say that X is *contained in* or is a *subset* of Y. We say that two sets X and Y are *equal,* written $X = Y$, if $X \subset Y$ and $Y \subset X$. It sometimes happens in manipulating sets that we end up with something that has no members. As a matter of notational convenience, we use $\varnothing$ to signify a set that has no elements and call $\varnothing$ the *empty set.*

We need a procedure for specifying sets. To obtain one, let $p(x)$ be a sentence containing a variable x. Then $\{x : p(x)\}$ will denote the set of objects x for which $p(x)$ is true. For example, consider the sentence "$x = 1$ or $x = 2$ or $x = 3$." Then $\{x : p(x)\}$ consists of the natural numbers 1, 2, and 3. This set is usually written $\{1, 2, 3\}$ for brevity.

EXAMPLE 2.1 Let $p(x)$ be the sentence "$x \in N$ and $x^2 \leq 40$." Then $\{x : p(x)\} = \{1, 2, 3, 4, 5, 6\}$. ■

EXAMPLE 2.2 If $a, b \in R$ with $a \leq b$, then we have the usual definitions of closed, open, and semiclosed intervals:

$$
\begin{aligned}
[a, b] &= \{x : x \in R \text{ and } a \leq x \leq b\} \\
(a, b) &= \{x : x \in R \text{ and } a < x < b\} \\
[a, b) &= \{x : x \in R \text{ and } a \leq x < b\} \\
(a, b] &= \{x : x \in R \text{ and } a < x \leq b\}.
\end{aligned}
$$

Infinite intervals are defined similarly; e.g.,

$$[a, +\infty) = \{x : x \in R \text{ and } x \geq a\}. \quad ■$$

For the remainder of this section, we will assume that we are dealing with a *universe* U. All objects under consideration will be members of U, and all sets will be subcollections of U.

Given $X \subset U$, the *complement* of X (relative to U), denoted by X^c, is defined by

$$X^c = \{x : x \notin X\}.$$

The set specified on the right should contain "$x \in U$" as part of its description, but this part is customarily omitted when it is understood that we are dealing with a fixed universe. It is easy to see that

$$\varnothing^c = U$$
$$U^c = \varnothing.$$

If X and Y are two subsets of U, the *union* of X and Y, denoted by $X \cup Y$, is defined by

$$X \cup Y = \{x : x \in X \text{ or } x \in Y\};$$

the *intersection* of X and Y, denoted by $X \cap Y$, is defined by

$$X \cap Y = \{x : x \in X \text{ and } x \in Y\}.$$

If $X \cap Y = \varnothing$, we say that X and Y are *mutually exclusive* or *disjoint.*

These concepts can be illustrated as follows. For U take the points inside a rectangle in a plane, and for a subset A of U take the points within and on a simple closed curve (e.g., a circle). If this is done for subsets $X, Y, Z, \ldots$ of U, the resulting picture is called a *Venn diagram.* The operations on sets defined above can be depicted as in Figure 2.1. Venn diagrams can be helpful for understanding set operations.

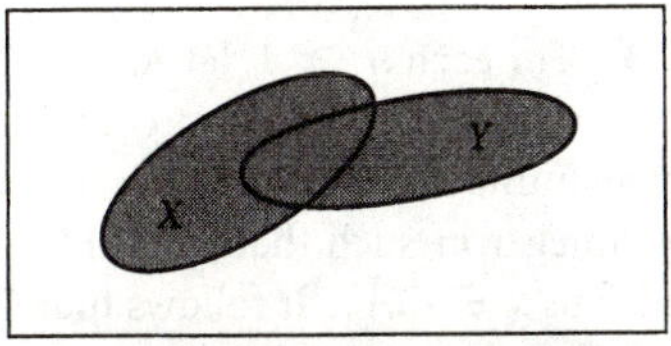

Shaded region: $X \cup Y$

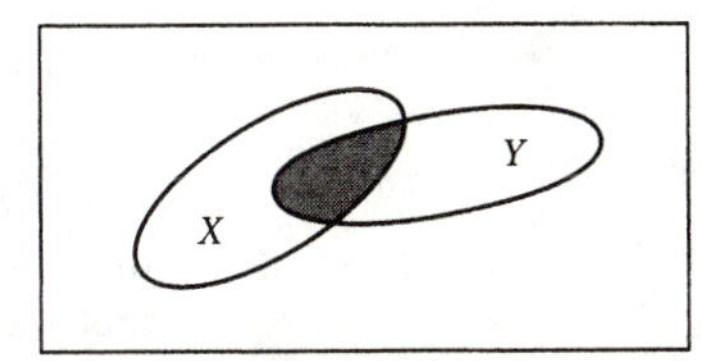

Shaded region: $X \cap Y$

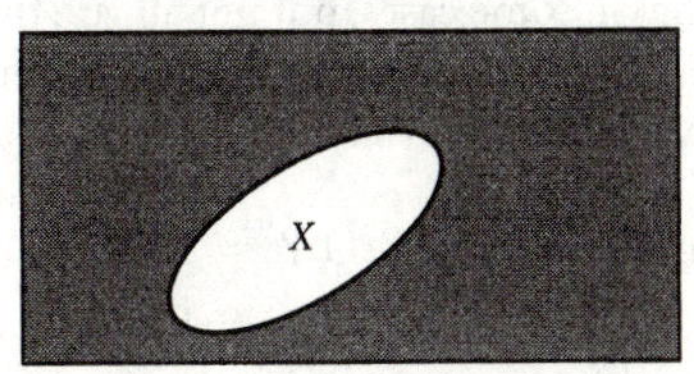

Shaded region: X^c

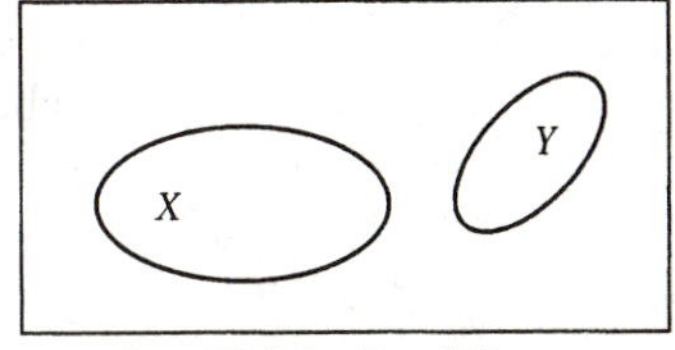

Disjoint X and Y

FIGURE 2.1 Venn diagrams illustrating set operations.

If $X_1, X_2, \ldots, X_n$ is a finite sequence of sets, their union is denoted and defined by

$$\bigcup_{i=1}^{n} X_i = \{x : x \in X_i \text{ for some } i = 1, \ldots, n\}$$

and their intersection by

$$\bigcap_{i=1}^{n} X_i = \{x : x \in X_i \text{ for all } i = 1, \ldots, n\}.$$

Similarly, if $X_1, X_2, \ldots$ is an infinite sequence of sets, then the union and intersection of the sets are defined by

$$\bigcup_{i=1}^{\infty} X_i = \{x : x \in X_i \text{ for some } i \geq 1\}$$

and

$$\bigcap_{i=1}^{\infty} X_i = \{x : x \in X_i \text{ for all } i \geq 1\},$$

respectively. As was the case with sums, rather than making the distinction between finite unions (intersections) and infinite unions (intersections), we usually just write $\cup X_i$ ($\cap X_i$) if the range on i is easily ascertained.

The following example requires the use of the *Archimedian property* of the real numbers, which states that if r is a real number, then there is a positive integer n such that $n > r$.

EXAMPLE 2.3 For each $n \geq 1$, let $A_n = [0, 1/n)$. Since $0 \in A_n$ for all $n \geq 1$, $0 \in \cap A_n$. Clearly, $\cap A_n$ cannot contain any negative numbers. But what about positive numbers? Assume $x > 0$. By the Archimedian property, there is a positive integer m such that $m > 1/x$. It follows that $x > 1/m$ so that $x \notin A_m$, and thus $x \notin \cap A_n$. It follows that $\cap A_n = \{0\}$. ■

The union, intersection, and complement operations on sets are subject to algebraic laws that in some cases, but not all, are the same as the algebraic laws for real numbers. Corresponding to the addition and multiplication of real numbers we have *commutative laws:*

$$X \cup Y = Y \cup X$$
$$X \cap Y = Y \cap X.$$

A proof of the commutative law for union requires that two things be proved; namely, that $X \cup Y \subset Y \cup X$ and $Y \cup X \subset X \cup Y$. Consider the first relation.

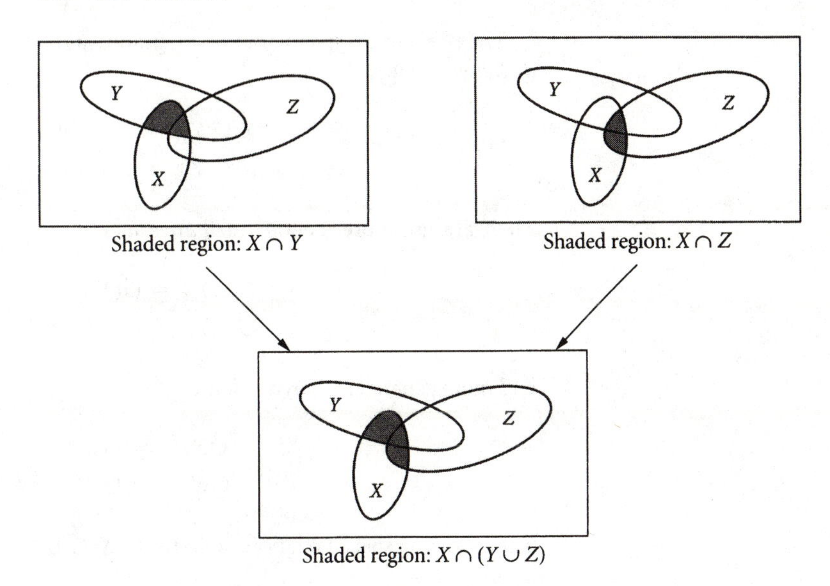

FIGURE 2.2 $X \cap (Y \cup Z) = (X \cap Y) \cup (X \cap Z)$.

Suppose $x \in X \cup Y$. Then $x \in X$ or $x \in Y$; but this statement is the same as $x \in Y$ or $x \in X$, and so $x \in Y \cup X$. Thus, $x \in X \cup Y$ implies $x \in Y \cup X$. At a crucial point in this argument, there is a claim that the statement "$x \in X$ or $x \in Y$" is equivalent to the statement "$x \in Y$ or $x \in X$." To justify this claim, we could move on to formal "truth tables," but we will not. The equivalence of the two statements is taken for granted as something from logic.

If X, Y, and Z are three sets, there are *associative laws:*

$$X \cup (Y \cup Z) = (X \cup Y) \cup Z$$
$$X \cap (Y \cap Z) = (X \cap Y) \cap Z.$$

The associative laws permit us to omit the parentheses altogether since they can be reinserted in any manner; e.g., $A \cup B \cup C \cup D = ((A \cup B) \cup C) \cup D = A \cup (B \cup (C \cup D))$.

There are also *distributive laws:*

$$X \cap (Y \cup Z) = (X \cap Y) \cup (X \cap Z)$$
$$X \cup (Y \cap Z) = (X \cup Y) \cap (X \cup Z).$$

A convincing, but not rigorous, argument that the first distributive law is true can be made by examining a Venn diagram for $X \cap (Y \cup Z)$ as in Figure 2.2. The two top shaded regions represent $X \cap Y$ and $X \cap Z$. Their union is the lower shaded region $X \cap (Y \cup Z)$.

The effect of complementation on unions and intersections is the subject of *de Morgan's laws:*

$$(X \cup Y)^c = X^c \cap Y^c$$
$$(X \cap Y)^c = X^c \cup Y^c.$$

There are also more general *distributive laws:*

$$X \cap \cup Y_n = \cup(X \cap Y_n)$$
$$X \cup \cap Y_n = \cap(X \cup Y_n)$$

and more general *de Morgan's laws:*

$$(\cup X_n)^c = \cap X_n^c$$
$$(\cap X_n)^c = \cup X_n^c.$$

The following special relations hold for all $X \subset U$:

$$X \cap \emptyset = \emptyset \qquad X \cup \emptyset = X$$
$$X \cup U = U \qquad X \cap U = X$$
$$X \cup X^c = U \qquad X \cap X^c = \emptyset.$$

Venn diagrams must be recognized for what they are—doodles. Equations relating sets cannot be proved using Venn diagrams. Such proofs require repeated applications of the laws defined above. Venn diagrams can be used legitimately to prove negative results, however.

EXAMPLE 2.4 Consider the equation $X \cap Y \cap Z = X \cap Y \cap (Y \cup Z)$. Is this equation true for all subsets X, Y, Z of a given U? The answer is no if we can construct a U and X, Y, Z for which the equation is not true. By constructing a Venn diagram for three sets, labeling the parts $1, 2, \ldots, 8$ as in Figure 2.3, and defining $U = \{1, 2, \ldots, 8\}, X = \{1, 4, 5, 7\}, Y = \{2, 5, 6, 7\}$, and $Z = \{3, 4, 6, 7\}$, we obtain $X \cap Y \cap Z = \{7\} \neq \{5, 7\} = X \cap Y \cap (Y \cup Z)$. We thus have a specific example of X, Y, and Z for which the above equation is not true, and therefore the equation is not always true. ■

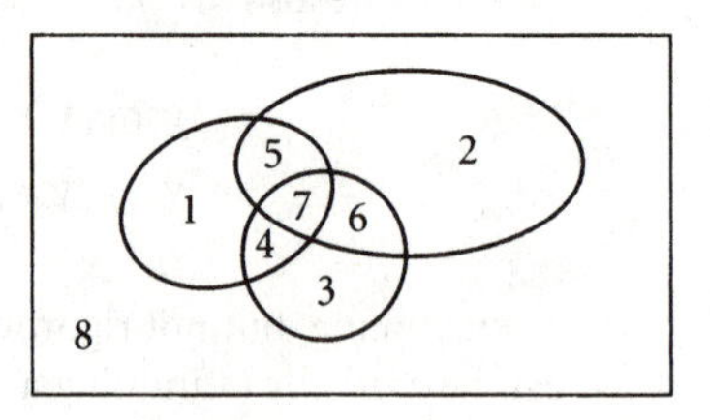

FIGURE 2.3 Counterexample.

Care must be taken when going beyond the relations listed above. For example, if $X \cup Y = X \cup Z$, there may be a temptation to conclude that $Y = Z$ because the analogous result is true in arithmetic. But the conclusion would not be valid. For example, let $U = \{1, 2, 3, 4\}$, $X = \{1, 2, 4\}$, $Y = \{2, 3\}$, and $Z = \{2, 3, 4\}$. Then $Y \neq Z$, but $X \cup Y = \{1, 2, 3, 4\} = X \cup Z$.

EXERCISES 2.2

1. Which of the following statements are correct? (a) $2 \in \{1, 2, 3\}$, (b) $2 \subset \{1, 2, 3\}$, (c) $\{2\} \in \{1, 2, 3\}$, (d) $\{2\} \subset \{1, 2, 3\}$.
2. Consider a universe U consisting of ordered pairs (i, j), $1 \leq i, j \leq 6$ where i represents the number of pips on a red die and j the number on a white die. Express the lay statement "the number of pips on the red die is greater than the number of pips on the white die" as a proposition concerning elements of U, and identify the set A specified by the proposition.
3. If $A_n = [0, 2^{1/n})$, $n \geq 1$, determine $\cap A_n$.
4. If $A_n = \{(x, y) : x \in R, y \in R, 0 \leq y \leq x^n, 0 \leq x < 1\}$, $n \geq 1$, determine $\cap A_n$.
5. If $A_n = \{(x, y) : x \in R, y \in R, 0 \leq y \leq x^n, 0 \leq x \leq 1\}$, $n \geq 1$, determine $\cap A_n$.
6. If A is any subset of the universe U, show that $(A^c)^c = A$.
7. If A and B are any two sets, show that $A \subset B$ if and only if $B^c \subset A^c$.
8. Is it true that $X \cap (Y \cup Z) = (X \cap Y) \cup Z$ for all subsets X, Y, and Z of the universe U? If not, give an example to show that the equation is not true in general.
9. Prove that $(X \cup Y) \cap (X \cap Y)^c = (X \cap Y^c) \cup (X^c \cap Y)$ for all subsets X, Y of the universe U.
10. If $A_n = [0, |\sin(n\pi/2)|]$, $n \geq 1$, determine $\cup_{n=1}^{\infty}(\cap_{k \geq n} A_k)$ and $\cap_{n=1}^{\infty}(\cup_{k \geq n} A_k)$.

2.3 COUNTABLE SETS

Let A and B be two nonempty sets and consider $A \times B$, the collection of all ordered pairs (x, y) with $x \in A$, $y \in B$. A *function* or *mapping* f from A to B is a subset $f \subset A \times B$ with the property that

$$(x, y) \in f \quad \text{and} \quad (x, z) \in f \text{ implies } y = z. \tag{2.1}$$

The *domain* of f is the set

$$\{x : x \in A \text{ and } (x, y) \in f \text{ for some } y \in B\}.$$

We will assume that A is chosen so that A is the domain of f. The *range* of f is the set

$$\{y : y \in B \text{ and } (x, y) \in f \text{ for some } x \in A\}.$$

If $(x, y) \in f$, then y is written $f(x)$ in the usual calculus terminology so that f consists of all pairs $(x, f(x))$ as x ranges over A. All of the above is condensed into a single symbol

$$f : A \to B.$$

EXAMPLE 2.5 Let $A = B = R$ and consider

$$f = \{(x, y) : x \in R, y \in R, -1 \le x \le 1, y \ge 0, x^2 + y^2 = 1\}.$$

Then f is a semicircle in the xy-plane. In the usual notation, $f(x) = \sqrt{1 - x^2}$ with domain $\{x : x \in R, -1 \le x \le 1\}$ and range $\{y : y \in R, 0 \le y \le 1\}$. ■

EXAMPLE 2.6 Let $A = B = R$ and consider

$$g = \{(x, y) : x \in R, y \in R, -1 \le x \le 1, x^2 + y^2 = 1\}.$$

In this case, for each x with $-1 < x < 1$, there are two values $g(x) = \pm\sqrt{1 - x^2}$ such that $(x, g(x)) \in g$. Therefore, g is not a function or mapping because 2.1 is not satisfied. ■

Finite and infinite sequences are specific examples of mappings. When we speak of a *finite sequence* of real numbers $\{a_k\}_{k=1}^n$ we are dealing with a collection of ordered pairs (k, a_k) where $k \in \{1, 2, \ldots, n\}$ and $a_k \in R$. If we let α be the collection of such pairs, then $\alpha : \{1, 2, \ldots, n\} \to R$. Similarly, an *infinite sequence* of real numbers is a mapping $\alpha : N \to R$; if we put $\alpha_k = \alpha(k)$, the usual notation for α is then $\alpha = \{\alpha_k\}_{k=1}^\infty$. Care must be taken to distinguish between the terms of an infinite sequence and the range of the sequence. For example, if $\alpha = \{(-1)^k\}_{k=1}^\infty$, then $-1, +1, -1, \ldots$ are the terms of the sequence, but the range is the set $\{-1, +1\}$.

We can use these concepts to make precise the commonly used term "finite." A set X is *finite* if for some $n \in N$ and some set B it is the range of a mapping $\alpha : \{1, 2, \ldots, n\} \to B$. A set is infinite if not finite.

The set X is *countable* if it is the range of an infinite sequence; i.e., the range of a mapping $\alpha : N \to B$ for some B containing X. We can always replace B by X. By definition, the empty set $\varnothing$ is countable. Finite sets are countable because if X is the range of the finite sequence $\{\alpha_k\}_{k=1}^n$, then it is the range of the infinite sequence $\{\alpha_k\}_{k=1}^\infty$ where $\alpha_k = \alpha_n$ for all $k \ge n$. The set of natural numbers $N = \{1, 2, \ldots\}$ is countable because N is the range of the map $I : N \to N$ where $I(n) = n, n \ge 1$. The set of even positive integers

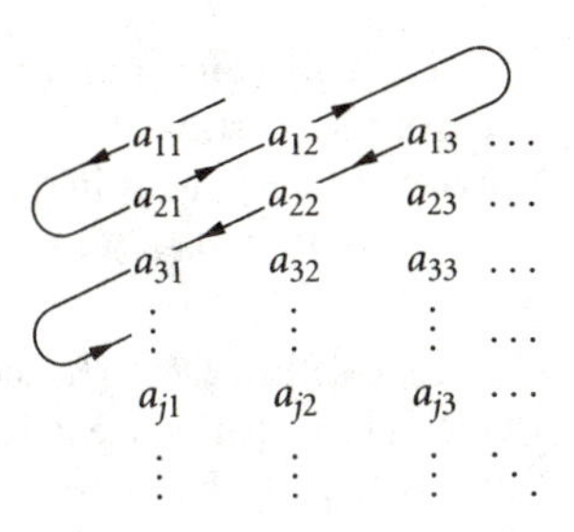

FIGURE 2.4 Countable union.

$\{2, 4, 6, \ldots\}$ is countable because it is the range of the map $\alpha : N \to N$ with $\alpha(n) = 2n, n \geq 1$. The set of negative integers $\{\ldots, -2, -1\}$ is countable because it is the range of the map $\alpha : N \to R$ with $\alpha(n) = -n, n \geq 1$.

A countable set can be finite or infinite. If we wish to exclude the finite case, we say that X is *countably infinite* if X is infinite and countable.

The union of two countable sets is again countable; in fact, the union of finitely many countable sets is again countable. The proof of the following theorem is more palatable if it is looked upon as a programing problem. An algorithm is given for each $n \geq 1$ for calculating the kth term of a sequence $\{x_{n,j}\}_{j=1}^{\infty}$, and we would like to define a single algorithm for listing all of the $x_{n,j}$.

Theorem 2.3.1 *The union of a countable collection of countable sets is countable.*

PROOF: We will assume that the collection is countably infinite. In this case, the collection is the range of a sequence $\{X_j\}_{j=1}^{\infty}$ of countable sets. We can assume that each of the X_j is the range of an infinite sequence by repeating one of its elements infinitely many times if this is not the case. Letting $X = \bigcup_{j=1}^{\infty} X_j$, we must show that there is a mapping $\alpha : N \to X$ having X as its range. For $j \geq 1$, let $X_j = \{a_{jk}\}_{k=1}^{\infty}$. The terms of X_j appear in the jth row of the array shown in Figure 2.4. An informal argument can be made for arranging the elements of this array as a sequence by following the path indicated in Figure 2.4. A map $\alpha : N \to X$ can be constructed using the same idea but following a diagonal from lower left to upper right, dropping down to the next diagonal, following the next diagonal from lower left to upper right, and so forth. We will illustrate the construction by using the identity

$$1 + 2 + \cdots + n = \frac{n(n+1)}{2} \tag{2.2}$$

to identify the element in the array corresponding to $\alpha(100)$. Note that

$$1 + 2 + \cdots + 13 = \frac{13 \cdot 14}{2} = 91,$$

which is the total number of elements in the array located in the first 13 diagonals. Starting at $a_{14,1}$, if we move 9 positions along the 14th diagonal, we arrive at the element in the array designated by $\alpha(100)$; it is easy to calculate that $\alpha(100) = a_{6,9}$. ■

EXAMPLE 2.7 The set of integers $Z = \{\ldots, -2, -1, 0, 1, 2, \ldots\}$ is countable. This is true since Z is the union of the countable sets $\{\ldots, -2, -1\}$, $\{0\}$, and $\{1, 2, \ldots\}$. ■

EXAMPLE 2.8 If $q \in N$, let $Z_q = \{\ldots, -2/q, -1/q, 0/q, 1/q, \ldots\}$. Then Z_q is countable since Z_q is the union of three sets $\{\ldots, -2/q, -1/q\}$, $\{0/q\}$, and $\{1/q, 2/q, \ldots\}$ each of which is easily seen to be countable. ■

EXAMPLE 2.9 The set Q of rational numbers p/q, where $q \in N$ and $p \in Z$, is countable. This follows from the fact that $Q = \cup_{q=1}^{\infty} Z_q$ and that each Z_q is countable and from Theorem 2.3.1. ■

Theorem 2.3.2 *If Y is countable and $X \subset Y$, then X is countable.*

PROOF: We can assume that $X \neq \emptyset$, because otherwise X is countable by definition. Since Y is countable, there is a mapping $\alpha : N \to Y$ with Y the range of the map. Let x_0 be a fixed element of X. Define $\beta : N \to X$ by putting

$$\beta(n) = \begin{cases} \alpha(n) & \text{if } \alpha(n) \in X \\ x_0 & \text{if } \alpha(n) \in Y \cap X^c \end{cases} \tag{2.3}$$

for $n \geq 1$. The range of β is then X and consequently X is countable. ■

The set R of real numbers is not countable. In view of the previous theorem, it suffices to show that $[0, 1)$ is not countable. This is done by using a method known as Cantor's diagonalization procedure. Each x in $[0, 1)$ has a decimal representation $x = .d_1d_2\cdots$ where $d_i \in \{0, 1, 2, \cdots, 9\}$, $i \geq 1$. But the representation is not unique. For example, $1/2 = .500\cdots = .499\cdots$. We will achieve uniqueness when this happens by using the representation that has all zeros beyond some point. Assume that $[0, 1)$ is countable. Then $[0, 1) = \{x_1, x_2, \ldots\}$. Suppose x_i has the unique decimal representation $x_i = .d_{i1}d_{i2}, \ldots, i \geq 1$, and consider the array

$$\begin{array}{cccc} .d_{11} & d_{12} & d_{13} & \ldots \\ .d_{21} & d_{22} & d_{23} & \ldots \\ .d_{31} & d_{32} & d_{33} & \ldots \\ \vdots & \vdots & \vdots & \ddots \end{array}$$

Consider the diagonal starting at d_{11}. For each $j \geq 1$, choose e_j different from d_{jj}, 0, and 9. Then $y = .e_1e_2\cdots$ represents a real number in $[0, 1)$ that is different from each x_i. But we assumed that the decimal representation of every real number in $[0, 1)$ appears in the above array, and we have a contradiction. Our assumption that $[0, 1)$ is countable leads to a contradiction.

EXERCISES 2.3

The last problem requires the use of the well-ordering property of the natural numbers N, which states that if $A \subset N$ and $A \neq \emptyset$, then A has a least element.

1. If $f = \{(x, y) : x \in R, y \in R, y = x^2, -1 \leq x \leq 1\}$, determine its domain and range.
2. If in the customary notation of the calculus $f(x) = \sqrt{1 - x^4}$, describe f as a subset of $R \times R$ and determine its domain and range.
3. If in the customary notation of the calculus $f(x) = 1/\sqrt{1 - x^2}$, describe f as a subset of $R \times R$ and determine its domain and range.
4. If $q \in N$ and $X = \{p/q : p \in N\}$, show that X is countable.
5. Let $X_1, X_2, \ldots, X_m$ be a finite sequence of countably infinite sets. Show that $X = X_1 \cup \cdots \cup X_m$ is countable.
6. Show that the set X of all infinite sequences of 0's and 1's is uncountable.
7. Show that $N \times N = \{(m, n) : m \in N, n \in N\}$ is countable by considering the collection of finite sets $A_k = \{(m, n) : m + n = k\}$.
8. Let A and B be countable sets. Show that $A \times B$ is countable.
9. Which of the following sets are countable?
   (a) The set of circles in the plane having centers with rational coordinates and rational radii.
   (b) The set of all polynomials $P(x) = a_nx^n + \cdots + a_1x + a_0$ having integer coefficients.
   (c) The set of all intervals $(a, b) \subset R$ having rational endpoints.
10. Let $X_1, X_2, \ldots$ be an infinite sequence of countably infinite sets. Show that $X = \cup_{n=1}^{\infty} X_n$ is countable.

2.4 AXIOMS

If A and B are disjoint collections of outcomes, we have seen that

$$P(A \cup B) = P(A) + P(B).$$

More generally, if $A_1, \ldots, A_n$ are disjoint, it follows from the empirical law that

$$P\left(\bigcup_{j=1}^{n} A_j\right) = \sum_{j=1}^{n} P(A_j). \tag{2.4}$$

If the total number of outcomes is finite, there is no more to be said in regard to Equation 2.4. But what if $\{A_j\}$ is an infinite sequence of disjoint collections? An experiment with an infinite number of outcomes was discussed in Chapter 1; namely, flipping a coin until a head appears for the first time. In this case, it is possible to have an infinite sequence $\{A_j\}$ of disjoint collections of outcomes, and so it makes sense to ask if

$$P\left(\bigcup_{j=1}^{\infty} A_j\right) = \sum_{j=1}^{\infty} P(A_j). \tag{2.5}$$

This is a moot question for all the examples with finitely many outcomes and has an affirmative answer for the single model just described. Since Equation 2.5 is compatible with every example we have considered, can we assume that Equation 2.5, in addition to Equations 1.1, 1.2, and 1.3, is valid in a general model for probability theory? We can, but we cannot have everything we would like to have. It turns out that we cannot assume that Equation 2.5 is valid for all sequences $\{A_j\}$ of disjoint collections of outcomes and at the same time assume that $P(A)$ is meaningful for all possible A. We must give up one of the two assumptions. We will give up the latter, and so $P(A)$ may not be meaningful for some A.

Let Ω denote the collection of all outcomes for a given experiment. The following definitions are needed to limit the A for which $P(A)$ will be defined.

Definition 2.1 *A collection $\mathcal{A}$ of subsets of Ω is an* algebra *if*

1. $A, B \in \mathcal{A}$ *implies* $A \cup B \in \mathcal{A}$.
2. $A \in \mathcal{A}$ *implies* $A^c \in \mathcal{A}$.
3. $\Omega \in \mathcal{A}$. ■

That is, $\mathcal{A}$ is an algebra of subsets of Ω if it is closed under the operations of union and complementation and $\Omega \in \mathcal{A}$. A mathematical induction argument can be used to show that an algebra $\mathcal{A}$ is closed under finite unions; i.e., if $A_1, A_2, \ldots, A_n \in \mathcal{A}$, then $\cup_{j=1}^{n} A_j \in \mathcal{A}$.

The important thing to remember about algebras is that by starting with a finite number of elements of the algebra and performing a finite number of union, intersection, and complementation operations on them, the result is still in the algebra.

EXAMPLE 2.10 Let $\mathcal{A}$ be an algebra of subsets of Ω and let A_1, A_2, A_3, A_4 be elements of the algebra with some or all having nonempty intersections. If we need to restructure the union $\cup_{i=1}^{4} A_i$ into a union of disjoint sets in $\mathcal{A}$, we could let $B_1 = A_1, B_2 = A_2 \cap A_1^c, B_3 = A_3 \cap (A_1 \cup A_2)^c$, and $B_4 = A_4 \cap (A_1 \cup A_2 \cup A_3)^c$. Then $B_j \subset A_j, 1 \le j \le 4$, the B_j are disjoint, and $\cup A_j = \cup B_j$. ■

We need to postulate more if we want to deal with infinite sequences.

Definition 2.2 *A collection $\mathcal{F}$ of subsets of Ω is a σ-*algebra *if*

1. *$\mathcal{F}$ is an algebra.*
2. *If $\{A_j\}$ is an infinite sequence in $\mathcal{F}$, then $\cup A_j \in \mathcal{F}$.* ■

The important thing to remember about σ-algebras is that by starting with a sequence of elements of the σ-algebra and performing countably many union, intersection, and complementation operations on them, the result is still in the σ-algebra.

EXAMPLE 2.11 Let Ω be a finite set of outcomes and let $\mathcal{A}$ be the collection of all subsets of Ω. Then $\mathcal{A}$ is an algebra. ■

EXAMPLE 2.12 Let Ω be any set and let $\mathcal{F}$ be the collection of all subsets of Ω. Then $\mathcal{F}$ is a σ-algebra. Clearly, $\Omega \in \mathcal{F}$ since $\Omega \subset \Omega$. If $\{A_n\}$ is a finite or infinite sequence in $\mathcal{F}$, then $\cup A_n$ is a subset of Ω and therefore is in $\mathcal{F}$. If $A \in \mathcal{F}$, then $A^c \subset \Omega$ and $A^c \in \mathcal{F}$. ■

If $\mathcal{B}$ is any collection of subsets of a set Ω, then there is a "smallest σ-algebra," denoted by $\sigma(\mathcal{B})$, that contains $\mathcal{B}$. In discussing probability, we began with objects ω that were used to form collections A that have now been used to form σ-algebras $\mathcal{F}$. This process has taken us through three hierarchical levels of set theory, and to prove the result just stated would require going to a fourth hierarchical level. This fourth level is left to more advanced texts. For the time being, we have all the concepts needed to describe a general probability model. Ω will be a fixed collection of outcomes.

Definition 2.3 *A* probability space *is a triple $(\Omega, \mathcal{F}, P)$ where $\mathcal{F}$ is a nonempty σ-algebra of subsets of Ω and P is a mapping from $\mathcal{F}$ to R satisfying*

1. $P(\Omega) = 1$.
2. $0 \leq P(A) \leq 1$ *for all* $A \in \mathcal{F}$.
3. *If $\{A_j\}$ is a finite or infinite disjoint sequence in $\mathcal{F}$, then*

$$P(\cup A_j) = \sum P(A_j). \quad \blacksquare$$

All the simple games of chance described in Chapter 1 for which Ω is finite, $\mathcal{F}$ is the collection of all subsets of Ω, and P is defined as described there result in probability spaces $(\Omega, \mathcal{F}, P)$.

Definition 2.4 *If $(\Omega, \mathcal{F}, P)$ is a probability space, elements of $\mathcal{F}$ are called* events. ■

We return to a model discussed in Section 1.5.

EXAMPLE 2.13 Let $\Omega = \{\omega_1, \omega_2, \ldots\}$ be countably infinite, $\mathcal{F}$ the σ-algebra of all subsets of Ω, and $p(\omega_i)$ a weight function as defined in Section 1.5. Define $P(A), A \in \mathcal{F}$, as in Section 1.5. Then $(\Omega, \mathcal{F}, P)$ is a probability space. To show this, we need only verify Item 3 of Definition 2.3. Let $\{A_j\}$ be a sequence of disjoint events in $\mathcal{F}$ and let $A = \cup A_j$. Suppose $A = \{\omega_{i_1}, \omega_{i_2}, \ldots\}$. The fact that the series $\sum_j p(\omega_{i_j})$ is a convergent series with sum $P(A)$ means that the terms of the series can be rearranged without affecting the sum of the series. This fact about absolutely convergent series is proved or at least discussed in most calculus books. We rearrange the terms of the series so that the terms $p(\omega_{i_j})$ with $\omega_{i_j} \in A_1$ come first, then the terms $p(\omega_{i_j})$ with $\omega_{i_j} \in A_2$ second, and so on, to obtain

$$\begin{aligned} P(A) &= \sum_j p(\omega_{i_j}) \\ &= \sum_{\omega_{i_j} \in A_1} p(\omega_{i_j}) + \sum_{\omega_{i_j} \in A_2} p(\omega_{i_j}) + \cdots \\ &= P(A_1) + P(A_2) + \cdots. \end{aligned}$$

Therefore,

$$P(\cup A_j) = \sum P(A_j),$$

and $(\Omega, \mathcal{F}, P)$ is a probability space. ■

We have previously encountered the following situation. Suppose a coin is flipped until a head appears for the first time with a maximum of n flips. Suppose n is a large positive integer. Let A be the event "the experiment terminates on the fifth flip." We can think of this experiment continuing through all n flips of the coin and simply ignoring what happens after the fifth flip. If $\omega \in A$, then the first four letters of ω are T, the fifth is H, and there are two choices for each of the remaining $n - 5$ letters. Thus, $|A| = 2^{n-5}$,

$$P(A) = \frac{2^{n-5}}{2^n} = \frac{1}{2^5},$$

and it appears that $P(A)$ does not depend upon n at all! This computation is based on Pascal's reasoning in which we think of the coin as continuing to be flipped beyond the fifth flip and simply ignoring everything beyond the fifth flip. Why bother to mention the number n at all? If we eliminate mentioning n at all, then we are confronted with a conceptual experiment in which a coin is continually flipped. We can, in fact, construct a probability space $(\Omega, \mathcal{F}, P)$ with Ω consisting of outcomes ω that are words of infinite length using the alphabet T, H, and probabilities for events such as A described above are calculated by fixing some large n. We will consider a more general model.

In the following example, Ω will denote the set of all infinite sequences $\{x_i\}_{i=1}^{\infty}$ where each x_i is a 1 or a 0. We can think of 1 and 0 as an encoding of H and T or S and F, respectively, where S stands for success and F stands for failure. This Ω is uncountable; i.e., not countable. (See Exercise 2.3.6.) Therefore, none of the models we have discussed pertain to Ω. The model depends upon a parameter p, called the *probability of success*, with $0 < p < 1$. The number $q = 1 - p$ is called the *probability of failure*. Whenever p and q appear, these conditions on p and q will be taken for granted without comment.

EXAMPLE 2.14 (Infinite Sequence of Bernoulli Trials) Fix $0 < p < 1$ and let $q = 1 - p$. Let Ω be the set described above and let $\mathcal{F}_0$ be the collection of subsets A of Ω of the form

$$A = \{\omega : \omega = \{x_i\}_{i=1}^{\infty}, x_{i_1} = \delta_1, \ldots, x_{i_n} = \delta_n\}, \tag{2.6}$$

where n is any positive integer, $1 \leq i_1 < i_2 < \cdots < i_n$, and each δ_i is a 0 or a 1. We think of the x_i as the results of successive trials. For $\mathcal{F}$ we take $\sigma(\mathcal{F}_0)$, the smallest σ-algebra containing $\mathcal{F}_0$. As an illustration of how probabilities are to be computed, consider the event "1 on the second trial, 0 on the fourth trial, and 1 on the eighth trial"; i.e., the event

$$A = \{\omega : \omega = \{x_i\}_{i=1}^{\infty}, x_2 = 1, x_4 = 0, x_8 = 1\}.$$

Then

$$P(A) = p^2 q = p^2(1 - p).$$

Note that

$$P(A) = p^{x_2+x_4+x_8} q^{3-(x_2+x_4+x_8)}.$$

For an event A of the type described in Equation 2.6, its probability is defined to be

$$P(A) = p^{(\sum_{j=1}^{n} \delta_j)} q^{(n - \sum_{j=1}^{n} \delta_j)}. \tag{2.7}$$

Note that $\sum_{j=1}^{n} \delta_j$ is the number of 1's in the trials numbered $i_1, i_2, \ldots, i_n$ and $n - \sum_{j=1}^{n} \delta_j$ is the number of 0's in the same trials. It cannot be done here, but it is possible to extend the definition of P so that $P(A)$ is defined for all $A \in \mathcal{F}$. Any set of outcomes that can be expressed in terms of events placing restrictions on only a finite number of trials will also be an event. Consider the event A described by "a 1 eventually appears in the outcome ω" ; i.e.,

$$A = \{\omega : \omega = \{x_j\}_{j=1}^{\infty}, \sum_{j=1}^{\infty} x_j \geq 1\}.$$

If we let A_j be the event "1 appears for the first time on the jth trial," then

$$A = \bigcup A_j \in \mathscr{F}$$

for the reasons just cited, and the A_j are disjoint. Since $P(A_j) = q^{j-1}p$ and the latter is the general term of a geometric series,

$$P(A) = \sum_{j=1}^{\infty} q^{j-1}p = p\sum_{j=1}^{\infty} q^{j-1} = \frac{p}{1-q} = 1. \blacksquare$$

There is no reason to limit the number of results of each trial to just the 0 and 1 of the preceding example. We can allow the possibility that each trial results in one of k possibilities $r_1, r_2, \ldots, r_k$ with associated weights $p_1, p_2, \ldots, p_k$, where $0 \leq p_i \leq 1, i = 1, \ldots, k$. Suppose $n \geq 1, 1 \leq i_1 < i_2 < \cdots < i_n$, and $\delta_1, \ldots, \delta_n \in \{r_1, \ldots, r_k\}$. For the event

$$A = \{\omega : \omega = \{x_j\}_{j=1}^{\infty}, x_{i_1} = \delta_1, \ldots, x_{i_n} = \delta_n\},$$

we can define

$$P(A) = p_1^{m_1} \times p_2^{m_2} \times \cdots \times p_k^{m_k},$$

where m_i is the number of trials resulting in $r_i, 1 \leq i \leq k$. This model is applicable to an unending sequence of throws of a die where the result of each throw is one of the integers 1, 2, 3, 4, 5, 6 with weight 1/6 associated with each.

EXERCISES 2.4

1. Using mathematical induction, write out a formal proof that an algebra $\mathscr{A}$ is closed under finite unions; i.e., for every $n \geq 1$,

$$A_1, \ldots, A_n \in \mathscr{A} \text{ implies } \bigcup_{j=1}^{n} A_j \in \mathscr{A}.$$

2. If $\mathscr{A}$ is an algebra of subsets of Ω, show that $\mathscr{A}$ is closed under finite intersections.
3. Let $\mathscr{F}$ be a σ-algebra of subsets of Ω. Show that $\mathscr{F}$ is closed under countable intersections; i.e., if $\{A_j\}$ is a finite or infinite sequence in $\mathscr{F}$, then $\cap A_j \in \mathscr{F}$.
4. Let Ω be an uncountable set and let $\mathscr{F}$ be the collection of subsets A of Ω such that either A is countable or A^c is countable. Show that $\mathscr{F}$ is a σ-algebra.
5. Consider an infinite sequence of Bernoulli trials with probability of success p. What is the probability that a success (or 1) will occur for the first time on an even-numbered trial?

6. An experiment consists of tossing a pair of dice until a score of 8 is observed for the first time, whereupon the experiment is terminated. What is the probability that it will terminate on an odd number of tosses of the dice?
7. A bowl contains w white chips, r red chips, and b black chips. Chips are successively selected at random from the bowl with replacement. What is the probability that a white chip will appear before a black chip?
8. If $\mathcal{F}$ is a σ-algebra of subsets of Ω and $\{A_j\}_{j=1}^{\infty}$ is an increasing sequence in $\mathcal{F}$ (i.e., $A_n \subset A_{n+1}$ for all $n \geq 1$), show that there is a disjoint sequence $\{B_j\}_{j=1}^{\infty}$ in $\mathcal{F}$ such that $\cup_{j=1}^{\infty} A_j = \cup_{j=1}^{\infty} B_j$.
9. If $\mathcal{F}$ is a σ-algebra of subsets of Ω, $\{A_j\}_{j=1}^{\infty}$ is a decreasing sequence in $\mathcal{F}$ (i.e., $A_{n+1} \subset A_n$ for all $n \geq 1$), and $A = \cap_{j=1}^{\infty} A_j$, show that there is a decreasing sequence $\{B_j\}_{j=1}^{\infty}$ in $\mathcal{F}$ such that $A_n = A \cup B_n, A \cap B_n = \emptyset$ for all $n \geq 1$ and $\cap_{j=1}^{\infty} B_j = \emptyset$.

2.5 PROPERTIES OF PROBABILITY FUNCTIONS

Throughout this section, $(\Omega, \mathcal{F}, P)$ will be a fixed probability space as described in Definition 2.3. We will now deduce several properties of the probability function P from the axioms listed in Definition 2.3.

Consider two events $A, B \in \mathcal{F}$. Since $\Omega = A \cup A^c$, intersecting both sides of this equation with B we obtain

$$B = B \cap \Omega = B \cap (A \cup A^c) = (B \cap A) \cup (B \cap A^c);$$

i.e., we can decompose B into two parts according to whether or not an outcome in B is in A or not in A. Since $B \cap A$ and $B \cap A^c$ are disjoint, by Item 3 of Definition 2.3,

$$P(B) = P(B \cap A) + P(B \cap A^c) \text{ for all } A, B \in \mathcal{F}. \tag{2.8}$$

If we put $B = \Omega$ and use Item 1 of Definition 2.3, then $1 = P(\Omega) = P(\Omega \cap A) + P(\Omega \cap A^c)$, so that

$$P(A^c) = 1 - P(A) \text{ for all } A \in \mathcal{F}. \tag{2.9}$$

In particular, $P(\emptyset) = 1 - P(\Omega) = 0$.

EXAMPLE 2.15 Consider n flips of a coin and let A be the event "the outcome ω has one or more heads." Calculating $P(A)$ directly is complicated, but calculating $P(A^c)$ is easily done because A^c consists of just one outcome having a label of n T's. Since each outcome has probability $1/2^n$, $P(A) = 1 - P(A^c) = 1 - (1/2^n)$. ■

Suppose now that $A, B \in \mathcal{F}, A \subset B$. Then $A \cap B = A$, and so Equation 2.8 becomes $P(B) = P(A) + P(B \cap A^c)$. Since $P(B \cap A^c) \geq 0$ by Item 2 of Definition 2.3,

$$P(A) \leq P(B) \text{ whenever } A, B \in \mathcal{F}, A \subset B. \tag{2.10}$$

If A and B are any two events, then $A \cup B = (A \cap B^c) \cup (A \cap B) \cup (A^c \cap B)$; i.e., $A \cup B$ can be split into three parts: (1) those outcomes in A but not in B, (2) those outcomes in both A and B, and (3) those outcomes in B but not in A. Thus,

$$P(A \cup B) = P(A \cap B^c) + P(A \cap B) + P(A^c \cap B).$$

Applying Equation 2.8 to the first and third terms on the right and simplifying, we obtain

$$P(A \cup B) = P(A) + P(B) - P(A \cap B). \tag{2.11}$$

EXAMPLE 2.16 A card is selected at random from a deck of 52 cards. What is the probability that the card selected will be a king or a spade? Let A be the event "the outcome ω is a king" and let B be the event "the outcome ω is a spade." The required probability is $P(A \cup B) = P(A) + P(B) - P(A \cap B) = 1/13 + 1/4 - 1/52 = 4/13$. ■

If A, B, and C are any three events, then

$$\begin{aligned} P(A \cup B \cup C) = {} & P(A) + P(B) + P(C) - P(A \cap B) \\ & - P(A \cap C) - P(B \cap C) + P(A \cap B \cap C). \end{aligned}$$

More generally, if $A_1, \ldots, A_N$ are any events, then

$$\begin{aligned} P(A_1 \cup \cdots \cup A_N) &= \sum_{i=1}^{N} P(A_i) - \sum_{1 \leq i_1 < i_2 \leq N} P(A_{i_1} \cap A_{i_2}) \\ &\quad + \sum_{1 \leq i_1 < i_2 < i_3 \leq N} P(A_{i_1} \cap A_{i_2} \cap A_{i_3}) \\ &\quad - \cdots \pm P(A_1 \cap A_2 \cap \cdots \cap A_N) \\ &= \sum_{r=1}^{N} (-1)^{r-1} \sum_{1 \leq i_1 < \cdots < i_r \leq N} P(A_{i_1} \cap \cdots \cap A_{i_r}). \end{aligned} \tag{2.12}$$

This result goes by the name *inclusion/exclusion principle* and can be proved using mathematical induction.

Returning to Equation 2.11,

$$P(A \cup B) \leq P(A) + P(B) \text{ for all } A, B \in \mathcal{F}$$

since $P(A \cap B) \geq 0$ by Item 2 of Definition 2.3. This inequality is a special case of a more general inequality whose proof will require the following lemma.

Lemma 2.5.1 *If $\{A_j\}_{j=1}^{\infty}$ is a sequence of events, then there is a disjoint sequence $\{B_j\}_{j=1}^{\infty}$ of events such that $B_j \subset A_j$ for all $j \geq 1$, $\cup_{j=1}^{n} A_j = \cup_{j=1}^{n} B_j$ for all $n \geq 1$, and $\cup B_j = \cup A_j$.*

PROOF: Let $B_1 = A_1$ and $B_j = A_j \cap (\cup_{i=1}^{j-1} A_i)^c$ for $j \geq 2$. Clearly, $B_j \subset A_j$ for all $j \geq 1$. For $1 \leq i \leq j-1$,

$$B_i \cap B_j \subset A_i \cap B_j \subset \left(\bigcup_{i=1}^{j-1} A_i\right) \cap B_j = \varnothing.$$

Thus, $B_i \cap B_j = \varnothing$ whenever $1 \leq i \leq j-1, j \geq 1$. This means that the B_j are disjoint. Clearly, $\cup_{j=1}^{n} B_j \subset \cup_{j=1}^{n} A_j$. Suppose $\omega \in \cup_{j=1}^{n} A_j$. Then there is a smallest integer $k \leq n$ such that $\omega \in A_k$. Thus, $\omega \in A_k \cap (\cup_{i=1}^{k-1} A_i)^c = B_k \subset \cup_{k=1}^{n} B_k$, and it follows that $\cup_{j=1}^{n} A_j \subset \cup_{j=1}^{n} B_j$ and therefore that the two are equal. The proof of the last assertion is essentially the same. ■

Theorem 2.5.2 (Boole's Inequality) *If $\{A_j\}$ is any sequence of events, then $P(\cup A_j) \leq \sum P(A_j)$.*

PROOF: By Lemma 2.5.1, there is a disjoint sequence of events $\{B_j\}$ such that $B_j \subset A_j, j \geq 1$, and $\cup B_j = \cup A_j$. By Inequality 2.10, $P(B_j) \leq P(A_j), j \geq 1$. Since the B_j are disjoint,

$$P(\cup A_j) = P(\cup B_j) = \sum P(B_j) \leq \sum P(A_j). \ ■$$

Theorem 2.5.3 *Let $\{A_j\}_{j=1}^{\infty}$ be a sequence of events.*

(i) If $A_1 \subset A_2 \subset \cdots$ is an increasing sequence and $A = \cup_{j=1}^{\infty} A_j$, then $P(A) = \lim_{n\to\infty} P(A_n)$.

(ii) If $A_1 \supset A_2 \supset \cdots$ is a decreasing sequence and $A = \cap_{j=1}^{\infty} A_j$, then $P(A) = \lim_{n\to\infty} P(A_n)$.

PROOF: (i) Let $\{A_j\}_{j=1}^{\infty}$ be an increasing sequence of events and let $A = \cup_{j=1}^{\infty} A_j \in \mathcal{F}$. Note that $\cup_{j=1}^{n} A_j = A_n$. By Lemma 2.5.1, there is a disjoint sequence of events $\{B_j\}_{j=1}^{\infty}$ such that $B_j \subset A_j, j \geq 1, \cup_{j=1}^{n} A_j = \cup_{j=1}^{n} B_j$, and $\cup A_j = \cup B_j$. By Item 3 of Definition 2.3,

$$\begin{aligned} P(A) &= P\left(\bigcup_{j=1}^{\infty} A_j\right) = P\left(\bigcup_{j=1}^{\infty} B_j\right) = \sum_{j=1}^{\infty} P(B_j) \\ &= \lim_{n\to\infty} \sum_{j=1}^{n} P(B_j) = \lim_{n\to\infty} P\left(\bigcup_{j=1}^{n} B_j\right) = \lim_{n\to\infty} P(A_n). \end{aligned}$$

(*ii*) Let $\{A_j\}_{j=1}^{\infty}$ be a decreasing sequence of events and let $A = \cap_{j=1}^{\infty} A_j \in \mathcal{F}$. Then $\{A_j^c\}_{j=1}^{\infty}$ is an increasing sequence of events, and $A^c = (\cap_{j=1}^{\infty} A_j)^c = \cup_{j=1}^{\infty} A_j^c$. By the first part of the proof,

$$\begin{aligned} 1 - P(A) = P(A^c) &= \lim_{n \to \infty} P(A_n^c) \\ &= \lim_{n \to \infty} (1 - P(A_n)) = 1 - \lim_{n \to \infty} P(A_n), \end{aligned}$$

and so $P(A) = \lim_{n \to \infty} P(A_n)$. ■

EXERCISES 2.5

1. In manufacturing brass cylindrical sleeves, 5 percent are defective because the outer diameter is too small and 3 percent are defective because the inner diameter is too large. What is the best you can say about the probability that a sleeve selected at random from a lot will be defective?
2. If A, B, and C are any three events, show that $P(A \cup B \cup C) = P(A)+P(B)+P(C)-P(A\cap B)-P(A\cap C)-P(B\cap C)+P(A\cap B\cap C)$.
3. Consider three events A, B, C for which $P(A) = 1/3$, $P(B) = 1/4$, $P(C) = 1/2$, $P(A \cap B) = 1/8$, $P(A \cap C) = 1/8$, $P(B \cap C) = 3/16$, and $P(A \cap B \cap C) = 1/32$. Calculate $P(A \cup B \cup C)$.
4. An integer is chosen at random between 0000 and 9999. (a) Use the inclusion/exclusion principle to calculate the probability that at least one 1 will appear in the number. (b) Calculate the same probability assuming that the experiment is that of four Bernoulli trials.
5. Show that the probability that one and only one of the events A and B will occur is
$$P(A) + P(B) - 2P(A \cap B).$$
6. The mid–seventeenth century gambler Chavalier de Méré thought that the probability of getting at least one ace with the throw of four dice is equal to the probability of getting at least one double ace in 24 throws of two dice. Was de Méré correct?
7. Consider an infinite sequence of Bernoulli trials with probability of success p. If ω_0 is any outcome, show that $P(\{\omega_0\}) = 0$. (Note: There is no significance to the fact that each outcome has probability 0 whereas the aggregate of all outcomes has probability 1! After all, points in the interval $[0, 1]$ have zero length, but the aggregate $[0, 1]$ has length1.)
8. If $P(A) = .8$ and $P(B) = .75$, show that $P(A \cap B) \geq .55$. More generally, show that if A and B are any two events, then
$$\min(P(A), P(B)) \geq P(A \cap B) \geq P(A) + P(B) - 1.$$

9. If $A_1, \ldots, A_n$ are any events, show that

$$P(A_1 \cap \cdots \cap A_n) \geq P(A_1) + \cdots + P(A_n) - (n-1).$$

2.6 CONDITIONAL PROBABILITY AND INDEPENDENCE

Conditional probabilities are defined for general probability spaces as in Equation 1.12.

Definition 2.5 *For $B \in \mathcal{F}$ with $P(B) > 0$ and $A \in \mathcal{F}$, define*

$$P(A|B) = \frac{P(A \cap B)}{P(B)}. \quad ■$$

Since $P(A|B)$ associates with each $A \in \mathcal{F}$ a real number, it is a function from $\mathcal{F}$ to R, which we denote by $P(\cdot|B)$ and call a *conditional probability function.* An immediate consequence of the definition is the equation

$$P(A \cap B) = P(A|B)P(B), \qquad A, B \in \mathcal{F}, P(B) > 0, \quad (2.13)$$

which is sometimes called the *law of compound probabilities.* If $P(B) = 0$, we usually define $P(A|B) = 0$, which is consistent with Equation 2.13 since $P(A \cap B) = 0$ whenever $P(B) = 0$.

It was pointed out at the end of Section 1.5 that some probability models are described not by specifying the probability of each outcome but rather by a combination of outcome probabilities and conditional probabilities as in the following example.

EXAMPLE 2.17 A bowl contains 10 red balls and 10 white balls. An experiment consists of selecting a ball at random from the bowl, replacing it by a ball of the other color, putting the replacement into the bowl, and then selecting a second ball at random from the bowl. There are four outcomes of the experiment: (R,R), (R,W), (W,R), and (W,W). Probabilities of these four outcomes are not given explicitly, but the model is described so that they can be determined. To do this, let R_1 denote the event "first ball selected is red" and let R_2 denote the event "second ball selected is red." The following numbers are the given data:

$$\begin{aligned} P(R_1) &= \frac{1}{2} \qquad & P(R_2^c|R_1) &= \frac{11}{20} \qquad & P(R_2|R_1) &= \frac{9}{20} \\ P(R_1^c) &= \frac{1}{2} \qquad & P(R_2|R_1^c) &= \frac{11}{20} \qquad & P(R_2^c|R_1^c) &= \frac{9}{20} \end{aligned}$$

As an illustration of these computations, consider $P(R_2|R_1)$. Given that the outcome is in R_1, at the time of the second selection there are 9 red and 11 white balls in the bowl, and so the probability that the second ball will be red is 9/20. Probabilities of individual outcomes can be calculated using Equation 2.13; e.g.,

$$P((R, R)) = P(R_1 \cap R_2) = P(R_2|R_1)P(R_1) = \frac{9}{20} \cdot \frac{1}{2} = \frac{9}{40}. \quad \blacksquare$$

All the theorems proved for probability functions in the previous section are true for conditional probability functions $P(\cdot|B)$ with a fixed $B \in \mathcal{F}, P(B) > 0$. To see this, define

$$\tilde{P}(A) = P(A|B) \text{ for } A \in \mathcal{F}.$$

Since $\tilde{P}(\Omega) = P(\Omega \cap B)/P(B) = 1$, Item 1 of Definition 2.3 is satisfied. Since for $A \in \mathcal{F}, 0 \leq \tilde{P}(A) = P(A|B) = P(A \cap B)/P(B) \leq 1$, Item 2 is satisfied. Let $\{A_j\}$ be a finite or infinite disjoint sequence in $\mathcal{F}$. Then

$$\tilde{P}(\cup A_j) = P(\cup A_j|B) = \frac{P((\cup A_j) \cap B)}{P(B)} = \frac{P(\cup(A_j \cap B))}{P(B)}.$$

Since the events $A_j \cap B$ are disjoint,

$$\tilde{P}(\cup A_j) = \frac{\sum P(A_j \cap B)}{P(B)} = \sum \frac{P(A_j \cap B)}{P(B)} = \sum P(A_j|B) = \sum \tilde{P}(A_j),$$

and Item 3 of Definition 2.3 is satisfied. Since the theorems proved in the previous section were consequences of Items 1, 2, and 3 in Definition 2.3, these same theorems are true for conditional probability functions $P(\cdot|B)$ for fixed $B \in \mathcal{F}, P(B) > 0$. For example, if $\{A_j\}_{j=1}^{\infty}$ is an increasing sequence of events with $A = \cup_{j=1}^{\infty} A_j$, it is not necessary to give a proof that

$$\lim_{n \to \infty} P(A_n|B) = P(A|B).$$

One of the most useful applications of conditional probabilities is known as *Bayes' rule.* Let $A_1, A_2, \ldots, A_n$ be a finite disjoint collection of events that exhausts Ω; i.e., $\Omega = \cup_{j=1}^{n} A_j$. We think of the $A_1, \ldots, A_n$ as a *stratification* of Ω. If B is any other event with $P(B) > 0, 1 \leq i \leq n$, then

$$P(A_i|B) = \frac{P(B \cap A_i)}{P(B)}.$$

Since $B = \cup_{j=1}^{n}(B \cap A_j)$ with the latter events disjoint,

$$P(B) = \sum_{j=1}^{n} P(B|A_j)P(A_j),$$

and so

$$P(A_i|B) = \frac{P(B|A_i)P(A_i)}{\sum_{j=1}^{n} P(B|A_j)P(A_j)}, \qquad i = 1, \dots, n. \qquad (2.14)$$

Note that all the probabilities $P(A_1), \dots, P(A_n), P(B|A_1), \dots, P(B|A_n)$ must be given data to apply Bayes' rule and that the $A_1, \dots, A_n$ are disjoint and exhaust Ω. It was tacitly assumed in this discussion that $P(A_i) > 0, 1 \le i \le n$. This is always true of at least one A_i, and the last equation is true assuming only that $P(B) > 0$.

EXAMPLE 2.18 A bowl contains three red balls and one white ball. A ball is selected at random from the bowl, replaced by a ball of the other color, and returned to the bowl. A second ball is then selected at random from the bowl. Given that the second ball is red, what is the probability that the first ball was red? Let R_i be the event "ith ball is red," $i = 1, 2$. Then R_1 and R_1^c are disjoint and exhaust Ω. We are given the data $P(R_1) = 3/4, P(R_1^c) = 1/4, P(R_2|R_1) = 1/2, P(R_2|R_1^c) = 1$. Thus,

$$P(R_1|R_2) = \frac{P(R_2|R_1)P(R_1)}{P(R_2|R_1)P(R_1) + P(R_2|R_1^c)P(R_1^c)} = \frac{3}{5}. \quad \blacksquare$$

The next application of Bayes' rule has to do with the settlement of paternity cases in a court of law and necessitates a crude review of genetics related to blood types.

In conceiving a child, each parent contributes one of the alleles O, A, or B to form one of the pairs OO, AO, AA, BO, BB, AB, called genotypes. Both A and B are dominant over O; neither A nor B is dominant over the other. The observed blood types, called phenotypes, of the child can be O, A, B, or AB. Figure 2.5 gives combinations of genotypes and phenotypes as well as the proportion of each combination in the general population.

Genotype	OO	OA	AA	OB	BB	AB
Phenotype	O	A	A	B	B	AB
Proportion	.479	.310	.050	.116	.007	.038

FIGURE 2.5 Frequencies of genotypes and phenotypes.

EXAMPLE 2.19 (Paternity Index) Jane of blood type A claims in court that Dick of blood type B is the father of her child of blood type B. The following calculations are made to support her claim. Consider an experiment in which a person is selected at random from the population of adult males. Let E be the event "The child of the person selected is of blood type B" and let F be the event "The person selected is Dick." The genotype of the child is either OB or BB. Since the mother has blood type A, her genotype can only be OA, and she passed on the O allele to her child. The genotype of Dick is unknown, but it must be either OB or BB. Let F_{OB} and F_{BB} be the events "Dick's genotype is OB" and "Dick's genotype is BB," respectively. Then

$$P(E|F) = \frac{P(E \cap F)}{P(F)} = \frac{P(E \cap F|F_{OB})P(F_{OB}) + P(E \cap F|F_{BB})P(F_{BB})}{P(F)}.$$

From Figure 2.5, $P(E \cap F|F_{OB}) = .116, P(F_{OB}) = .5, P(E \cap F|F_{BB}) = 1, P(F_{BB}) = .007,$ and $P(F) = .123$. Therefore,

$$P(E|F) = \frac{(.116)(.5) + .007}{.123} = .528.$$

We now calculate $P(E|F^c)$, the probability that the child is of blood type B given that someone other than Dick is the father. Since that someone must have blood type B and therefore genotype OB or BB, and in the first case there is a 50–50 chance that the B allele will be passed to the child,

$$P(E|F^c) = (.116)(.5) + .007 = .065.$$

The quantity

$$I = \frac{P(E|F)}{P(E|F^c)} = \frac{1}{.123} = 8.13$$

is called the *paternity index* and is interpreted to mean that a person of blood type B is eight times more likely to be the father than some other person. The paternity index is just as applicable to one man of blood type B as it is to any other man of the same blood type. This is a useful index, but it does not give us the probability that Dick is the father of the child with blood type B. By Bayes' rule,

$$P(F|E) = \frac{P(E|F)P(F)}{P(E|F)P(F) + P(E|F^c)P(F^c)}.$$

We can use the calculations above to obtain the two conditional probabilities on the right, but to complete the computation we need to know $P(F)$. Jane claims that this number should be 1 and Dick claims that it should be 0. In this situation it is customary to compromise by using the figure $P(F) = .5$, in which case $P(F|E) = .89$. ■

The reader might question the applicability of Bayes' rule in paternity cases but not Bayes' rule itself. The basic premise in this example is that Jane chose an adult male at random from the population of adult males and the chosen person fathered her child.

As in Chapter 1, we will interpret $P(A|B)$ as the probability of A given the partial information that the outcome is in B. It sometimes happens that the partial information is irrelevant as far as the event A is concerned; i.e., $P(A|B) = P(A)$ or, using Equation 2.13, $P(A \cap B) = P(A)P(B)$. In this case, the events A and B are said to be independent. We will reformulate the definition so that it is not required that $P(B) > 0$.

Definition 2.6 *The events $A, B \in \mathcal{F}$ are* independent events *if*

$$P(A \cap B) = P(A)P(B). \quad ■$$

The definition is now symmetric in A and B.

EXAMPLE 2.20 Consider a roll of two dice, one red and one white. Let $R_i, i = 1, \ldots, 6$ be the event "i pips on the red die" and let $W_j, j = 1, \ldots, 6$ be the event "j pips on the white die." Any pair R_i and W_j are independent events since

$$P(R_i \cap W_j) = P((i,j)) = \frac{1}{36} = \frac{1}{6} \cdot \frac{1}{6} = P(R_i)P(W_j). \quad ■$$

Generally speaking, any event specified solely by conditions on a red die will be independent of any event specified solely by conditions on a white die. Let A be the event "even number of pips on the red die" and let B be the event "odd number of pips on the white die." By examining Figure 1.2, $P(A \cap B) = 1/4 = 1/2 \cdot 1/2 = P(A)P(B)$.

Theorem 2.6.1 *If A and B are independent events, then each of the pairs A and B^c, A^c and B, A^c and B^c are independent.*

PROOF: Consider the pair A and B^c. By Equation 2.8, $P(A \cap B^c) = P(A) - P(A \cap B) = P(A) - P(A)P(B) = P(A)(1 - P(B)) = P(A)P(B^c)$, and so A and B^c are independent. Similarly for the other two pairs. ■

If A, B, and C are any three events, independence of the three could be taken to mean that the three pairs A and B, A and C, and B and C are independent pairs. This type of independence is called *pairwise independence.* In some models there is a stronger built-in independence.

Definition 2.7 *The events $A_1, \ldots, A_n \in \mathcal{F}$ are* mutually independent *if*

$$\begin{aligned} P(A_{i_1} \cap A_{i_2}) &= P(A_{i_1})P(A_{i_2}), \qquad i_1 \neq i_2, 1 \leq i_1, i_2 \leq n \\ P(A_{i_1} \cap A_{i_2} \cap A_{i_3}) &= P(A_{i_1})P(A_{i_2})P(A_{i_3}) \\ &\quad \text{for } i_1, i_2, i_3 \text{ distinct}, 1 \leq i_1, i_2, i_3 \leq n \\ &\vdots \\ P(A_1 \cap A_2 \cap \cdots \cap A_n) &= P(A_1)P(A_2) \times \cdots \times P(A_n). \ \blacksquare \end{aligned}$$

The total number of conditions imposed in this definition is easily calculated using Equation 1.10 and is $2^n - n - 1$. It is possible for events to be pairwise independent but not mutually independent.

EXAMPLE 2.21 Suppose a pair of dice are rolled, one red and one white. Let A be the event "odd number of pips on the red die," B the event "odd number of pips on the white die," and C the event "the score is odd." Checking Figures 1.2 and 1.3, it is easy to see that A, B, and C are pairwise independent, but $P(A \cap B \cap C) = 0 \neq (1/2)^3 = P(A)P(B)P(C)$ since $A \cap B \cap C = \emptyset$. ■

Caveat: Independent and mutually exclusive are not the same.

Theorem 2.6.2 *If $A_1, A_2, \ldots, A_n$ are mutually independent events, then $B_1, B_2, \ldots, B_n$ are also mutually independent where each B_j is A_j or A_j^c.*

In rolling a pair of dice, the numbers of pips on each die constitute independent events. Coin flipping also has built-in independence.

EXAMPLE 2.22 Consider an infinite sequence of Bernoulli trials with probability of success p. Suppose $\delta_1, \delta_2, \ldots, \delta_n \in \{0, 1\}$ are given and $1 \leq i_1 < i_2 < \cdots < i_n$. For $j = 1, 2, \ldots, n$, let

$$A_{i_j} = \{\omega : \omega = \{x_i\}_{i=1}^{\infty}, x_{i_j} = \delta_j\};$$

i.e., if $\omega \in A_{i_j}$, then the result of the i_j trial is δ_j and nothing else is known about the results of the other trials. The events $A_{i_1}, \ldots, A_{i_n}$ are then mutually independent events. According to Equation 2.7, $P(A_{i_j}) = p^{\delta_j} q^{1-\delta_j}$, and since $A_{i_1} \cap \cdots \cap A_{i_n} = \{\omega : \omega = \{x_i\}_{i=1}^{\infty}, x_{i_1} = \delta_1, \ldots, x_{i_n} = \delta_n\}$,

$$\begin{aligned} P(A_{i_1} \cap \cdots \cap A_{i_n}) &= p^{\sum_{i=1}^n \delta_i} q^{(n - \sum_{i=1}^n \delta_i)} \\ &= \prod_{i=1}^n (p^{\delta_i} q^{1-\delta_i}) \\ &= P(A_{i_1}) \times \cdots \times P(A_{i_n}). \end{aligned}$$

Since this is true for any set of integers $1 \leq i_1 < i_2 < \cdots < i_n$, the $2^n - n - 1$ conditions for mutual independence are fulfilled. ■

Theorem 2.6.3 *Let $A_1, A_2, \ldots, A_n$ be mutually independent events and let $I = \{i_1, \ldots, i_k\}$, $J = \{j_1, \ldots, j_{n-k}\}$ be nonempty disjoint subsets of $\{1, 2, \ldots, n\}$. Then any event constructed from the $A_{i_1}, \ldots, A_{i_k}$ is independent of any event constructed from the $A_{j_1}, \ldots, A_{j_{n-k}}$.*

This theorem is proved in more advanced texts. For now we must be satisfied with verifying it in specific cases in the exercises.

EXERCISES 2.6

Bayes' rule is needed for some of the following problems.

1. Let A, B, and C be mutually independent events. Show that A, B^c, and C are mutually independent.
2. Let $A_1, A_2, \ldots, A_n$ be mutually independent events. Show that $B_1, B_2, \ldots, B_n$ are mutually independent events where B_i is A_i or A_i^c, $i = 1, 2, \ldots, n$.
3. (a) If A, B, and C are any three events, show that $P(A \cap B \cap C) = P(A|B \cap C)P(B|C)P(C)$ provided the conditional probabilities are defined. (b) State a generalization for events $A_1, A_2, \ldots, A_n$.
4. A bowl contains 10 red balls and 10 white balls. A ball is selected at random from the bowl, replaced by a ball of the other color, and returned to the bowl. This procedure is repeated two more times. An outcome is defined to be an ordered triple (i, j, k) where i is the number of red balls in the bowl after the first return, j is the number after the second return, and k is the number after the third return. Determine the probability of each outcome.
5. A coin is flipped twice in succession. Let A be the event "head on the first flip," B the event "head on the second flip," and C the event "the two flips match." (a) Are the events A, B, and C pairwise independent? (b) Are they mutually independent? In both cases, justify your answer.
6. Binary digits 0 and 1 are transmitted over a communications channel. If a 1 is sent, it will be received as a 1 with probability .95 and as a 0 with probability .05; if a 0 is sent, it will be received as a 0 with probability .99 and as a 1 with probability .01. If the probabilities that a 0 or 1 is sent are equal, what is (a) the probability that a 1 was sent given that a 1 was received and (b) the probability that a 0 was sent given that a 0 was received?
7. If, in the previous problem, three successive binary digits are transmitted with independence between digits, what is the probability that 111 was sent given that 101 was received?
8. There are three chests with two drawers each, and each drawer contains a gold coin or a silver coin. Chest 1 contains two gold coins, Chest 2

contains a gold coin and a silver coin, and Chest 3 contains two silver coins. (Gold coins? A very old problem.) A chest is selected at random, and then one of its two drawers is selected at random and opened. If a gold coin is observed, what is the probability that the other drawer contains a gold coin?

9. Let A, B, C, and D be mutually independent events. Show that $A \cup B$ and $C \cap D$ are independent events.

10. Consider an event A and an infinite sequence of disjoint events $\{A_j\}_{j=1}^{\infty}$ such that A and A_j are independent for each $j \geq 1$. Show that A and $\cup A_j$ are independent events.

11. A mechanical system consists of components A, B_1, B_2, C, and D as indicated in the diagram. The system will function if there is a path from α to β along which all components are functioning. (a) If in a specified period of time, A, C, and D will each malfunction with probability .05 while B_1 and B_2 will each malfunction with probability .2, all independently of one another, what is the probability that the system will function during the period? (b) If B_3 is added to the system in parallel with B_1 and B_2 and with the same probability of malfunction, what is the probability that the system will function?

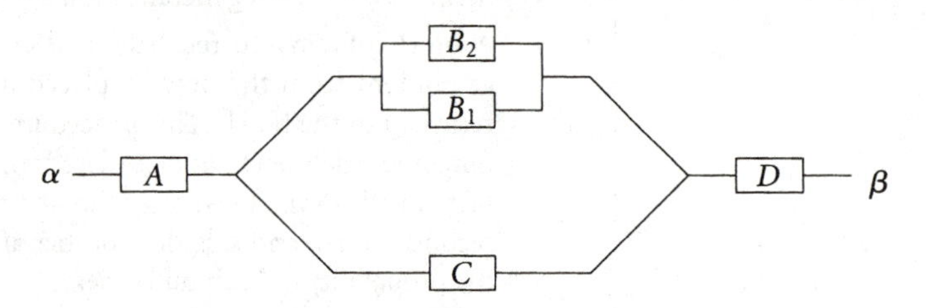

The following problem does not require mathematical software such as Mathematica or Maple V, but using a hand calculator is a bit tedious.

12. Suppose in the previous exercise that the components B_1 and B_2 are replaced by $B_1, \ldots, B_m$ connected in parallel and C is replaced by $C_1, \ldots, C_n$ connected in parallel. Assume that each B_i will malfunction with probability .4 and each C_j will malfunction with probability .6. If the B_i cost \$100 each and the C_j cost \$80 each, how many of the B_i and C_j components are required to ensure that the total system will function with probability at least .88 and will minimize the cost?

2.7 SOME APPLICATIONS

The first application will deal with such questions as "How secure is your remotely operated garage door opener? Your computer password? Your answering machine access number? Your telephone calling card number?" Such applications require an extension of the definition of mutually independent events to countable collections. Fix the probability space $(\Omega, \mathcal{F}, P)$.

Definition 2.8 *The events $A_1, A_2, \ldots$ are* mutually independent *if every finite subcollection consists of mutually independent events.* ■

EXAMPLE 2.23 Consider an infinite sequence of Bernoulli trials with probability of success p. For each $j \geq 1$, let $\delta_j \in \{0, 1\}$ and define

$$A_j = \{\omega : \omega = \{x_i\}_{i=1}^{\infty}, x_j = \delta_j\}, \qquad j \geq 1.$$

If $\{A_{i_1}, \ldots, A_{i_n}\}$ is any subcollection of the A_i, we saw in the previous section that the $A_{i_1}, \ldots, A_{i_n}$ are mutually independent events. Thus, the events $A_1, A_2, \ldots$ are mutually independent. ■

Consider any sequence of events $\{A_j\}_{j=1}^{\infty}$ and an outcome ω. What is to be meant by the statement that ω belongs to infinitely many of the A_j? It should mean that no matter how far out you go in the sequence, the ω should belong to an A_j out beyond that point; i.e., for every $k \geq 1$, ω belongs to some A_j with $j \geq k$ or, in the language of set theory, $\omega \in \cup_{j \geq k} A_j$. Since this is true for every $k \geq 1$, $\omega \in \cap_{k \geq 1} \cup_{j \geq k} A_j$.

Definition 2.9 *If $\{A_j\}_{j=1}^{\infty}$ is any sequence of events, we define*

$$\{A_n \ i.o.\} = \cap_{k \geq 1} \cup_{j \geq k} A_j. \ ■$$

Note that $\{A_n \ i.o.\}$, read "A_n infinitely often," is an event, because $\mathcal{F}$ is closed under countable unions and intersections.

What is the complement of $\{A_n \ i.o.\}$? If ω is in the complement, then it is not true that $\omega \in A_j$ for infinitely many j; i.e., $\omega \in A_j$ for at most finitely many A_j. Formally, by de Morgan's laws,

$$\{A_n \ i.o.\}^c = \cup_{k \geq 1} \cap_{j \geq k} A_j^c.$$

This brings us to a famous theorem that for some reason is called a lemma.

Lemma 2.7.1 (Borel-Cantelli) *Let $\{A_j\}_{j=1}^{\infty}$ be an infinite sequence of events.*

(i) If $\sum_{j=1}^{\infty} P(A_j)$ converges, then $P(\{A_n \ i.o.\}) = 0$.

(ii) If the A_j are mutually independent events and $\sum_{j=1}^{\infty} P(A_j)$ diverges, then $P(\{A_n \ i.o.\}) = 1$.

PROOF: (*i*) Assume that the series $\sum_{j=1}^{\infty} P(A_j)$ converges. Since the sequence $\{\cup_{j \geq k} A_j\}_{k=1}^{\infty}$ is a decreasing sequence and $\{A_n\ i.o.\} = \cap_{k=1}^{\infty} \cup_{j \geq k} A_j$,

$$P(\{A_n\ i.o.\}) = \lim_{k \to \infty} P(\cup_{j \geq k} A_j)$$

by Theorem 2.5.3. By Theorem 2.5.2,

$$P(\cup_{j \geq k} A_j) \leq \sum_{j=k}^{\infty} P(A_j).$$

Thus,

$$0 \leq P(\{A_n\ i.o.\}) \leq \sum_{j=k}^{\infty} P(A_j)$$

for all $k \geq 1$. Since the series $\sum_{j=1}^{\infty} P(A_j)$ converges, the sum on the right has the limit zero as $k \to \infty$. Since $P(\{A_n\ i.o.\})$ does not depend upon k, $P(\{A_n\ i.o.\}) = 0$.

(*ii*) It is easily checked using calculus that the graph of the equation $y = 1 - x$ lies below the graph of the equation $y = e^{-x}$ for $x \geq 0$; i.e., $1 - x \leq e^{-x}$ for all $x \geq 0$. Therefore,

$$P(A_j^c) = 1 - P(A_j) \leq e^{-P(A_j)}, \qquad j \geq 1. \tag{2.15}$$

Consider $\{A_n\ i.o.\}^c = \cup_{k \geq 1} \cap_{j \geq k} A_j^c$. Since $\{\cap_{j=k}^{r} A_j^c\}_{r=k}^{\infty}$ is a decreasing sequence and

$$\cap_{j \geq k} A_j^c \subset \cap_{j=k}^{r} A_j^c, \qquad \text{for all } r \geq k,$$

it follows that

$$P(\cap_{j \geq k} A_j^c) \leq \lim_{r \to \infty} P(\cap_{j=k}^{r} A_j^c).$$

Since $A_k, \ldots, A_r$ are mutually independent events, $A_k^c, \ldots, A_r^c$ are mutually independent events and

$$\begin{aligned} P(\cap_{j \geq k} A_j^c) &\leq \lim_{r \to \infty} \prod_{j=k}^{r} P(A_j^c) \\ &\leq \lim_{r \to \infty} \prod_{j=k}^{r} e^{-P(A_j)} \\ &= \lim_{r \to \infty} e^{-\sum_{j=k}^{r} P(A_j)} \end{aligned}$$

by Inequality 2.15. Since the series $\sum_{j=k}^{\infty} P(A_j)$ diverges to $+\infty$ for each $k \geq 1$, the limit on the right is zero, and so $P(\cap_{j \geq k} A_j^c) = 0$ for all $k \geq 1$. Therefore,

$$0 \leq P(\{A_n \ i.o.\}^c) = P(\cup_{k \geq 1} \cap_{j \geq k} A_j^c) \leq \sum_{k=1}^{\infty} P(\cap_{j \geq k} A_j^c) = 0$$

and $P(\{A_n \ i.o.\}^c) = 0$. Thus, $P(\{A_n \ i.o.\}) = 1$, as was to be proved. ■

Lemma 2.7.2 *If* $\{A_{i_1}, A_{i_2}, \ldots\}$ *is any subcollection of the collection* $\{A_1, A_2, \ldots\}$, *then* $\{A_{i_n} \ i.o.\} \subset \{A_n \ i.o.\}$.

PROOF: If ω is in infinitely many of the A_{i_j}, then it is in infinitely many of the A_j. ■

EXAMPLE 2.24 (Password Problem) Consider an infinite sequence of Bernoulli trials with probability of success p. Consider the four-letter word 1001. You may substitute the binary representation of your social security number (which may require up to 30 binary digits 0 and 1), computer password, answering machine access number, or telephone calling card number for this number. What is the probability that the word 1001 will appear infinitely often in an outcome $\omega = \{x_j\}_{j=1}^{\infty}$? For each $j \geq 1$, let

$$B_j = \{\omega : \omega = \{x_i\}_{i=1}^{\infty}, x_j = 1, x_{j+1} = 0, x_{j+2} = 0, x_{j+3} = 1\}.$$

If $\omega \in B_1$, then ω looks like 1001.... If $\omega \in \{B_n \ i.o.\}$, then 1001 appears in ω infinitely often. Although the events $B_1, B_2, \ldots$ are not mutually independent, the events $B_1, B_5, B_9, \ldots$ are mutually independent because they are based on nonoverlapping sets of four trials. Since each B_j has probability p^2q^2 and $\sum_{j=0}^{\infty} P(B_{1+4j}) = \sum_{j=1}^{\infty} p^2q^2$ diverges, $P(\{B_{1+4n} \ i.o.\}) = 1$ by (ii) of the Borel-Cantelli lemma. Since $B_1, B_5, B_9, \ldots$ is a subcollection of the collection $B_1, B_2, \ldots$ and $\{B_{1+4n} \ i.o.\} \subset \{B_n \ i.o.\}$ by the preceding lemma, $P(\{B_n \ i.o.\}) = 1$. ■

How safe is your remotely operated garage door opener, computer password, answering machine access number, or telephone calling card number? It depends upon how long it would take a random generator of 1's and 0's to hit the electronic combination. The question should not be "Can it be violated?" but rather "How long will it take?" But that is another mathematical problem to which we will return in Chapter 4.

To illustrate the inclusion/exclusion principle, consider a deck of cards that are numbered $1, 2, \ldots, N$. Suppose the deck is thoroughly shuffled and the cards are dealt one by one onto positions numbered $1, 2, \ldots, N$. A match occurs at position j if the card numbered j is at that position. If all $N!$ arrangements of the deck are equally likely, what is the probability that there will be at least one match? Let A_j be the event "there is a match at the jth position." The answer to the question lies in calculating $P(A_1 \cup \cdots \cup A_N)$. This probability can be calculated using the inclusion/exclusion principle given by Equation 2.12:

$$\begin{aligned} P(A_1 \cup \cdots \cup A_N) &= \sum_{i=1}^{N} P(A_i) - \sum_{1 \le i_1 < i_2 \le N} P(A_{i_1} \cap A_{i_2}) \\ &\quad + \sum_{1 \le i_1 < i_2 < i_3 \le N} P(A_{i_1} \cap A_{i_2} \cap A_{i_3}) \\ &\quad - \cdots \pm P(A_1 \cap A_2 \cap \cdots \cap A_N) \\ &= \sum_{r=1}^{N} (-1)^{r-1} \sum_{1 \le i_1 < \cdots < i_r \le N} P(A_{i_1} \cap \cdots \cap A_{i_r}). \end{aligned}$$

Consider a typical term $P(A_{i_1} \cap \cdots \cap A_{i_r})$ where $1 \le i_1 < i_2 < \cdots i_r \le N$. For an outcome to be in $A_{i_1} \cap \cdots \cap A_{i_r}$ there must be matches at the $i_1, \ldots, i_r$ positions. The number of outcomes with such matches is $(N - r)!$. Thus,

$$P(A_{i_1} \cap \cdots \cap A_{i_r}) = \frac{(N-r)!}{N!}$$

and

$$\sum_{1 \le i_1 < i_2 < \cdots < i_r \le N} P(A_{i_1} \cap \cdots \cap A_{i_r}) = \binom{N}{r} \frac{(N-r)!}{N!}$$

since the sum on the left has $\binom{N}{r}$ terms corresponding to the number of ways of choosing a subset $\{i_1, \ldots, i_r\}$ from $\{1, \ldots, N\}$. Therefore,

$$\begin{aligned} P(A_1 \cup \cdots \cup A_N) &= \sum_{r=1}^{N} (-1)^{r-1} \binom{N}{r} \frac{(N-r)!}{N!} \\ &= \sum_{r=1}^{N} \frac{(-1)^{r-1}}{r!} \\ &= -\sum_{r=1}^{N} \frac{(-1)^r}{r!}. \end{aligned}$$

The last sum is a partial sum for the Maclaurin series expansion $e^{-x} = \sum_{r=0}^{\infty} ((-1)^r/r!) x^r$ with $x = 1$ except for a missing $r = 0$ term. If N is large, the sum $\sum_{r=1}^{N} (-1)^r/r!$ can be approximated by $e^{-1} - 1$. Thus, for large N, the probability of at least one match is

$$P(A_1 \cup \cdots \cup A_N) \approx 1 - \frac{1}{e},$$

and the probability of no match is approximately $1/e$. Actually, the approximation of the probability of no match by $1/e$ is quite good even for N as small as 6, the error in this case being on the order of .0002.

The following example appears in many guises.

EXAMPLE 2.25 (Coupon Collector Problem) Any one of N different coupons (e.g., baseball cards) is included in a commercial product. Assume independence between purchases and that the coupons are equally likely to appear in a product. If a collector purchases the product n times, $n \geq N$, what is the probability that a complete set of coupons will be collected? Suppose the coupons are labeled $1, 2, \ldots, N$. Let A_j be the event that coupon j does not appear among the n purchases, $j = 1, \ldots, N$. The probability that a complete set is not collected in n purchases is then $P(A_1 \cup \cdots \cup A_N)$. By the inclusion/exclusion principle,

$$\begin{aligned} P(A_1 \cup \cdots \cup A_N) &= \sum_{r=1}^{N} (-1)^{r-1} \sum_{1 \leq i_1 < \cdots < i_r \leq N} P(A_{i_1} \cap \cdots \cap A_{i_r}) \\ &= \sum_{i=1}^{N} P(A_i) - \sum_{1 \leq i < j \leq N} P(A_i \cap A_j) \\ &\quad + \cdots + (-1)^{N-1} P(A_1 \cap \cdots A_N). \end{aligned}$$

Note that the last term $P(A_1 \cap \cdots \cap A_N) = 0$ since it is impossible for no coupon to appear. Consider A_i. The probability that coupon i will not appear with a particular purchase is $1 - (1/N)$, and the probability that it will not appear in n purchases is $(1 - (1/N))^n$. Thus,

$$\sum_{i=1}^{N} P(A_i) = N\left(1 - \frac{1}{N}\right)^n = \binom{N}{1}\left(1 - \frac{1}{N}\right)^n.$$

Consider A_i and $A_j, i \neq j$. The probability that coupons i and j will not appear with a particular purchase is $(1 - (2/N))$, and the probability that they will not appear in n purchases is $(1 - (2/N))^n$. Since there are $\binom{N}{2}$ choices of i and j with $1 \leq i < j \leq N$,

$$\sum_{1 \leq i < j \leq N} P(A_i \cap A_j) = \binom{N}{2}\left(1 - \frac{2}{N}\right)^n.$$

Similarly,

$$\sum_{1 \leq i_1 < \cdots < i_r \leq N} P(A_{i_1} \cap \cdots \cap A_{i_r}) = \binom{N}{r}\left(1 - \frac{r}{N}\right)^n.$$

Therefore,

$$P(A_1 \cup \cdots \cup A_N) = \sum_{r=1}^{N-1} (-1)^{r-1} \binom{N}{r} \left(1 - \frac{r}{N}\right)^n.$$

For example, if $N = 6$ and $n = 25$, then $P(A_1 \cup \cdots \cup A_6) = .062$ and the probability of collecting a complete set of 6 coupons with 25 purchases is .938. ■

EXERCISES 2.7

1. A deck of cards numbered $1, 2, \ldots, 10$ is shuffled and the cards are dealt one by one onto positions $1, 2, \ldots, 10$. Calculate the exact probability of at least one match.
2. Consider an infinite number of Bernoulli trials with probability of success $p \neq 1/2, 0 < p < 1$. An *equalization* occurs as of some trial if there is an equal number of heads and tails. Equalization can occur only on an even number of trials. (a) If A_{2n} is the event "Equalization occurs on the $2n$ trial," show that

$$P(A_{2n}) = \binom{2n}{n} p^n (1-p)^n.$$

(b) What is the probability that an infinite number of equalizations will occur?

The next two problems relate to an infinite sequence of Bernoulli trials with probability of success 1/2. A *run* of length r beginning on the nth trial occurs if there are 1's on the n through $(n + r - 1)$ trials followed immediately by a 0. For integers $n, r \geq 1$, let $A_{n,r}$ be the event consisting of those outcomes for which there is a run of length greater than or equal to r beginning on the nth trial.

3. If r is a fixed positive integer, determine $P(A_{n,r}\ i.o.)$.
4. If for $n \geq 1$ and $\delta > 0$, $r_n = (1 + \delta) \log_2 n$, determine $P(A_{n,r_n}\ i.o.)$.

The following problems pertain to the game of craps, which is played with two dice according to the following rules:

- You win on the first roll if you roll a score of 7 or 11.
- You lose on the first roll if you roll a score of 2, 3, or 12.
- If you do not roll a 2, 3, 7, 11, or 12 on the first roll, the score becomes your "point" for subsequent rolls.
- You win on subsequent rolls if you roll your point without having rolled a 7 and lose if you roll a 7 without having rolled your point.

5. Assuming independence between trials, describe appropriate Ω, $\mathcal{F}$, and P.

6. What is the probability that you will win with a point of 8?
7. What is the probability that you will win at craps?
8. What is the probability that the game will terminate?

The following problems require mathematical software, such as Mathematica or Maple V, or much patience.

9. A deck of 52 cards numbered $1, 2, \ldots, 52$ is shuffled and the cards are dealt one by one onto positions $1, 2, \ldots, 52$. Calculate the probability of at least one match without using the approximation $1 - 1/e$.
10. A commercial product includes a coupon that can be either a worthless coupon or one of eight collectible coupons. If 30 percent of the coupons are worthless and the collectible coupons occur in equal proportions, how many products must be purchased to be 95 percent confident of obtaining a complete set of collectible coupons, assuming that the coupons are inserted randomly into the product?

SUPPLEMENTAL READING LIST

R. W. Hamming (1991). *The Art of Probability for Scientists and Engineers.* Redwood City, Calif.: Addison-Wesley.

CHAPTER 3

RANDOM VARIABLES

3.1 INTRODUCTION

The score obtained upon rolling two dice and the number of heads in n flips of a coin are examples of random variables. It is possible to forgo the apparatus of the first two chapters and deal directly with a primitive concept of random variables by specifying certain probability statements about the random variables. Eventually, however, the study of algebraic and limiting properties of random variables would lead to the considerations of the first two chapters.

One of the problems we will study in this chapter is the gambler's ruin problem, which apparently appeared in print for the first time in a paper by Huygens around the beginning of the eighteenth century. This problem was solved by James Bernoulli in a paper published posthumously in 1713. A more modern method, the method of difference equations, will be used to solve the ruin problem. Another important methodology for solving probability problems involves generating functions, which were introduced by de Moivre around 1740 and treated exhaustively by Laplace at the end of the eighteenth century. The reader wanting to learn more applications of generating functions, or of probability theory in general, would be well advised to read the book by Feller listed at the end of the chapter.

3.2 RANDOM VARIABLES

Unless otherwise specified, $(\Omega, \mathcal{F}, P)$ will be a fixed probability space. At this juncture, we are not going to give the most general definition of a random variable but will keep things as simple as possible.

Definition 3.1 *A map $X : \Omega \to R$ is a* random variable *if the range of X is a countable set $\{x_1, x_2, \ldots\}$, finite or infinite, and $\{\omega : X(\omega) = x_j\} \in \mathcal{F}$ for all $j \geq 1$.* ■

A random variable as just defined is customarily called a "discrete random variable," but the prefix "discrete" will be dropped because no other type of random variable will be considered until much later. The definition will be extended to allow the possibility that X can take on the value $+\infty$.

In all the probability models considered so far, except for an infinite sequence of Bernoulli trials, $\mathcal{F}$ consists of all subsets of Ω. When this is the case, $\{\omega : X(\omega) = x_j\}$ is just another subset of Ω and therefore is in $\mathcal{F}$. In most cases, showing that X is a random variable simply amounts to verifying that the range of X is countable.

The notation for the event $\{\omega : X(\omega) = x_j\}$ will be compressed to $(X = x_j)$ by suppressing the ω. The same is true for other events; e.g., the event $\{\omega : a < X(\omega) \leq b\}$ will be compressed to $(a < X \leq b)$.

Definition 3.2 *Let X be a random variable with range $\{x_1, x_2, \ldots\}$. The* density function f_X *is the real-valued function on the range of X defined by*

$$f_X(x_j) = P(X = x_j), \qquad j = 1, 2, \ldots \quad ■$$

The range of X can be finite or infinite. The density function f_X will be denoted simply by f if the meaning is clear from the context. It is important to keep in mind that the domains of f_X and P are not the same. P is a function on $\mathcal{F}$ with values in R, whereas f_X is a function on the range $\{x_1, x_2, \ldots\}$ of the random variable X with values in R.

EXAMPLE 3.1 Let X be the score upon rolling two dice. The function p defined on $\{2, 3, \ldots, 12\}$ as in Figure 1.3 is the density function of X; e.g., $f_X(7) = p(7) = 6/36$. ■

EXAMPLE 3.2 (Binomial Density) Consider n Bernoulli trials with probability of success p. Let X be the number of successes in n trials; i.e., if $\omega = \{x_j\}_{j=1}^n$ with $x_j \in \{0, 1\}$, then $X(\omega) = \sum_{j=1}^n x_j$. Since the range of X is $\{0, 1, \ldots, n\}$, X is a random variable. If $\sum_{j=1}^n x_j = k$, then there are k successes in ω and $n - k$ failures, and so ω has probability $p^k q^{n-k}$. But there are $\binom{n}{k}$ outcomes with this property, and so

$$f_X(k) = \binom{n}{k} p^k q^{n-k}, \qquad k = 0, 1, \ldots, n$$

is the density function. This density function is called the *binomial density function* with parameters n and p and is denoted by $b(\cdot; n, p)$. ■

EXAMPLE 3.3 Between a source S and a collector C there is an absorption medium as indicated below.

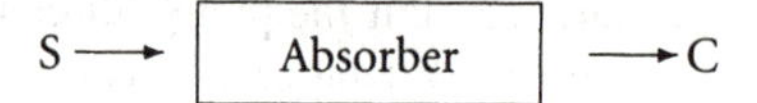

The probability that a given particle emitted from S is not absorbed is p, and the probability that it is absorbed is $1 - p, 0 < p < 1$. Assume that the particles are absorbed independently of each other. If n particles are emitted, the probability that exactly k particles will reach C is given by the binomial density

$$b(k; n, p) = \binom{n}{k} p^k (1-p)^{n-k}, \qquad k = 0, \ldots, n. \blacksquare$$

EXAMPLE 3.4 (Geometric Density Function) Consider an infinite sequence of Bernoulli trials with probability of success p as described in Example 2.14. If X is defined as the first trial at which success occurs, then we have a small problem in that $X(\omega_0)$ is not defined for the outcome $\omega_0 = (0, 0, \ldots)$ consisting of all 0's. It was shown in Example 2.14 that a success will eventually occur with probability 1. We can therefore define $X(\omega_0)$ however we choose, the result having no bearing on the computation of probabilities. We choose to define $X(\omega_0) = +\infty$. The range of X is then $N \cup \{+\infty\}$, which is countable. For $k \in N$,

$$(X = k) = \{\omega : \omega = \{x_j\}_{j=1}^{\infty}, x_1 = 0, \cdots, x_{k-1} = 0, x_k = 1\} \in \mathcal{F},$$

and since $(X = +\infty) = (\cup_{k=1}^{\infty}(X = k))^c \in \mathcal{F}$, X is a random variable. We saw in Section 2.4 that

$$f_X(k) = P(X = k) = pq^{k-1}, \qquad k = 1, 2, \ldots$$

This density function is called a *geometric density function* with parameters p and $q = 1 - p, 0 < p < 1$. ■

Note that the definition of a random variable has been extended in this example because Definition 3.1 requires that the values of X be real numbers and R does not contain $+\infty$. If the value $+\infty$ is allowed, the random variable is called an *extended real-valued random variable.* The only situation in which a random variable will be allowed to take on the value $+\infty$ is that in which X measures some waiting time as in the previous example. In either case, the criterion is still the same, because if $(X = x_j) \in \mathcal{F}$ for all real values x_j of

X, then $(X = +\infty) = (\cup_j (X = x_j))^c \in \mathcal{F}$, since $\mathcal{F}$ is a σ-algebra. In most instances, the original definition is applicable.

The geometric density is applicable to physical systems for which aging is not a factor. For example, the waiting time, in discrete units, for a radioactive atom to decay has a geometric density in which the parameter p can be determined from the half-life of the atom.

EXAMPLE 3.5 (Negative Binomial Density) Consider an infinite sequence of Bernoulli trials with probability of success p. Fix a positive integer r and let T_r be the trial at which the rth success occurs for the first time. If for the outcome ω the rth success never occurs, then we put $T_r(\omega) = +\infty$. The range of T_r is $\{r, r+1, \ldots, +\infty\}$. If $x \geq r$ is a positive integer, it is easy to see that $(T_r = x)$ is a condition on just finitely many trials, and so T_r is an extended real-valued random variable. Each outcome in the event $(T_r = x)$ has probability $p^r q^{x-r}$. Since $r - 1$ of the trials preceding the xth trial must be successes, the number of outcomes in this event is $\binom{x-1}{r-1}$. Thus,

$$P(T_r = x) = \binom{x-1}{r-1} p^r q^{x-r}, \qquad x = r, r+1, \ldots$$

Changing the scale by replacing x by $x + r$,

$$\begin{aligned} P(T_r - r = x) &= \binom{x+r-1}{r-1} p^r q^x \\ &= \binom{x+r-1}{x} p^r q^x, \qquad x = 0, 1, 2, \ldots \end{aligned}$$

By Exercise 1.3.10,

$$P(T_r - r = x) = \binom{-r}{x} p^r (-q)^x, \qquad x = 0, 1, 2, \ldots$$

It follows that $T_r - r$ has the density function

$$f(x) = \binom{-r}{x} p^r (-q)^x, \qquad x = 0, 1, 2, \ldots$$

which is called the *negative binomial density* with parameters r and p. The name arises from the fact that

$$(1-q)^{-r} = \sum_{x=0}^{\infty} \binom{-r}{x} (-q)^x$$

by the generalized binomial theorem (see Exercise 1.3.4). Accordingly,

$$\sum_{x=0}^{\infty} \binom{-r}{x} p^r(-q)^x = p^r(1-q)^{-r} = 1,$$

and it follows that $P(T_r < +\infty) = P(T_r \in \{r, r+1, \ldots\}) = 1$. This means that the rth success will eventually occur with probability 1. ■

Caveat: In the discussions that follow, assume that all random variables are real-valued unless explicitly stated otherwise.

The next density, the Poisson density, can be obtained as a limiting case of the binomial density as follows. Consider a sequence of experiments described by a binomial density for which the probability of success depends upon n; that is, consider a sequence of binomial densities $b(\cdot; n, p_n)$, $n \geq 1$. Assume that as n increases, p_n varies in such a way that $np_n \to \lambda > 0$ for some fixed λ. Fix an integer $k \geq 0$. Then

$$\lim_{n\to\infty} b(k; n, p_n) = \lim_{n\to\infty} \binom{n}{k} p_n^k (1-p_n)^{n-k}.$$

For large n, $p_n \approx \lambda/n$, $1 - p_n \approx 1 - (\lambda/n)$, and

$$\begin{aligned}
&\binom{n}{k} p_n^k (1-p_n)^{n-k} \\
&\approx \frac{(n)_k}{k!}\left(\frac{\lambda}{n}\right)^k \left(1 - \frac{\lambda}{n}\right)^{n-k} \\
&= \frac{\lambda^k}{k!} \frac{n(n-1) \times \cdots \times (n-k+1)}{n^k} \left(1 - \frac{\lambda}{n}\right)^n \left(1 - \frac{\lambda}{n}\right)^{-k} \\
&= \frac{\lambda^k}{k!}\left(1 - \frac{1}{n}\right)\left(1 - \frac{2}{n}\right) \times \cdots \times \left(1 - \frac{k-1}{n}\right)\left(1 - \frac{\lambda}{n}\right)^n \left(1 - \frac{\lambda}{n}\right)^{-k}.
\end{aligned}$$

Since k is fixed, there are a fixed number of factors in the last product, and the limit of the product is the product of the limits. Since $\lim_{n\to\infty}(1 - (j/n)) = 1$ for $j = 1, \ldots, k-1$, $\lim_{n\to\infty}(1 - (\lambda/n))^n = e^{-\lambda}$ from the calculus, and $\lim_{n\to\infty}(1 - (\lambda/n))^{-k} = 1$,

$$\lim_{n\to\infty} \binom{n}{k} p_n^k (1-p_n)^{n-k} = \frac{\lambda^k e^{-\lambda}}{k!}.$$

Therefore,

$$\lim_{n\to\infty} b(k; n, p_n) = \frac{\lambda^k e^{-\lambda}}{k!}, \qquad k \geq 0.$$

Can the function $f(k) = (\lambda^k e^{-\lambda})/k!, k \geq 0$, serve as the density of a random variable X?

According to Definition 3.1, the domain of the density function f_X is the range $\{x_1, x_2, \ldots\}$ of X. Whenever it is convenient to do so, we will define $f_X(x) = 0$ for $x \notin \{x_1, x_2, \ldots\}$. With this convention in mind, a density function has the following properties:

$$0 \leq f(x) \leq 1 \qquad \text{for all } x \in R. \tag{3.1}$$

There is a countable set $\{x_1, x_2, \ldots\}$ such that $\sum_j f(x_j) = 1$ and $f(x) = 0$ whenever $x \notin \{x_1, x_2, \ldots\}$. (3.2)

Conversely, given such a function, we can construct a probability space $(\Omega, \mathcal{F}, P)$ by taking $\Omega = \{x_1, x_2, \ldots\}$, $\mathcal{F}$ the collection of all subsets of Ω, and defining P using the weight function $f(x_j)$ as in Section 1.5. The random variable X defined on Ω by $X(x_j) = x_j$ then has f as its density function. To show that this construction can be applied to the function $f(k)$ of the previous paragraph, we need the Maclaurin series expansion

$$e^\lambda = \sum_{k=0}^{\infty} \frac{\lambda^k}{k!}. \tag{3.3}$$

EXAMPLE 3.6 (Poisson Density Function) Fix a positive number λ and let

$$f(k) = \begin{cases} (\lambda^k e^{-\lambda})/k! & \text{if } k = 0, 1, \ldots \\ 0 & \text{otherwise.} \end{cases}$$

Clearly, $f(x) \geq 0$ for all $x \in R$. Since $\{0, 1, \ldots\}$ is a countable set, $f(x) = 0$ for all $x \notin \{0, 1, \ldots\}$, and $\sum_{k=0}^{\infty} (\lambda^k e^{-\lambda})/k! = 1$ by Equation 3.3, the function f satisfies 3.1 and 3.2. Thus, there is a probability space $(\Omega, \mathcal{F}, P)$ and a random variable X having f as its density function. This density function is called a *Poisson density function* with parameter λ and is usually denoted by $p(\cdot; \lambda)$. ■

The Poisson density is usually applied in situations in which there are a large number n of trials with a small probability p that an event will occur in each trial and with $\lambda = np$ moderate in magnitude.

EXAMPLE 3.7 An electronic system has a periodic operating cycle of 0.01 second. In each of the cycles, an event can occur with probability .001. What is the probability of observing fewer than 15 events in a 100-second time interval? During 100 seconds, 10,000 cycles will be observed. Letting $\lambda = 10{,}000(.001) = 10$, the probability that k events will be observed is given by the Poisson density $p(k; 10)$. The required probability is

$$\sum_{k=0}^{14} p(k; 10) = \sum_{k=0}^{14} \frac{10^k e^{-10}}{k!} \approx .9165. \blacksquare$$

EXAMPLE 3.8 (Uniform Density Function) For fixed $n \in N$, let $\Omega = \{1, 2, \ldots, n\}$ and let $f(k) = 1/n$ for $k = 1, 2, \ldots, n$. Then f satisfies 3.1 and 3.2, and there is a random variable X having f as its density function. $\blacksquare$

If A is any set of real numbers and X is a random variable, then

$$(X \in A) = \{\omega : X(\omega) \in A\} = \cup_{x_j \in A}\{\omega : X(\omega) = x_j\}$$

belongs to $\mathcal{F}$ since each set $\{\omega : X(\omega) = x_j\} \in \mathcal{F}$. Since the events in the union are disjoint,

$$P(X \in A) = \sum_{x_j \in A} P(X = x_j) = \sum_{x_j \in A} f_X(x_j). \tag{3.4}$$

This equation allows us to compute probabilities related to the random variable X.

EXAMPLE 3.9 Suppose X has a geometric density with parameters p and $q = 1 - p$. Then

$$P(X \geq 10) = P(X \in [10, \infty)) = \sum_{j=10}^{\infty} pq^{j-1} = q^9. \blacksquare$$

Let X be any random variable and let φ be a real-valued function on R. Given any $\omega \in \Omega$, it makes sense to form the composite function $\varphi(X(\omega))$. This composite map is denoted by $\varphi(X)$. The range of $\varphi(X)$ is the countable set $\{\varphi(x_1), \varphi(x_2), \ldots\}$. Let y be in the range of $\varphi(X)$. To show that $(\varphi(X) = y) \in \mathcal{F}$, let $x_{i_1}, x_{i_2}, \ldots$ be those values of X for which $\varphi(x_{i_j}) = y$. Then $(\varphi(X) = y) = \cup_j (X = x_{i_j}) \in \mathcal{F}$, since X is a random variable. This shows that $\varphi(X)$ is a random variable.

EXAMPLE 3.10 If X is a random variable, then X^2 is the random variable defined for each ω by $X^2(\omega) = (X(\omega))^2$, $\sin X$ is the random variable defined for each ω by $(\sin X)(\omega) = \sin(X(\omega))$, $|X|$ is the random variable defined for each ω by $|X|(\omega) = |X(\omega)|$, and so forth. $\blacksquare$

Given a random variable X and a real-valued function φ on R, how do we determine the density function f_Z of the random variable $Z = \varphi(X)$? There is no algorithm for generating the density function f_Z.

EXAMPLE 3.11 Let X be a random variable having the geometric density

$$f_X(x) = pq^{x-1}, \qquad x = 1, 2, \ldots$$

and let $Y = \min(X, 5)$. The range of Y is the set $\{1, 2, 3, 4, 5\}$. If $X(\omega)$ is 1, 2, 3, or 4, then $Y(\omega) = X(\omega)$ and $f_Y(x) = f_X(x), x = 1, 2, 3, 4$. If $X(\omega) \geq 5$, then $Y(\omega) = \min(X(\omega), 5) = 5$ so that $(Y = 5) = (X \geq 5) = \cup_{x=5}^{\infty}(X = x)$. Therefore,

$$f_Y(5) = \sum_{x=5}^{\infty} pq^{x-1} = q^4.$$

Therefore,

$$f_Y(y) = \begin{cases} pq^{y-1} & \text{if } y = 1, 2, 3, 4 \\ q^4 & \text{if } y = 5 \\ 0 & \text{otherwise.} \end{cases}$$ ■

Consider now two random variables X and Y on the same probability space with ranges $\{x_1, x_2, \ldots\}$ and $\{y_1, y_2, \ldots\}$, respectively, and let ψ be a real-valued function of two variables. Then for each $\omega \in \Omega$, $\psi(X(\omega), Y(\omega))$ defines a new map from Ω to R that is denoted by $\psi(X, Y)$. The range of $\psi(X, Y)$ is the set $\{\psi(x_i, y_j) : i \geq 1, j \geq 1\}$, which is a subset of the set $\cup_{i \geq 1}\{\psi(x_i, y_j) : j \geq 1\}$, which is countable by Theorem 2.3.1 and is therefore countable. Let z be any value of $\psi(X, Y)$ and let $(x_{i_1}, y_{j_1}), (x_{i_2}, y_{j_2}), \ldots$ be those ordered pairs for which $\psi(x_{i_k}, y_{j_k}) = z$. Then

$$\begin{aligned}(\psi(X, Y) = z) &= \bigcup_k \{\omega : X(\omega) = x_{i_k}, Y(\omega) = y_{j_k}\} \\ &= \bigcup_k ((X = x_{i_k}) \cap (Y = y_{j_k})) \in \mathcal{F}\end{aligned}$$

since X and Y are random variables. Thus, $Z = \psi(X, Y)$ is a random variable.

EXAMPLE 3.12 If X and Y are random variables, then $X + Y$, $X - Y$, XY, $X^2 + Y^2$, $\max(X, Y)$, $\min(X, Y)$, $\sin XY$, and so forth, are all random variables. ■

More generally, if $X_1, \ldots, X_n$ are n random variables and ψ is a real-valued function of n variables, then $\psi(X_1, \ldots, X_n)$ defined by

$$\psi(X_1, \ldots, X_n)(\omega) = \psi(X_1(\omega), \ldots, X_n(\omega))$$

for $\omega \in \Omega$ is a random variable. Finding the density function of $Z = \psi(X_1, \ldots, X_n)$ can be difficult, depending upon ψ. We will show how this can be done when $n = 2$ in some cases using the joint density of two random variables.

Definition 3.3 *Let X and Y be random variables with ranges $\{x_1, x_2, \ldots\}$ and $\{y_1, y_2, \ldots\}$, respectively. The* joint density function $f_{X,Y}$ *of X and Y is defined on the set of ordered pairs $\{(x_i, y_j) : i \geq 1, j \geq 1\}$ by*

$$f_{X,Y}(x_i, y_j) = P(X = x_i, Y = y_j), \qquad i, j \geq 1. \blacksquare$$

Equation 3.4 can be extended to two or more random variables. Let X and Y be two random variables with ranges $\{x_1, x_2, \ldots\}$ and $\{y_1, y_2, \ldots\}$, respectively, and let A be any subset of $R \times R$. Since $((X, Y) \in A) = \cup_{(x_i, y_j) \in A}(X = x_i, Y = y_j)$,

$$P((X, Y) \in A) = \sum_{(x_i, y_j) \in A} P(X = x_i, Y = y_j) = \sum_{(x_i, y_j) \in A} f_{X,Y}(x_i, y_j). \quad (3.5)$$

In calculating probabilities pertaining to the pair X, Y, there is some latitude in the choice of A. The set A can usually be defined by replacing X and Y by typical values x_i and y_j, respectively.

EXAMPLE 3.13 Suppose two dice are rolled, one red and one white. Let X be the number of pips on the red die, let Y be the number on the white die, and let Z be the maximum of the two numbers; i.e., $Z = \max(X, Y)$. The joint density of X and Y is

$$f_{X,Y}(x, y) = \frac{1}{36}, \qquad x, y = 1, 2, \ldots, 6.$$

Suppose we want to find the joint density of X and Z. Both X and Z have range $\{1, 2, \ldots, 6\}$. Let x and z be typical values of X and Z, respectively. Since $Z \geq X$, $f_{X,Z}(x, z) = P(X = x, Z = z) = 0$ if $z < x$. If $z = x$, then we must have $Y \leq x$. Hence, $(X = x, Z = x) = (X = x, Y \leq x)$. To put the event $(X = x, Y \leq x)$ into the form $((X, Y) \in A)$, formally replace X by i and Y by j in the first event to define $A = \{(i, j) : i = x, j \leq x\}$. Thus, by Equation 3.5,

$$f_{X,Z}(x, z) = P(X = x, Z = x) = P((X, Y) \in A) = \sum_{i = x, 1 \leq j \leq x} \frac{1}{36} = \frac{x}{36}$$

whenever $z = x$. If $z > x$, the event $(X = x, Z = z)$ can occur only if $Y = z$; i.e., $(X = x, Z = z) = (X = x, Y = z)$ whenever $z > x$, and thus $f_{X,Z}(x, z) = P(X = x, Y = z) = 1/36$ whenever $z > x$. In summary,

$$f_{X,Z}(x,z) = \begin{cases} 0 & \text{if } z < x \\ x/36 & \text{if } z = x \\ 1/36 & \text{if } z > x. \ \blacksquare \end{cases} \tag{3.6}$$

Definition 3.4 *The* joint density function $f_{X_1,\ldots,X_n}$ *of the random variables* $X_1, \ldots, X_n$ *is the real-valued function*

$$f_{X_1,\ldots,X_n}(x_1, \ldots, x_n) = P(X_1 = x_1, \ldots, X_n = x_n). \ \blacksquare$$

Of course, if any x_i is not in the range of X_i, then the probability on the right is zero. If it is clear from context, the joint density will be denoted simply by $f(x_1, \ldots, x_n)$, keeping in mind that the order of the variables in the argument of f corresponds to the order of the random variables.

If the joint density $f_{X_1,\ldots,X_n}$ of $X_1, \ldots, X_n$ is known, then the joint density of any subcollection of the X_j's can be determined. For example, suppose we want to determine the joint density of $X_1, \ldots, X_{n-1}$. Let $\{x_{n1}, x_{n2}, \ldots\}$ be the range of X_n. Then

$$\Omega = \bigcup_k (X_n = x_{nk}).$$

Intersecting both sides of this equation by $(X_1 = x_1, \ldots, X_{n-1} = x_{n-1})$,

$$(X_1 = x_1, \ldots, X_{n-1} = x_{n-1}) = \bigcup_k (X_1 = x_1, \ldots, X_{n-1} = x_{n-1}, X_n = x_{nk}).$$

Since the events on the right are disjoint,

$$f_{X_1,\ldots,X_{n-1}}(x_1, \ldots, x_{n-1}) = \sum_k f_{X_1,\ldots,X_n}(x_1, \ldots, x_{n-1}, x_{nk}).$$

Alternatively,

$$f_{X_1,\ldots,X_{n-1}}(x_1, \ldots, x_{n-1}) = \sum_{x_n} f_{X_1,\ldots,X_n}(x_1, \ldots, x_{n-1}, x_n) \tag{3.7}$$

since $f_{X_1,\ldots,X_n}(x_1, \ldots, x_{n-1}, x_n) = 0$ whenever x_n is not one of the values of X_n. This procedure can be repeated as often as necessary to obtain the joint density of any subcollection.

EXAMPLE 3.14 A point with integer coordinates (X, Y) is chosen at random from the triangle with vertices at $(1, 1)$, $(n, 1)$, and (n, n) where n is a fixed positive integer. Since the total number of points with integer coordinates (x, y) in the triangle is $1 + 2 + \cdots + n = (n(n+1))/2$, the density function of the pair (X, Y) is

$$f(x, y) = \begin{cases} 2/(n(n+1)) & \text{if } 1 \le y \le x, x = 1, 2, \ldots, n \\ 0 & \text{otherwise.} \end{cases}$$

For $x = 1, 2, \ldots, n$, the density $f_X(x)$ is given by

$$f_X(x) = \sum_{y=1}^{x} \frac{2}{n(n+1)} = \frac{2x}{n(n+1)}.$$

Thus,

$$f_X(x) = \begin{cases} 2x/(n(n+1)) & \text{if } x = 1, 2, \ldots, n \\ 0 & \text{otherwise.} \end{cases}$$

For $y = 1, 2, \ldots, n$, the density $f_Y(y)$ is given by

$$f_Y(y) = \sum_{x=y}^{n} \frac{2}{n(n+1)} = \frac{2(n-y+1)}{n(n+1)}.$$

Thus,

$$f_Y(y) = \begin{cases} (2(n-y+1))/(n(n+1)) & \text{if } y = 1, 2, \ldots, n \\ 0 & \text{otherwise.} \end{cases} \quad \blacksquare$$

EXERCISES 3.2

1. A bus tour operator uses a bus with a capacity of 45 passengers but sells 50 tickets. If one person out of 12 is a no-show, what is the probability that everyone who shows up for the tour will be accommodated?
2. What is the maximum number of tickets the bus tour operator should sell to be able to accommodate all that show up with probability at least equal to .90?
3. An electronic system has an operating cycle of 0.01 second. During successive time intervals of length 5×10^{-6}, an event may occur with probability $p = .0005$. What is the approximate probability that fewer than 8 events will occur during 10 cycles?
4. The random variables X and Y have the joint density function

$$f_{X,Y}(x, y) = \begin{cases} 2/(n(n+1)) & \text{if } 1 \le y \le -x + n + 1, 1 \le x \le n \\ 0 & \text{otherwise} \end{cases}$$

where x and y are positive integers. Find $f_X(x)$ and $f_Y(y)$.

5. Suppose a pair of dice are rolled and Z is the larger of the number of pips on each. What is the density of Z?

6. Denote the general term of the binomial density with parameters n and p by

$$b(k; n, p) = \binom{n}{k} p^k q^{n-k}, \qquad k = 0, 1, \ldots, n.$$

Find a recursion formula for calculating $b(k; n, p)$ from $b(k-1; n, p)$, and put the ratio of the two in the form $1 + \cdots$. For what value or values of k is $b(k; n, p)$ a maximum?

7. The two random variables X and Y have the joint density

$$f_{X,Y}(x, y) = \frac{\alpha^x \beta^y e^{-(\alpha+\beta)}}{x!y!}, \qquad x, y = 0, 1, 2, \ldots$$

where $\alpha, \beta > 0$. What are the densities of X and Y?

8. Suppose the random variables X and Y have ranges $\{x_1, x_2, \ldots\}$ and $\{y_1, y_2, \ldots\}$, respectively, and their joint density has the form

$$f_{X,Y}(x_i, y_j) = f(x_i)g(y_j), \qquad i, j = 1, 2, \ldots$$

Express the densities of X and Y in terms of f and g.

9. The random variables X and Y have the joint density function tabulated below. Find the densities of X and Y and calculate $P(Y \geq X)$.

	$y = 1$	$y = 2$	$y = 3$	$y = 4$	$y = 5$	
$x = 1$	.03	.07	.04	.06	.05	
$x = 2$	.01	.01	.08	.09	.06	
$x = 3$	.07	.06	.06	.03	.03	
$x = 4$	.01	.13	.02	.07	.02	

10. The cards of a deck are numbered $1, 2, \ldots, 50$. The deck is thoroughly shuffled and then two cards are dealt. Let X and Y be the numbers on the first and second cards, respectively. Use Equation 3.5 to calculate $P(|X - Y| \geq 2)$.

The following problems require mathematical software such as Mathematica or Maple V.

11. A jumbo jet with a capacity of 365 passengers is oversold by 10 tickets. If one person out of 25 is a no-show, what is the probability that all of those who show up will board the jet?

12. What is the maximum number of tickets the airline should sell to be able to accommodate all who show up with probability .99?

3.3 INDEPENDENT RANDOM VARIABLES

The discussion of the empirical law in Chapter 1 does not include the phrase "under identical conditions," which is usually a part of the discussion. The phrase means that an experiment should be repeated in such a way that the outcome of a repetition should not be influenced by the outcome of previous repetitions; i.e., it should be independent of previous outcomes of the experiment. If $X_1, X_2, \ldots$ denote the outcomes of successive repetitions, then the events $(X_1 = x_1), (X_2 = x_2), \ldots$ should be independent events. The formal definition will be given in terms of joint densities.

Definition 3.5 *The random variables* $X_1, \ldots, X_n$ *are* independent *if*

$$f_{X_1,\ldots,X_n}(x_1, \ldots, x_n) = f_{X_1}(x_1) \times f_{X_2}(x_2) \times \cdots \times f_{X_n}(x_n)$$

for all $x_1, \ldots, x_n \in R$. ■

That is, $X_1, \ldots, X_n$ are independent if their joint density is the product of their individual densities. We frequently use the fact that if $X_1, \ldots, X_n$ are independent random variables and $\{i_1, \ldots, i_k\}$ is a subset of $\{1, 2, \ldots, n\}$, then $X_{i_1}, \ldots, X_{i_k}$ are independent random variables. Consider, for example, $X_1, \ldots, X_{n-1}$. By Equation 3.7,

$$\begin{aligned}
f_{X_1,\ldots,X_{n-1}}(x_1, \ldots, x_{n-1}) &= \sum_{x_n} f_{X_1,\ldots,X_n}(x_1, \ldots, x_{n-1}, x_n) \\
&= \sum_{x_n} f_{X_1}(x_1) \times \cdots \times f_{X_{n-1}}(x_{n-1}) f_{X_n}(x_n) \\
&= f_{X_1}(x_1) \times \cdots \times f_{X_{n-1}}(x_{n-1}) \sum_{x_n} f_{X_n}(x_n) \\
&= f_{X_1}(x_1) \times \cdots \times f_{X_{n-1}}(x_{n-1}).
\end{aligned}$$

Thus, $X_1, \ldots, X_{n-1}$ are independent random variables.

EXAMPLE 3.15 Consider the roll of two dice, one red and one white. Let X be the number of pips on the red die and Y the number on the white die. Since $f_X(x) = P(X = x) = 1/6$ for $x = 1, \ldots, 6$, $f_Y(y) = P(Y = y) = 1/6$ for $y = 1, \ldots, 6$, and $f_{X,Y}(x, y) = 1/36$ for $1 \le x, y \le 6$,

$$f_{X,Y}(x,y) = f_X(x)f_Y(y) \qquad \text{for all } x, y \in R.$$

Thus, X and Y are independent random variables. ■

Independence of random variables X and Y implies much more. Suppose A and B are any sets of real numbers and the ranges of X and Y are $\{x_1, x_2, \ldots\}$ and $\{y_1, y_2, \ldots\}$, respectively. Then

$$P(X \in A, Y \in B) = P(X \in A)P(Y \in B).$$

This follows from independence, since

$$\begin{aligned} P(X \in A, Y \in B) &= \sum_{x_i \in A, y_j \in B} f_{X,Y}(x_i, y_j) \\ &= \sum_{x_i \in A, y_j \in B} f_X(x_i)f_Y(y_j) \\ &= \left(\sum_{x_i \in A} f_X(x_i)\right)\left(\sum_{y_j \in B} f_Y(y_j)\right) \\ &= P(X \in A)P(Y \in B). \end{aligned}$$

For example,

$$P(X \geq x, Y \geq y) = P(X \geq x)P(Y \geq y).$$

More generally, if $X_1, X_2, \ldots, X_n$ are independent random variables and $A_1, A_2, \ldots, A_n$ are any sets of real numbers, then

$$P(X_1 \in A_1, \ldots, X_n \in A_n) = P(X_1 \in A_1) \times \cdots \times P(X_n \in A_n).$$

EXAMPLE 3.16 Consider n Bernoulli trials with probability of success p. For $j = 1, \ldots, n$, let $X_j = 1$ if there is a success on the jth trial and let $X_j = 0$ if there is a failure on the jth trial. Note that the set $(X_j = x_j)$ involves a condition imposed solely on the jth trial. It follows that the events $(X_1 = x_1), \ldots, (X_n = x_n)$ are mutually independent, and therefore

$$P(X_1 = x_1, \ldots, X_n = x_n) = P(X_1 = x_1) \times \cdots \times P(X_n = x_n);$$

that is,

$$f_{X_1, \ldots, X_n}(x_1, \ldots, x_n) = f_{X_1}(x_1) \times \cdots \times f_{X_n}(x_n),$$

and therefore $X_1, \ldots, X_n$ are independent random variables. ■

Example 3.16 is a special case of a more general model. Instead of allowing each trial to have just two outcomes 0 and 1, we could allow r outcomes on each trial. For example, in eight repeated tosses of a die, the outcome of each toss is one of the integers $1, 2, \ldots, 6$. A typical outcome might look like $\omega = (3, 5, 2, 1, 2, 3, 3, 4)$. As in the Bernoulli case, we can define random variables $X_1, X_2, \ldots, X_8$ to specify outcomes of individual tosses; i.e., for this outcome, $X_1(\omega) = 3, X_2(\omega) = 5, \ldots, X_8(\omega) = 4$. In the Bernoulli case, we also counted the number of successes; in tossing a die, we could count the number of times Y_i the outcome i appears among the eight tosses of the die. For the above outcome, $Y_1(\omega) = 1, Y_2(\omega) = 2, Y_3(\omega) = 3, Y_4(\omega) = 1, Y_5(\omega) = 1, Y_6(\omega) = 0$.

EXAMPLE 3.17 (Multinomial Density Function) Consider a basic experiment with r outcomes that we choose to label as $1, 2, \ldots, r$ having probabilities $p_1, \ldots, p_r$, respectively. Consider the compound experiment of n independent repetitions of this basic experiment. An outcome ω of the compound experiment is an ordered n-tuple $\omega = (i_1, \ldots, i_n)$ where each $i_j \in \{1, 2, \ldots, r\}$. We associate with each $\omega = (i_1, \ldots, i_n)$ the weight $p(\omega) = p_{i_1} \times \cdots \times p_{i_n}$. Letting Ω be the collection of all such outcomes, Ω is finite and we can take for $\mathcal{F}$ the collection of all subsets of Ω. For $j = 1, \ldots, n$, we can define a random variable X_j by putting $X_j(\omega) = i_j$ whenever $\omega = (i_1, \ldots, i_n)$. Probabilities have been defined so that

$$P(X_1 = i_1, \ldots, X_n = i_n) = p_{i_1} \times \cdots \times p_{i_n}$$

for $i_j \in \{1, \ldots, r\}, j = 1, \ldots, n$. Since $\sum_{j=1}^{r} p_j = 1$,

$$\begin{aligned}
P(X_1 = i_1) &= \sum_{i_2,\ldots,i_n=1}^{r} P(X_1 = i_1, X_2 = i_2, \ldots, X_n = i_n) \\
&= \sum_{i_2,\ldots,i_n=1}^{r} p_{i_1} \times \cdots \times p_{i_n} \\
&= \sum_{i_2=1}^{r} \cdots \sum_{i_{n-1}=1}^{r} \left(\sum_{i_n=1}^{r} p_{i_1} \times \cdots \times p_{i_n} \right) \\
&= \sum_{i_2=1}^{r} \cdots \sum_{i_{n-1}=1}^{r} p_{i_1} \times \cdots \times p_{i_{n-1}} \left(\sum_{i_n=1}^{r} p_{i_n} \right) \\
&= \sum_{i_2=1}^{r} \cdots \sum_{i_{n-1}=1}^{r} p_{i_1} \times \cdots \times p_{i_{n-1}} \\
&\ \ \vdots \\
&= p_{i_1} \sum_{i_2=1}^{r} p_{i_2} \\
&= p_{i_1}.
\end{aligned}$$

Similarly, $P(X_j = i_j) = p_{i_j}, j = 1, \ldots, n$. Thus,

$$P(X_1 = i_1, \ldots, X_n = i_n) = P(X_1 = i_1) \times \cdots \times P(X_n = i_n),$$

and $X_1, \ldots, X_n$ are independent random variables. For $1 \leq k \leq r$, let $Y_k(\omega)$ be the number of k's in the outcome $\omega = (i_1, \ldots, i_n)$. Note that $Y_1 + \cdots + Y_r = n$. Let $n_1, \ldots, n_r$ be nonnegative integers with $n = n_1 + \cdots + n_r$. Any outcome ω for which the number of 1's is n_1, the number of 2's is n_2, and so on, has probability $p_1^{n_1} \times \cdots \times p_r^{n_r}$. Since there are many outcomes fitting this description,

$$P(Y_1 = n_1, \ldots, Y_r = n_r) = C(n; n_1, \ldots, n_r) p_1^{n_1} \times \cdots \times p_r^{n_r}$$

where $C(n; n_1, \ldots, n_r)$ is the number of such outcomes. We can calculate this constant as follows. The number of ways of selecting n_1 positions out of the n positions to be filled with 1's is $\binom{n}{n_1}$; having done this, the number of ways of selecting n_2 positions out of the remaining $n - n_1$ to be filled with 2's is $\binom{n - n_1}{n_2}$, and so forth. Thus,

$$\begin{aligned} &P(Y_1 = n_1, \ldots, Y_r = n_r) \\ &= \binom{n}{n_1}\binom{n - n_1}{n_2} \times \cdots \times \binom{n - n_1 - \cdots - n_{r-1}}{n_r} p_1^{n_1} \times \cdots \times p_r^{n_r}. \end{aligned}$$

Expressing the binomial coefficients in terms of factorials and simplifying,

$$P(Y_1 = n_1, \ldots, Y_r = n_r) = \frac{n!}{n_1! n_2! \cdots n_r!} p_1^{n_1} \times \cdots \times p_r^{n_r}.$$

This joint density of $Y_1, \ldots, Y_r$ is called the *multinomial density*. ■

EXAMPLE 3.18 Suppose a die is tossed 12 times in succession. The probability that there will be two 1's, one 2, four 3's, one 4, three 5's, and one 6 is

$$\begin{aligned} P(Y_1 = 2, Y_2 = 1, Y_3 = 4, Y_4 = 1, Y_5 = 3, Y_6 = 1) &= \frac{12!}{2!1!4!1!3!1!} \frac{1}{6^{12}} \\ &\approx .00076. \end{aligned}$$ ■

Definition 3.6 *The random variables of the sequence $X_1, X_2, \ldots$ are* independent *if for every $n \geq 1, X_1, X_2, \ldots, X_n$ are independent random variables.* ■

Definition 3.7 *The sequence of random variables $X_1, X_2, \ldots$ is called an* infinite sequence of Bernoulli random variables *with probability of success p if they are independent, $P(X_j = 1) = p$, and $P(X_j = 0) = 1 - p$ for all $j \geq 1$.* ■

EXAMPLE 3.19 Consider an infinite sequence of Bernoulli trials with probability of success p as described in Example 2.14. For $j \geq 1$, let $X_j = 1$ if there is a success on the jth trial and let $X_j = 0$ if there is a failure on the jth trial. It was shown in Example 2.22 that the events $(X_1 = x_1), \ldots, (X_n = x_n)$ are mutually independent for every $n \geq 1$. Thus, the random variables of the sequence $X_1, X_2, \ldots$ are independent. It was shown in Example 2.22 that

$$P(X_1 = x_1, \ldots, X_n = x_n) = p^{\sum_{i=1}^n x_i} q^{(n - \sum_{i=1}^n x_i)}.$$

The probability $P(X_2 = x_1, \ldots, X_{n+1} = x_n)$ is also equal to the product on the right side of this equation. More generally, we have the following property of the joint density of n consecutive X_j's:

$$f_{X_1, \ldots, X_n}(x_1, \ldots, x_n) = f_{X_{k+1}, \ldots, X_{k+n}}(x_1, \ldots, x_n) \tag{3.8}$$

for all $n, k \geq 1$; i.e., the joint density of $X_k, X_{k+1}, \ldots, X_{k+n}$ is independent of k. In particular, the probability of getting r successes in the first n trials is the same as the probability of getting r successes in any n trials. ■

Theorem 3.3.1 *Let X and Y be independent random variables with ranges $\{x_1, x_2, \ldots\}$ and $\{y_1, y_2, \ldots\}$, respectively, and let $Z = X + Y$. Then*

$$f_Z(z) = \sum_i f_X(x_i) f_Y(z - x_i) = \sum_j f_X(z - y_j) f_Y(y_j). \tag{3.9}$$

PROOF: For fixed $z \in R$, $f_Z(z) = P(Z = z) = P(X + Y = z)$. Now stratify the event $(X + Y = z)$ according to the values of X; i.e., write

$$(X + Y = z) = \bigcup_i (X + Y = z, X = x_i).$$

Since $(X + Y = z, X = x_i) = (Y = z - x_i, X = x_i)$,

$$(X + Y = z) = \bigcup_i (X = x_i, Y = z - x_i).$$

Since the events on the right are disjoint,

$$P(X + Y = z) = \sum_i P(X = x_i, Y = z - x_i).$$

By independence of X and Y,

$$f_Z(z) = \sum_i f_{X,Y}(x_i, z - x_i) = \sum_i f_X(x_i) f_Y(z - x_i).$$

A similar argument applies to the second assertion. ■

If the random variables of Theorem 3.3.1 are of a particular type, the formula takes on a simpler form.

Definition 3.8 *The random variable X is* nonnegative integer-valued *if $f_X(x) = 0$ whenever $x \notin \{0, 1, 2, \ldots\}$.* ■

Theorem 3.3.2 *If X and Y are independent nonnegative integer-valued random variables and $Z = X + Y$, then*

$$f_Z(z) = \begin{cases} \sum_{x=0}^{z} f_X(x) f_Y(z - x) & \text{if } z = 0, 1, 2, \ldots \\ 0 & \text{otherwise.} \end{cases}$$

PROOF: Note that Z is also nonnegative integer-valued. By Theorem 3.3.1, for $z = 0, 1, 2, \ldots,$

$$f_Z(z) = \sum_{x=0}^{\infty} f_X(x) f_Y(z - x).$$

Note that $f_Y(z - x) = 0$ whenever $x > z$ and the infinite limit can be replaced by z. ■

EXAMPLE 3.20 Let X and Y be independent random variables having Poisson densities with parameters λ_1 and λ_2, respectively. Let $Z = X + Y$. Consider any $z \in \{0, 1, 2, \ldots\}$. By Theorem 3.3.2 and the binomial theorem,

$$\begin{aligned} f_Z(z) &= \sum_{x=0}^{z} \frac{\lambda_1^x e^{-\lambda_1}}{x!} \frac{\lambda_2^{z-x} e^{-\lambda_2}}{(z - x)!} \\ &= \frac{e^{-(\lambda_1+\lambda_2)}}{z!} \sum_{x=0}^{z} \binom{z}{x} \lambda_1^x \lambda_2^{z-x} \\ &= \frac{e^{-(\lambda_1+\lambda_2)}}{z!} (\lambda_1 + \lambda_2)^z. \end{aligned}$$

It follows that Z has a Poisson density with parameter $\lambda_1 + \lambda_2$. ■

EXAMPLE 3.21 Let X and Y be independent random variables having binomial densities $b(\cdot; m, p)$ and $b(\cdot; n, p)$, respectively. The range of $Z =$

$X + Y$ is then $\{0, 1, \ldots, m + n\}$. Suppose z is in the range of Z. By Theorem 3.3.2 and Equation 1.11,

$$\begin{aligned}
f_Z(z) &= \sum_{x=0}^{z} b(x; m, p)b(z - x; n, p) \\
&= \sum_{x=0}^{z} \binom{m}{x} p^x q^{m-x} \binom{n}{z-x} p^{z-x} q^{n-z+x} \\
&= p^z q^{(m+n)-z} \sum_{x=0}^{z} \binom{m}{x}\binom{n}{z-x} \\
&= \binom{m+n}{z} p^z q^{(m+n)-z}.
\end{aligned}$$

Thus, Z has the binomial density $b(\cdot; m + n, p)$. ■

The last two examples have extensions to sums of finitely many random variables. The following lemma will be needed. Recall that a density function $f_X(x)$ is zero when x is not in the range of X.

Lemma 3.3.3 *If $X_1, X_2, \ldots, X_n$ are independent random variables, then $X_1 + X_2 + \cdots + X_{n-1}$ and X_n are independent random variables.*

PROOF: ($n = 4$ case) Let α and β be values of $X_1 + X_2 + X_3$ and X_4, respectively. Then

$$\begin{aligned}
&f_{X_1+X_2+X_3,X_4}(\alpha, \beta) \\
&= P(X_1 + X_2 + X_3 = \alpha, X_4 = \beta) \\
&= \sum_{x_3} P(X_1 + X_2 + X_3 = \alpha, X_3 = x_3, X_4 = \beta) \\
&= \sum_{x_3} P(X_1 + X_2 = \alpha - x_3, X_3 = x_3, X_4 = \beta) \\
&= \sum_{x_3}\sum_{x_2} P(X_1 = \alpha - x_3 - x_2, X_2 = x_2, X_3 = x_3, X_4 = \beta).
\end{aligned}$$

Since X_1, X_2, X_3, and X_4 are independent random variables,

$$\begin{aligned}
&f_{X_1+X_2+X_3,X_4}(\alpha, \beta) \\
&= \sum_{x_3}\sum_{x_2} P(X_1 = \alpha - x_3 - x_2)P(X_2 = x_2)P(X_3 = x_3)P(X_4 = \beta).
\end{aligned}$$

Using the fact that X_1, X_2, and X_3 are independent random variables,

$$\begin{aligned}
&f_{X_1+X_2+X_3,X_4}(\alpha,\beta)\\
&= \sum_{x_3}\sum_{x_2} P(X_1 = \alpha - x_3 - x_2, X_2 = x_2, X_3 = x_3)P(X_4 = \beta)\\
&= \left(\sum_{x_3}\sum_{x_2} P(X_1 + X_2 + X_3 = \alpha, X_2 = x_2, X_3 = x_3)\right)P(X_4 = \beta).
\end{aligned}$$

Applying Equation 3.7 two times in succession,

$$f_{X_1+X_2+X_3,X_4}(\alpha,\beta) = f_{X_1+X_2+X_3}(\alpha)f_{X_4}(\beta). \blacksquare$$

This result is a special case of a more general result. Consider a collection $X_1, \ldots, X_n, X_{n+1}, \ldots, X_{n+m}$ of independent random variables. Let ϕ and ψ be real-valued functions of n and m variables, respectively. Then $\phi(X_1, \ldots, X_n)$ and $\psi(X_{n+1}, \ldots, X_{n+m})$ are independent random variables. Any potential for gaining insight into probability theory by proving this result would be overwhelmed by the cumbersome notation required at this stage.

Theorem 3.3.4 *Let $X_1, X_2, \ldots, X_k$ be independent random variables.*

(i) If each X_i has a binomial density with parameters n_i and p, then $X_1 + \cdots + X_k$ has a binomial density with parameters $n_1 + \cdots + n_k$ and p.

(ii) If each X_i has a Poisson density with parameter λ_i, then $X_1 + \cdots + X_k$ has a Poisson density with parameter $\lambda_1 + \cdots + \lambda_k$.

(iii) If each X_i has a negative binomial density with parameters r_i and p, then $X_1 + \cdots + X_k$ has a negative binomial density with parameters $r_1 + \cdots + r_k$ and p.

Assertions (i) and (ii) were proved in the $n = 2$ case in the last two examples; the general cases use these results and a mathematical induction argument. Assertion (iii) is left as an exercise in the $n = 2$ case, the general case again being an easy application of mathematical induction.

The problem of finding the density function of $Z = \phi(X_1, \ldots, X_k)$ can be difficult. Sometimes independence makes it possible, as in the following example.

EXAMPLE 3.22 Let X and Y be independent random variables each of which has a uniform density on $\{1, 2, \ldots, n\}$ and let $Z = \max(X, Y)$. The range of Z is then $\{1, 2, \ldots, n\}$. Suppose z is in the range. Rather than calculating $f_Z(z) = P(Z = z)$, we calculate $P(Z \le z)$ for reasons that will become apparent.

$$\begin{aligned}
P(Z \le z) &= P(\max(X, Y) \le z)\\
&= P(X \le z, Y \le z)
\end{aligned}$$

$$= \sum_{x=1}^{z}\sum_{y=1}^{z} P(X = x, Y = y)$$

$$= \sum_{x=1}^{z}\sum_{y=1}^{z} P(X = x)P(Y = y)$$

$$= \frac{z^2}{n^2}.$$

By Equation 2.8, for $2 \leq z \leq n$,

$$\begin{aligned} P(Z = z) &= P((Z \leq z) \cap (Z \leq z-1)^c) \\ &= P(Z \leq z) - P((Z \leq z) \cap (Z \leq z-1)) \\ &= P(Z \leq z) - P(Z \leq z-1) \\ &= \frac{2z-1}{n^2}. \end{aligned}$$

It is easy to see that this result holds for $z = 1$ also. Thus,

$$f_Z(z) = \begin{cases} (2z-1)/n^2 & \text{if } z = 1, 2, \ldots, n \\ 0 & \text{otherwise.} \end{cases} \qquad (3.10)$$

■

EXERCISES 3.3

Problem 1 requires the following fact from the calculus. If $\sum_{n=0}^{\infty} a_n x^n$ and $\sum_{n=0}^{\infty} b_n x^n$ are power series, the product of their sums can be written $\sum_{n=0}^{\infty} c_n x^n$, where $c_n = \sum_{k=0}^{n} a_k b_{n-k}$ for $n \geq 1$, on the common interval of convergence.

1. If a and b are any real numbers and z is a nonnegative integer, show that

$$\binom{a+b}{z} = \sum_{x=0}^{z} \binom{a}{x}\binom{b}{z-x}.$$

2. Let X and Y be independent random variables having negative binomial densities with parameters r and p and s and p, respectively. Derive the density of $Z = X + Y$.
3. Let X and Y be independent random variables each having a uniform density on $\{1, 2, \ldots, n\}$. Calculate $P(X \geq Y)$ and $P(X = Y)$.
4. Let N be a random variable having a Poisson density with parameter $\lambda > 0$. Given that $N = n$, n Bernoulli trials are performed, the number X of successes is counted, and the number Y of failures is counted. Show that X and Y are independent random variables.

5. Let X and Y be independent random variables having geometric densities with the same parameter p. Calculate $P(X \geq Y)$ and $P(X = Y)$.
6. Let X and Y be as in Problem 5. Find the density of $Z = X + Y$.
7. Let X and Y be independent random variables having uniform densities on $\{1, 2, \ldots, n\}$ and let $Z = X + Y$. Find the density of Z.
8. Let X and Y be independent random variables and let ϕ and ψ be two real-valued functions on R. Show that $\phi(X)$ and $\psi(Y)$ are independent random variables.

Solving the following problem without the benefit of Mathematica or Maple V software would be tedious.

9. Let X and Y be independent random variables with X having a binomial density $b(\cdot; 10, 1/2)$ and Y having a uniform density on $\{1, 2, 3\}$. Find the density of $Z = X + Y$ accurate to three decimal places.

3.4 GENERATING FUNCTIONS

In some instances, the problem of finding the density function of a sum of two random variables can be transformed into a purely algebraic problem using generating functions.

Definition 3.9 *Let $\{a_j\}_{j=0}^{\infty}$ be a sequence of real numbers. If the power series $\sum_{j=0}^{\infty} a_j t^j$ has $(-t_0, t_0)$ as its interval of convergence for some $t_0 > 0$, then the function $A(t) = \sum_{j=0}^{\infty} a_j t^j$ is called its* generating function. ■

EXAMPLE 3.23 If $a_j = 1$ for all $j \geq 0$, then $A(t) = \sum_{j=0}^{\infty} t^j$ has $(-1, 1)$ as its interval of convergence and $A(t) = 1/(1-t)$. If $a_j = 1/j!$ for all $j \geq 0$, then $A(t) = \sum_{j=0}^{\infty} t^j/j!$ has $(-\infty, \infty)$ as its interval of convergence and $A(t) = e^t$. If $a_0 = a_1 = 0$ and $a_j = 1$ for all $j \geq 2$, then $A(t) = \sum_{j=2}^{\infty} t^j$ has $(-1, 1)$ as its interval of convergence and $A(t) = t^2/(1-t)$. ■

Returning to the notation of Definition 3.9, if there is an $M \in R$ such that $|a_j| \leq M$ for all $j \geq 0$, then the series $\sum_{j=0}^{\infty} a_j t^j$ converges absolutely at least for $-1 < t < 1$, since the general term of the series $\sum_{j=0}^{\infty} |a_j t^j|$ is dominated by the general term of the series $\sum_{j=0}^{\infty} M|t|^j$, which is known to converge for $|t| < 1$, and thus the interval of convergence of $\sum_{j=0}^{\infty} a_j t^j$ contains the interval $(-1, 1)$.

An important result about generating functions is the fact that if a function can be represented as the sum of a power series on an open interval containing 0, then that representation is unique; i.e., if

$$f(t) = \sum_{j=0}^{\infty} a_j t^j = \sum_{j=0}^{\infty} b_j t^j$$

on an open interval containing 0, then $a_j = b_j$ for all $j \geq 0$.

EXAMPLE 3.24 Suppose we have found that a generating function is given by

$$A(t) = \frac{1}{1-t^2}, \qquad |t| < 1.$$

What is the sequence $\{a_j\}_{j=0}^{\infty}$? We can interpret $1/(1-t^2)$ as the sum of the geometric series $\sum_{j=0}^{\infty}(t^2)^j = \sum_{j=0}^{\infty} t^{2j}$. Thus,

$$A(t) = \sum_{j=0}^{\infty} a_j t^j = \sum_{j=0}^{\infty} t^{2j},$$

and corresponding coefficients of t^j are equal. Noting that coefficients of even powers of t on the right are equal to 1 and coefficients of odd powers of t on the right are 0, $a_j = (1/2)(1-(-1)^{j+1}), j = 0, 1, 2, \ldots$. ■

Another important property of power series is the following. If $\{a_j\}_{j=0}^{\infty}$ and $\{b_j\}_{j=0}^{\infty}$ are sequences of real numbers and $c_j = a_0 b_j + \cdots + a_j b_0, j \geq 0$, then the power series $\sum_{j=0}^{\infty} c_j t^j$ converges absolutely on the common interval of convergence of the series $\sum_{j=0}^{\infty} a_j t^j$ and $\sum_{j=0}^{\infty} b_j t^j$, and

$$\sum_{j=0}^{\infty} c_j t^j = \left(\sum_{j=0}^{\infty} a_j t^j\right)\left(\sum_{j=0}^{\infty} b_j t^j\right).$$

It is important to remember that the method of generating functions applies only to nonnegative integer-valued random variables.

Definition 3.10 *If X is a nonnegative integer-valued random variable with density f_X, its* generating function *is the function $\hat{f}_X$ on $[-1, 1]$ defined by*

$$\hat{f}_X(t) = \sum_{x=0}^{\infty} f_X(x) t^x, \qquad -1 \leq t \leq 1. \ ■$$

Note that the generating function of X is the same as the generating function of the sequence $\{f_X(x)\}_{x=0}^{\infty}$. Since $|f_X| \leq 1$ for all $x \geq 0$, $\hat{f}_X(t)$ is certainly defined on $(-1, 1)$. But since $\sum_{x=0}^{\infty} f_X(x) = 1$, the power series converges absolutely when $|t| = 1$. Thus, $\hat{f}_X$ is defined on $[-1, 1]$.

EXAMPLE 3.25 Let X have a geometric density with parameter p, $0 < p < 1$. Then

$$\hat{f}_X(t) = \sum_{x=1}^{\infty} pq^{x-1}t^x = pt\sum_{x=1}^{\infty} (qt)^{x-1} = pt\sum_{y=0}^{\infty} (qt)^y = \frac{pt}{1-qt}. \quad \blacksquare$$

EXAMPLE 3.26 Let X have a binomial density with parameters n and p. Then

$$\begin{aligned}\hat{f}_X(t) &= \sum_{x=0}^{\infty} f_X(x)t^x \\ &= \sum_{x=0}^{n} f_X(x)t^x \\ &= \sum_{x=0}^{n} \binom{n}{x} p^x q^{n-x} t^x \\ &= (pt+q)^n. \quad \blacksquare\end{aligned}$$

EXAMPLE 3.27 Let X have a Poisson density with parameter $\lambda > 0$. Then

$$\hat{f}_X(t) = \sum_{x=0}^{\infty} \frac{\lambda^x e^{-\lambda}}{x!} t^x = e^{-\lambda} \sum_{x=0}^{\infty} \frac{(\lambda t)^x}{x!} = e^{-\lambda} e^{\lambda t} = e^{\lambda(t-1)}. \quad \blacksquare$$

EXAMPLE 3.28 Let X have a negative binomial density with parameters r and p. Then

$$\hat{f}(t) = \sum_{x=0}^{\infty} \binom{-r}{x} p^r (-q)^x t^x = p^r \sum_{x=0}^{\infty} \binom{-r}{x} (-qt)^x.$$

By the generalized binomial theorem,

$$\hat{f}_X(t) = p^r(1-qt)^{-r} = \frac{p^r}{(1-qt)^r}. \quad \blacksquare$$

The following theorem is one justification for introducing generating functions.

Theorem 3.4.1 *If X and Y are independent nonnegative integer-valued random variables and $Z = X + Y$, then $\hat{f}_Z(t) = \hat{f}_X(t)\hat{f}_Y(t)$ for all $t \in [-1, 1]$.*

PROOF: First note that

$$\begin{aligned}\hat{f}_X(t)\hat{f}_Y(t) &= \left(\sum_{x=0}^{\infty} f_X(x)t^x\right)\left(\sum_{y=0}^{\infty} f_Y(y)t^y\right)\\ &= \sum_{z=0}^{\infty} c_z t^z\end{aligned}$$

where $c_z = \sum_{x=0}^{z} f_X(x)f_Y(z-x)$. By Theorem 3.3.2,

$$\sum_{z=0}^{\infty} c_z t^z = \sum_{z=0}^{\infty} f_Z(z)t^z = \hat{f}_Z(t).$$

Thus, $\hat{f}_Z(t) = \hat{f}_X(t)\hat{f}_Y(t)$. ■

EXAMPLE 3.29 Let X be a random variable taking on the values $1, 2, 3$ with probabilities $.02, .53, .45$, respectively, and let Y be a random variable taking on the values $1, 2, 3, 4$ with equal probabilities. What is the density of $Z = X + Y$, assuming that X and Y are independent? The generating function of X is $f_X(t) = .02t + .53t^2 + .45t^3$, the generating function of Y is $F_Y(t) = .25t + .25t^2 + .25t^3 + .25t^4$, and the generating function of Z is

$$\begin{aligned}f_Z(t) &= f_X(t)f_Y(t)\\ &= (.02t + .53t^2 + .45t^3)(.25t + .25t^2 + .25t^3 + .25t^4)\\ &= .005t^2 + .1375t^3 + .25t^4 + .25t^5 + .245t^6 + .1125t^7.\end{aligned}$$

Therefore, $f_Z(2) = .005, f_Z(3) = .1375, f_Z(4) = f_Z(5) = .25, f_Z(6) = .245$, and $f_Z(7) = .1125$. ■

Corollary 3.4.2 *If $X_1, \ldots, X_n$ are independent nonnegative integer-valued random variables and $Z = X_1 + \cdots + X_n$, then $\hat{f}_Z(t) = \prod_{i=1}^{n} \hat{f}_{X_i}(t)$.*

PROOF: The statement is trivially true when $n = 1$. Assume it is true for $n - 1$. Since $X_1 + \cdots + X_{n-1}$ and X_n are independent by Lemma 3.3.3,

$$\hat{f}_Z(t) = \hat{f}_{X_1+\cdots+X_{n-1}}(t) \cdot \hat{f}_{X_n}(t)$$

by Theorem 3.4.1. By the induction hypothesis, $\hat{f}_{X_1+\cdots+X_{n-1}}(t) = \prod_{j=1}^{n-1} \hat{f}_{X_j}(t)$. Therefore, $\hat{f}_Z(t) = \prod_{j=1}^{n} \hat{f}_{X_j}(t)$. ■

This corollary provides an alternative proof of the three assertions in Theorem 3.3.4.

EXAMPLE 3.30 Let $X_1, \ldots, X_k$ be independent random variables and let $Z = X_1 + \cdots + X_k$. If each X_i has a negative binomial density with parameters r_i and p, then

$$\hat{f}_Z(t) = \prod_{i=1}^{k} \frac{p^{r_i}}{(1 - qt)^{r_i}} = \frac{p^r}{(1 - qt)^r}$$

where $r = r_1 + \cdots + r_k$. But by the generalized binomial theorem,

$$p^r(1 - qt)^{-r} = \sum_{z=0}^{\infty} \binom{-r}{z} p^r(-qt)^z.$$

Therefore,

$$\hat{f}_Z(t) = \sum_{z=0}^{\infty} \binom{-r}{z} p^r(-q)^z tz,$$

and it follows that

$$f_Z(z) = \binom{-r}{z} p^r(-q)^z, \qquad z = 0, 1, 2, \ldots;$$

i.e., Z has a negative binomial density with parameters $r = r_1 + \cdots + r_k$ and p. ■

Having tediously calculated the probabilities $p(x)$ of getting a score of x upon rolling three dice in Exercise 1.4.3, the reader will appreciate the ease with which these probabilities can be calculated using generating functions. Let $X_1, X_2,$ and X_3 denote the number of pips on each of the dice and let $X = X_1 + X_2 + X_3$. The generating function of each X_j is

$$\hat{f}_{X_j}(t) = \frac{t}{6} + \frac{t^2}{6} + \cdots + \frac{t^6}{6} = \frac{t}{6}\frac{1 - t^6}{1 - t}.$$

Since $X_1, X_2,$ and X_3 are independent, by Corollary 3.4.2,

$$\hat{f}_X(t) = \left(\frac{t}{6} + \frac{t^2}{6} + \cdots + \frac{t^6}{6}\right)^3.$$

Expanding the expression on the right side using mathematical software,

$$\begin{aligned}\hat{f}_X(t) &= \frac{1}{216}t^3 + \frac{1}{72}t^4 + \frac{1}{36}t^5 + \frac{5}{108}t^6 + \frac{5}{72}t^7 + \frac{7}{72}t^8 \\ &+ \frac{25}{216}t^9 + \frac{1}{8}t^{10} + \frac{1}{8}t^{11} + \frac{25}{216}t^{12} + \frac{7}{72}t^{13} + \frac{5}{72}t^{14} \\ &+ \frac{5}{108}t^{15} + \frac{1}{36}t^{16} + \frac{1}{72}t^{17} + \frac{1}{216}t^{18}.\end{aligned}$$

Since $f_X(x)$ is just the coefficient of t^x, the probabilities can be read off; e.g., $P(X = 13) = 7/72$.

Generating functions are particularly useful for solving difference equations, as in the following example.

EXAMPLE 3.31 Consider an infinite sequence of Bernoulli trials with probability of success p. For each $n \geq 1$, let p_n be the probability of the event E_n "an even number of successes in the first n trials." We will use the fact that the probability of an even number of successes in trials $1, \ldots, n-1$ is the same as the probability of an even number of successes in trials $2, \ldots, n$. If the first trial results in a failure, in order for an outcome to be in E_n there must be an even number of successes in trials $2, \ldots, n$, and if the first trial results in success, there must be an odd number of successes in trials $2, \ldots, n$. Thus, $p_n = qp_{n-1} + p(1 - p_{n-1})$.

Decomposing E_n in this way makes sense only for $n \geq 2$. In one trial there is only one way to get an even number of successes, namely none at all, by that trial resulting in failure. Thus, $p_1 = q$. If the equation above is to hold when $n = 1$, we must have $q = p_1 = qp_0 + p(1 - p_0)$, and so p_0 must be taken equal to 1. Therefore, the p_n satisfy the difference equation

$$p_n = qp_{n-1} + p(1 - p_{n-1}), \qquad n \geq 1 \tag{3.11}$$

and the initial condition $p_0 = 1$. To solve the equation subject to the initial condition, let $P(t)$ be the generating function of the sequence $\{p_n\}_{n=0}^{\infty}$; i.e., $P(t) = \sum_{n=0}^{\infty} p_n t^n$. Multiplying both sides of Equation 3.11 by t^n and summing $n = 1, 2, \ldots$,

$$\begin{aligned}\sum_{n=1}^{\infty} p_n t^n &= qt\sum_{n=1}^{\infty} p_{n-1}t^{n-1} + pt\sum_{n=1}^{\infty} t^{n-1} - pt\sum_{n=1}^{\infty} p_{n-1}t^{n-1} \\ &= qt\sum_{n=0}^{\infty} p_n t^n + pt\sum_{n=0}^{\infty} t^n - pt\sum_{n=0}^{\infty} p_n t^n.\end{aligned}$$

Since $\sum_{n=1}^{\infty} p_n t^n = P(t) - p_0 = P(t) - 1$ and $\sum_{n=0}^{\infty} t^n = 1/(1-t)$,

$$P(t) - 1 = qtP(t) + \frac{pt}{1-t} - ptP(t).$$

Solving for $P(t)$,

$$P(t) = \frac{1}{1 - qt + pt} + \frac{pt}{(1-t)(1 - qt + pt)}.$$

Applying the method of partial fraction expansions to the second term on the right,

$$P(t) = \frac{1}{1 - qt + pt} + \frac{p}{1 - q + p}\frac{1}{1 - t} - \frac{p}{1 - q + p}\frac{1}{1 - qt + pt}.$$

Since $1 - q + p = 2p$,

$$P(t) = \frac{1}{2}\frac{1}{1 - t} + \frac{1}{2}\frac{1}{1 - qt + pt}$$

and

$$2P(t) = \frac{1}{1 - t} + \frac{1}{1 - qt + pt}.$$

Regarding the two terms on the right as sums of geometric series,

$$\sum_{n=0}^{\infty} 2p_n t^n = \sum_{n=0}^{\infty} t^n + \sum_{n=0}^{\infty}(q - p)^n t^n = \sum_{n=0}^{\infty}(1 + (q - p)^n)t^n.$$

Equating coefficients of t^n, we obtain

$$p_n = \frac{1}{2}(1 + (q - p)^n), \qquad n \geq 1.$$

This solution for p_n is much more enlightening than the solution

$$p_n = b(0; n, p) + b(2; n, p) + b(4; n, p) + \cdots + b(2m; n, p)$$

where m is the greatest integer such that $2m \leq n$. ■

It is often necessary to interchange the order of summation of two infinite series. The essential facts will be presented; proofs are contained in the appendix at the end of the chapter.

A map $a : N \times N \to R$ is called a *double sequence*, and its value at (i, j) is denoted by $a_{i,j}$. We also write $a = \{a_{i,j}\} = \{a_{i,j}\}_{i,j=1}^{\infty}$. The double sequence $\{a_{i,j}\}$ *converges* and has *limit* L if for every $\epsilon > 0$ there are integers $M, N \geq 1$ such that

$$|a_{i,j} - L| < \epsilon \qquad \text{for all } m \geq M, n \geq N.$$

In this case we write $\lim_{i,j \to \infty} a_{i,j} = L$.

Given a double sequence $\{a_{i,j}\}$, the formal expression

$$\sum_{i,j=1}^{\infty} a_{i,j}$$

can be formed and is called a *double series*. For each $m \geq 1, n \geq 1$, the following partial sum can be formed:

$$S_{m,n} = \sum_{\substack{1 \leq i \leq m \\ 1 \leq j \leq n}} a_{i,j}.$$

EXAMPLE 3.32 Consider the double sequence $\{a_{i,j}\}$ defined by $a_{i,j} = (-1)^{i+j}(1/j^i)$. Since $S_{2,3}$ is a sum of finitely many terms, the terms can be added in any order. Fixing j and summing over i,

$$\begin{aligned} S_{2,3} &= \sum_{j=1}^{3}(-1)^{1+j}\frac{1}{j} + \sum_{j=1}^{3}(-1)^{2+j}\frac{1}{j^2} \\ &= 1 - \frac{1}{2} + \frac{1}{3} - 1 + \frac{1}{4} - \frac{1}{9}. \end{aligned}$$

On the other hand, fixing i and summing over j,

$$\begin{aligned} S_{2,3} &= \sum_{i=1}^{2}(-1)^{i+1}\frac{1}{1^i} + \sum_{i=1}^{2}(-1)^{i+2}\frac{1}{2^i} + \sum_{i=1}^{2}(-1)^{i+3}\frac{1}{3^i} \\ &= 1 - 1 - \frac{1}{2} + \frac{1}{4} + \frac{1}{3} - \frac{1}{9}. \quad \blacksquare \end{aligned}$$

Definition 3.11 *The double series $\sum_{i,j=1}^{\infty} a_{i,j}$ is said to converge and have sum S if $\lim_{i,j\to\infty} S_{i,j} = S$; i.e., if for each $\varepsilon > 0$ there are integers $M, N \geq 1$ such that*

$$|S_{i,j} - S| < \varepsilon \qquad \text{for all } i \geq M, j \geq N. \quad \blacksquare$$

If the $a_{i,j} \geq 0$ for all $i, j \geq 1$, we say that $\sum_{i,j=1}^{\infty} a_{i,j}$ diverges to $+\infty$ if for every $L \in R$ there are integers $M, N \geq 1$ such that $S_{i,j} \geq L$ for all $i \geq M, j \geq N$.

Given the double series $\sum_{i,j=1}^{\infty} a_{i,j}$ we can form the iterated sums

$$\sum_{i=1}^{\infty}\left(\sum_{j=1}^{\infty} a_{i,j}\right) \text{ and } \sum_{j=1}^{\infty}\left(\sum_{i=1}^{\infty} a_{i,j}\right).$$

Proofs often depend upon showing that the latter two iterated series are equal; i.e., the order of summation can be interchanged.

Theorem 3.4.3 *If $\sum_{i,j=1}^{\infty} a_{i,j}$ is a double series with $a_{i,j} \geq 0$ for all $i, j \geq 1$, then*

$$\sum_{i,j=1}^{\infty} a_{i,j} = \sum_{i=1}^{\infty}\left(\sum_{j=1}^{\infty} a_{i,j}\right) = \sum_{j=1}^{\infty}\left(\sum_{i=1}^{\infty} a_{i,j}\right)$$

even if any sum is $+\infty$.

The double series $\sum_{i,j=1}^{\infty} a_{i,j}$ converges absolutely if the double series $\sum_{i,j=1}^{\infty} |a_{i,j}|$ converges.

Theorem 3.4.4 *If the double series $\sum_{i,j=1}^{\infty} a_{i,j}$ converges absolutely, then*

$$\sum_{i,j=1}^{\infty} a_{i,j} = \sum_{i=1}^{\infty}\left(\sum_{j=1}^{\infty} a_{i,j}\right) = \sum_{j=1}^{\infty}\left(\sum_{i=1}^{\infty} a_{i,j}\right).$$

The next application of generating functions has to do with the sum of a random number of random variables. Consider any infinite sequence of random variables $\{X_j\}_{j=1}^{\infty}$ and let N be any random variable taking on values in $\{1, 2, \ldots\}$. An outcome ω determines an infinite sequence $X_1(\omega), X_2(\omega), \ldots$ as well as a positive integer $N(\omega)$, and we can form the sum of the first $N(\omega)$ terms of the infinite sequence, which is denoted by

$$S_N(\omega) = X_1(\omega) + X_2(\omega) + \cdots + X_{N(\omega)}(\omega).$$

S_N is called the sum of a *random number of random variables.* If N is a constant n, then S_n is the sum of a fixed number of random variables. That S_N is a random variable follows from the fact that

$$(S_N = s) = \bigcup_{n=1}^{\infty}(S_N = s, N = n) = \bigcup_{n=1}^{\infty}((N = n) \cap (S_n = s))$$

and the fact that each S_n is a random variable.

Theorem 3.4.5 *If $\{X_j\}_{j=1}^{\infty}$ is an infinite sequence of independent nonnegative integer-valued random variables all having the same density function f, N is a positive integer-valued random variable, and $N, X_1, X_2, \ldots$ are independent, then*

$$\hat{f}_{S_N}(t) = \hat{f}_N(\hat{f}_{X_1}(t)).$$

PROOF: By stratifying the event $(S_N = s)$ according to the values of N and using the fact that $S_N = S_n$ on the event $(N = n)$,

$$\begin{aligned}\hat{f}_{S_N}(t) &= \sum_{s=0}^{\infty} f_{S_N}(s)t^s \\ &= \sum_{s=0}^{\infty} P(S_N = s)t^s \\ &= \sum_{s=0}^{\infty}\sum_{n=1}^{\infty} P(S_N = s, N = n)t^s \\ &= \sum_{s=0}^{\infty}\sum_{n=1}^{\infty} P(S_n = s, N = n)t^s.\end{aligned}$$

Since $N, X_1, \ldots, X_n$ are independent random variables, $X_1 + \cdots + X_n$ and N are independent random variables by Lemma 3.3.3. By Theorem 3.4.4,

$$\begin{aligned}\hat{f}_{S_N}(t) &= \sum_{s=0}^{\infty}\sum_{n=1}^{\infty} P(S_n = s)P(N = n)t^s \\ &= \sum_{n=1}^{\infty}\left(\sum_{s=0}^{\infty} P(S_n = s)t^s\right)P(N = n) \\ &= \sum_{n=0}^{\infty} \hat{f}_{S_n}(t)f_N(n) \\ &= \sum_{n=0}^{\infty} (\hat{f}_{X_1}(t))^n f_N(n)\end{aligned}$$

where the terms corresponding to $n = 0$ in the last two expressions can be included since $f_N(0) = 0$. Since $\hat{f}_N(t) = \sum_{n=1}^{\infty} t^n f_N(n)$,

$$\hat{f}_{S_N}(t) = \hat{f}_N(\hat{f}_{X_1}(t)). \quad \blacksquare$$

EXAMPLE 3.33 Suppose the wind carries N seeds onto a given plot of land where N has a Poisson density with parameter $\lambda > 0$ and each seed has probability p of germinating, independently of the number of seeds and independently of the other seeds. Let $X_1, X_2, \ldots$ be an infinite sequence of Bernoulli random variables with $P(X_j = 1) = p$. The number of germinating seeds is then $S_N = X_1 + \cdots + X_N$. Since $\hat{f}_{X_1}(t) = pt + q$ and $\hat{f}_N(t) = e^{\lambda(t-1)}$,

$$\begin{aligned}\hat{f}_{S_N}(t) &= \hat{f}_N(pt + q) \\ &= e^{\lambda p(t-1)}.\end{aligned}$$

It follows that S_N has a Poisson density with parameter λp. ■

EXERCISES 3.4

1. Let X be a random variable having a uniform density on $\{1, 2, \ldots, n\}$. Find the generating function of X.
2. The sequence of real numbers $\{a_j\}_{j=1}^{\infty}$ has the generating function $A(t) = 1 - (1 - t^2)^{1/2}$. Find a formula for the a_j.
3. If the random variable X has each of the following generating functions, what is the corresponding density function?
   (a) $\hat{f}_X(t) = t/(3 - 2t)$
   (b) $\hat{f}_X(t) = e^{(t-1)/4}$
   (c) $\hat{f}_X(t) = t/(8 - 7t)^5$
4. A die is rolled to determine how many times a coin is to be flipped and then the coin is flipped that many times. Let X be the number of heads so obtained. Find the generating function of X.
5. Consider the generating function

$$\hat{f}(t) = \frac{1}{6}\left(\left(\frac{t}{3} + \frac{2}{3}\right) + \left(\frac{t}{3} + \frac{2}{3}\right)^2 + \cdots + \left(\frac{t}{3} + \frac{2}{3}\right)^6\right).$$

   Describe a compound experiment and an associated random variable having this generating function.
6. If the random variable X has the generating function

$$\hat{f}_X(t) = e^{2(t^2-1)},$$

   what is the density function of X?
7. Consider an infinite sequence of Bernoulli random variables $X_1, X_2, \ldots$ with probability of success p. Let E_n be the event "there are an even number of successes in the first n trials." Express E_n in terms of the random variables $X_1, X_2, \ldots$ and use the theorems of probability theory to derive Equation 3.11 by stratifying E_n according to the values of X_1; i.e., $p_n = P(E_n) = P(E_n \cap (X_1 = 0)) + P(E_n \cap (X_1 = 1))$, and so forth.

The following problems require software such as Mathematica or Maple V.

8. Consider an infinite sequence of Bernoulli trials with probability of success $p = 1/2$. For $n \geq 1$, let q_n be the probability that the pattern 11 will not appear in the first n trials (i.e., the probability that there will not be two consecutive 1's). Derive a difference equation for the q_n, specify initial conditions, and find a formula for the q_n.

9. If 10 dice are tossed simultaneously, what is the probability of getting a score of 42?

10. If X has the binomial density $b(\cdot\,; 5, .5)$, Y has a uniform density on $\{1, 2, \ldots, 6\}$, and X, Y are independent random variables, find the density of $Z = X + Y$.

11. For $j = 1, \ldots, 10$, the random variable X_j takes on the values 1 and 0 with probabilities p_j and $1 - p_j$, respectively, where $p_j = (.95)^j/2$. If $X = X_1 + \cdots + X_{10}$, find the density of X assuming that $X_1, \ldots, X_{10}$ are independent.

3.5 GAMBLER'S RUIN PROBLEM

Suppose a gambler and an opponent have a combined capital of a units and the gambler has x units of capital where $1 \leq x \leq a - 1$. The gambler wagers one unit on successive plays of a game in which the probability that he will win one unit is p and that he will lose one unit is $q = 1 - p$, where $0 < p < 1$. The gambler is ruined if his capital ever reaches zero units; his opponent is ruined if the gambler's capital ever reaches a units. What is the probability that the gambler will be ruined eventually? Since it is conceivable that the wagers could go on forever and neither be ruined, it is of interest to also find the probability that the opponent will be ruined.

A more immediate question concerns a probability model for which these questions make sense. Let $\{X_j\}$ be an infinite sequence of Bernoulli trials with probability of success p so that the X_j's are independent, $P(X_j = 1) = p$, and $P(X_j = 0) = q$. If for each $j \geq 1$ we let $Y_j = 2X_j - 1$, then the Y_j's are independent random variables with $P(Y_j = 1) = p$ and $P(Y_j = -1) = q$. For $j \geq 1$, Y_j represents the gambler's gain on the jth play of the game. His capital as of the jth play is then $S_j = x + Y_1 + \cdots Y_j$, $j \geq 1$. The gambler is ruined if $S_1 = 0$ or $0 < S_1 < a, \ldots, 0 < S_{j=1} < a, S_j = 0$ for some $j \geq 2$. The probability of eventual ruin q_x, which depends upon x, is given by

$$q_x = P(S_1 = 0 \text{ or } 0 < S_1 < a, \ldots, 0 < S_{j-1} < a, S_j = 0 \text{ for some } j \geq 2).$$

Since the indicated events are mutually exclusive,

$$q_x = P(S_1 = 0) + P(0 < S_1 < a, \ldots, 0 < S_{j-1} < a, S_j = 0 \text{ for some } j \geq 2).$$

Suppose $1 < x < a - 1$. Then ruin cannot occur on the first wager, and

$$q_x = P(0 < S_1 < a, \ldots, 0 < S_{j-1} < a, S_j = 0 \text{ for some } j \geq 2).$$

We now show that

$$q_x = pq_{x+1} + qq_{x-1}, \qquad 1 < x < a - 1.$$

A "probabilistic argument" can be made as follows. The first wager can result in winning one unit, with probability p, whereupon the gambler's capital becomes $x + 1$ and the probability of subsequent ruin is q_{x+1}; since an event determined solely by the first wager and an event determined by subsequent wagers are independent, the probability of winning the first wager and then being ruined is pq_{x+1}. Similarly, the probability of losing the first wager and then being ruined is qq_{x-1}. Since these two possibilities are mutually exclusive,

$$q_x = pq_{x+1} + qq_{x-1}, \qquad \text{for } 1 < x < a - 1.$$

The same argument applies when $x = 1$, with the exception that the probability of losing the first wager and then being ruined is $q \cdot 1$, since ruin has already occurred on the first wager. Thus, $q_1 = pq_2 + q$, and if the equation above is to hold when $x = 1$, we must have $q_0 = 1$. Similarly, $q_{a-1} = qq_{a-2}$, and we must have $q_a = 0$. The q_x must then satisfy the difference equation

$$q_x = pq_{x+1} + qq_{x-1}, \qquad 1 \le x \le a - 1 \tag{3.12}$$

subject to the boundary conditions

$$q_0 = 1, q_a = 0. \tag{3.13}$$

One way of solving such a problem is to try known functions successively until we come across a solution; e.g., $q_x = A, q_x = Bx, q_x = Cx^2, \ldots, q_x = D\lambda^x$, and so forth, where $A, B, C, D, \ldots$ are constants. It is easy to check that if A is any constant, then $q_x = A$ satisfies the difference equation but does not satisfy the boundary conditions. Trying $q_x = B\lambda^x$ results in a quadratic equation in λ that has two roots $\lambda = 1$ and $\lambda = q/p$. At this point, we must consider two cases according to whether $p \neq q$ or $p = q = 1/2$, since there is only one solution in the latter case. Suppose first that $p \neq q$ so that there are two distinct roots of the quadratic equation. In this case there are two solutions $q_x = A$ and $q_x = B(q/p)^x$, but neither satisfies both boundary conditions. Noting that the difference equation has the property that if $q_x^{(1)}$ and $q_x^{(2)}$ are two solutions, then $q_x^{(1)} + q_x^{(2)}$ is also a solution, we might try

$$q_x = A + B\left(\frac{q}{p}\right)^x.$$

In this case, A and B can be chosen so that both boundary conditions are satisfied and satisfy the equations

$$A + B = 1$$
$$A + B\left(\frac{q}{p}\right)^a = 0.$$

Solving for A and B,

$$q_x = \frac{(q/p)^a - (q/p)^x}{(q/p)^a - 1}, \qquad 1 \le x \le a - 1 \tag{3.14}$$

provided $p \neq q$. Suppose now that $p = q = 1/2$. Again $q_x = A$ is a solution of the difference equation

$$q_x = \frac{1}{2}q_{x+1} + \frac{1}{2}q_{x-1}, \qquad 1 \le x \le a - 1$$

but does not satisfy both boundary conditions. This time $q_x = Bx$ satisfies the difference equation but not the boundary conditions. The function $q_x = A + Bx$ will satisfy all conditions and leads to the solution

$$q_x = 1 - \frac{x}{a}, \qquad 1 \le x \le a - 1 \tag{3.15}$$

provided $p = q = 1/2$.

Equations 3.14 and 3.15 provide the answer to the first of the two questions originally raised about the probability of eventual ruin. What about the second question pertaining to the probability p_x that the gambler will wipe out his adversary? It is not necessary to repeat the arguments given above, since we can interpret p_x as the probability of ruin for the adversary, in which case x is replaced by $a - x$ and p by q in the equations above. In the $p \neq q$ case,

$$p_x = \frac{(p/q)^a - (p/q)^{a-x}}{(p/q)^a - 1}, \qquad 1 \le x \le a - 1 \tag{3.16}$$

and in the $p = q = 1/2$ case,

$$p_x = \frac{x}{a}, \qquad 1 \le x \le a - 1. \tag{3.17}$$

Returning to the probability of ruin, we have found a solution to the problem, depending upon whether $p \neq q$ or $p = q = 1/2$. How do we know that the q_x is the real solution to our problem? Perhaps there is some other solution q_x that satisfies the difference equation and the boundary conditions. It is a question of the uniqueness of the solution. Suppose $q_x^{(1)}$ and $q_x^{(2)}$ are two solutions. Then $u_x = q_x^{(1)} - q_x^{(2)}$ will satisfy the equation

$$u_x = pu_{x+1} + qu_{x-1}, \qquad 1 \le x \le a - 1 \tag{3.18}$$

and the boundary conditions

$$u_0 = 0, u_a = 0. \tag{3.19}$$

Assume that $u_x \not\equiv 0$. By replacing u_x by $-u_x$, if necessary, we can assume that $u_y > 0$ for some $y \in \{0, 1, \ldots, a\}$. Consider the finite set of numbers $\{u_0, u_1, \ldots, u_a\}$. There is some m for which u_m is the largest of the numbers in this set; i.e., $u_m \geq u_x$ for $x = 1, 2, \ldots, a - 1$ and $u_m > 0$. If there is more than one such m, we can assume by the well-ordering principle that m is the smallest integer with this property. Then $u_{m-1} < u_m$. But $u_m = pu_{m+1} + qu_{m-1} < pu_{m+1} + qu_m \leq pu_m + qu_m = u_m$, a contradiction. The assumption that $u_x \not\equiv 0$ leads to a contradiction and therefore $u_x \equiv 0$; i.e., $q_x^{(1)} = q_x^{(2)}$ for $x = 1, 2, \ldots, a - 1$. Thus, the q_x given by Equation 3.14 or 3.15 is the only solution of the difference equation satisfying the boundary conditions.

What happens if the gambler decides to wager one-half unit each time instead of one unit? Will this improve his chances of avoiding eventual ruin? The effect of this change is to double the number of units. In the $p = q = 1/2$ case, the probability of eventual ruin is

$$1 - \frac{2x}{2a} = 1 - \frac{x}{a}, \tag{3.20}$$

and there is no change in the probability of eventual ruin. Suppose $p \neq q$. In this case the probability of eventual ruin is

$$\frac{(q/p)^{2a} - (q/p)^{2x}}{(q/p)^{2a} - 1} = q_x \cdot \frac{(q/p)^a + (q/p)^x}{(q/p)^a + 1}. \tag{3.21}$$

In the usual situation in which the game is unfair to the gambler $q > p$, the second factor on the right is greater than 1 so that wagering half a unit instead of a whole unit actually increases the probability of eventual ruin.

EXAMPLE 3.34 Suppose the gambler has an initial capital of \$100 and he decides in advance to continue placing wagers of \$10 on a game with $p = .45$ until he has increased his capital by \$10 or has been ruined. He then has 10 units to wager. By Equation 3.14, the probability of eventual ruin is $q_{10} = .204$. Thus, there is a probability of .796 of achieving the goal of increasing his capital by \$10. Of course, if upon winning the \$10 the gambler gets greedy and continues to play against an adversary who for all practical purposes is infinitely rich, then it is simply a question of how long it will take for ruin to occur. But that is another mathematical problem. ■

EXERCISES 3.5

1. What is the probability that the wagering will eventually terminate?
2. If $q > p$, what is the gambler's probability of eventual ruin against an infinitely rich adversary?
3. If $\{i_1, \ldots, i_m\}$ and $\{j_1, \ldots, j_n\}$ are disjoint sets of positive integers, it is known that events of the type $(Y_{i_1} = \delta_1, \ldots, Y_{i_m} = \delta_m)$ and $(Y_{j_1} = \epsilon_1, \ldots, Y_{j_n} = \epsilon_n)$ are independent. Show that the events $(Y_1 = 1)$ and $(Y_2 + \cdots + Y_j = y$ for some $j \geq 2)$ are independent.

4. Modify the gambler's ruin problem by allowing the possibility of a tie on each play of the game so that there are positive numbers α, β, γ, with $\alpha + \beta + \gamma = 1$ such that $P(Y_j = 1) = \alpha, P(Y_j = 0) = \beta$, and $P(Y_j = -1) = \gamma$, and let q_x be the probability of eventual ruin for the gambler. Derive a difference equation for the q_x and appropriate boundary conditions. Solve for the q_x and draw conclusions.

3.6 APPENDIX

Proof of Theorem 3.4.3 We need only deal with the first equation because the second can be obtained by interchanging the role of i and j. Suppose first that $\sum_{i,j=1}^{\infty} a_{i,j}$ converges and has sum S. Clearly, $S_{i,j} \leq S$ for all $i,j \geq 1$. Since $\lim_{i,j\to\infty} S_{i,j} = S$, given $\varepsilon > 0$ there are integers $M, N \geq 1$ such that

$$S - \varepsilon < \sum_{\substack{1\leq j\leq m\\ 1\leq i\leq n}} a_{i,j} \leq S \qquad \text{for all } m \geq M, n \geq N.$$

Since

$$\sum_{\substack{1\leq j\leq m\\ 1\leq i\leq n}} a_{i,j} = \sum_{i=1}^{m}\sum_{j=1}^{n} a_{i,j}$$

is valid for finite sums, $S - \varepsilon < \sum_{i=1}^{m}\sum_{j=1}^{n} a_{i,j} \leq S$ for all $m \geq M, n \geq N$. Since $\sum_{j=1}^{n} a_{i,j}$ is an increasing sequence for each i, with m fixed we can take the limit as $n \to \infty$ to obtain $S - \varepsilon < \sum_{i=1}^{m}\sum_{j=1}^{\infty} a_{i,j} \leq S$. Since the middle expression increases with m and is bounded above by S, the series $\sum_{i=1}^{\infty}(\sum_{j=1}^{\infty} a_{i,j})$ converges and

$$S - \varepsilon < \sum_{i=1}^{\infty}\left(\sum_{j=1}^{\infty} a_{i,j}\right) \leq S.$$

Since ε is arbitrary,

$$\sum_{i=1}^{\infty}\left(\sum_{j=1}^{\infty} a_{i,j}\right) = S.$$

Suppose now that $\sum_{i=1}^{\infty}(\sum_{j=1}^{\infty} a_{i,j})$ converges to S in R. Given $\varepsilon > 0$, there is an $M \geq 1$ such that $S - \varepsilon < \sum_{i=1}^{m}(\sum_{j=1}^{\infty} a_{i,j}) < S + \varepsilon$ for all $m \geq M$, from which it follows that for each $i = 1, \ldots, m$, the series $\sum_{j=1}^{\infty} a_{i,j}$ converges. Thus, $S - \varepsilon < \lim_{n\to\infty} \sum_{i=1}^{m}\sum_{j=1}^{n} a_{i,j} < S + \varepsilon$, and there is an $N \geq 1$ such that $S - \varepsilon < \sum_{i=1}^{m} \sum_{j=1}^{n} a_{i,j} = S_{m,n} < S + \varepsilon$ for all $m \geq M, n \geq N$. This shows that $\sum_{i,j=1}^{\infty} a_{i,j}$ converges and has sum $S = \sum_{i=1}^{\infty}(\sum_{j=1}^{\infty} a_{i,j})$.

Assume that the double series $\sum_{i,j=1}^{\infty} a_{i,j}$ diverges to $+\infty$. Given any $L \in R$, there are integers $M, N \geq 1$ such that

$$\sum_{i=1}^{m} \sum_{j=1}^{n} a_{i,j} = \sum_{\substack{1 \leq j \leq m \\ 1 \leq i \leq n}} a_{i,j} > L \qquad \text{for all } m \geq M, n \geq N.$$

Thus, $\sum_{i=1}^{m}(\sum_{j=1}^{\infty} a_{i,j}) > L$ for all $m \geq M$. Thus, the sequence

$$\sum_{i=1}^{m}\left(\sum_{j=1}^{\infty} a_{i,j}\right)$$

diverges to $+\infty$, and so

$$\sum_{i=1}^{\infty}\left(\sum_{j=1}^{\infty} a_{i,j}\right) = \sum_{i,j=1}^{\infty} a_{i,j} = +\infty.$$

Finally, suppose that the series $\sum_{i=1}^{\infty}(\sum_{j=1}^{\infty} a_{i,j})$ diverges to $+\infty$. To deal with this case, note that $\lim_{n\to\infty} S_{n,n}$ exists as a real number or $\lim_{n\to\infty} S_{n,n} = +\infty$ since $\{S_{n,n}\}_{n=1}^{\infty}$ is an increasing sequence of real numbers. In the latter case, given $L \in R$ there is an $M \geq 1$ such that $S_{n,n} > L$ for all $n \geq M$, and therefore $S_{m,n} > L$ for all $m \geq M, n \geq M$; i.e., $\lim_{m,n\to\infty} S_{m,n} = \sum_{i,j=1}^{\infty} a_{i,j} = +\infty$. On the other hand, if $\lim_{n\to\infty} S_{n,n} = S \in R$, then it is easy to see that $\lim_{m,n\to\infty} S_{m,n} = S$, and by the first part of the proof $\sum_{i=1}^{\infty}(\sum_{j=1}^{\infty} a_{i,j})$ converges to $S \in R$, a contradiction. Therefore, $\lim_{m,n\to\infty} S_{m,n} = \sum_{i,j=1}^{\infty} a_{i,j} = +\infty$. ■

The following functions will be needed for the next proof. For $x \in R$, let

$$x^+ = \max(x, 0)$$
$$x^- = \max(-x, 0).$$

Then it is easy to see, by considering the two cases $x \geq 0$ and $x \leq 0$ separately, that

$$x = x^+ - x^-$$
$$|x| = x^+ + x^-$$
$$0 \leq x^+ \leq |x|$$
$$0 \leq x^- \leq |x|.$$

Proof of Theorem 3.4.4: Since $0 \le a_{i,j}^{\pm} \le |a_{i,j}|$, the double series $\sum_{i,j=1}^{\infty} a_{i,j}^{\pm}$ converges. By Theorem 3.4.3,

$$\sum_{i,j=1}^{\infty} a_{i,j}^{\pm} = \sum_{i=1}^{\infty}\left(\sum_{j=1}^{\infty} a_{i,j}^{\pm}\right) = \sum_{j=1}^{\infty}\left(\sum_{i=1}^{\infty} a_{i,j}^{\pm}\right).$$

Taking the difference between the + and − versions results in the conclusions of the theorem. ■

SUPPLEMENTAL READING LIST

W. Feller (1957). *An Introduction to Probability Theory and Its Applications*, 2nd ed. New York: Wiley.

CHAPTER 4

EXPECTATION

4.1 INTRODUCTION

The concept of expectation was first formalized in print by Huygens in the middle of the seventeenth century, and it has played an essential role in probability theory ever since. The expected value of a random variable is a number that summarizes information about a random variable. From the time of its inception until the middle of the twentieth century, the concept of expected value developed along two paths: the discrete and continuous cases. Although it is possible to treat both paths simultaneously, we will stay with the discrete for the time being.

Among other things, we will determine the expected duration of play in the gambler's ruin problem, discuss prediction and filtering theory, and look briefly at some applications to communication theory.

4.2 EXPECTED VALUE

Unless specified otherwise in examples, $(\Omega, \mathcal{F}, P)$ will be a fixed probability space. The idea behind expected value is very simple. If a gambler wagers on a game in which he can win one unit with probability $p = 3/4$ and lose two units with probability $q = 1/4$ and he plays 100 games, then according to the empirical law for relative frequencies, the gambler would expect to win about 75 games and lose about 25 games. Thus, he would expect to win about $75 \cdot 1$ units and lose about $25 \cdot 2$ units with a net gain of $75 \cdot 1 - 25 \cdot 2$ units. Putting this on a per-game basis, he would expect a net gain per game of

$$\frac{75}{100}\cdot 1 - \frac{25}{100}\cdot 2 = \frac{3}{4}(1) + \frac{1}{4}(-2) = \frac{1}{4}$$

where the coefficients 3/4 and 1/4 are the probabilities of winning 1 and -2 units, respectively.

Definition 4.1 *Let X be a random variable with range $\{x_1, x_2, \ldots\}$, finite or infinite. The* expected value *of X, denoted by $E[X]$, is defined as the real number*

$$E[X] = \sum_i x_i f_X(x_i)$$

provided that the series converges absolutely and X is said to have finite expectation; *if $P(X \geq 0) = 1$ and the series diverges, $E[X]$ is defined as $+\infty$.* ■

If the range of X is finite, the series on the right is a finite sum and there is no question of convergence, absolute or otherwise. The question of absolute convergence is appropriate only when the series is infinite. Recall that if a series converges but not absolutely, then it is conditionally convergent. In the case of a conditionally convergent series, a rearrangement of the terms of the series can alter the sum of the series. If the sum in the series above is conditionally convergent, then one person listing the values of X in one order might arrive at a different sum than would some other person listing the values in another order. Under absolute convergence, the order in which the values of X are listed does not matter.

EXAMPLE 4.1 Let X have a uniform density on $\{0, 1, \ldots, n\}$ so that $f_X(x) = 1/(n+1), x = 0, 1, \ldots, n$. By Exercise 1.2.6,

$$E[X] = \sum_{j=0}^{n} j\frac{1}{n+1} = \frac{1}{n+1}\sum_{j=1}^{n} j = \frac{1}{n+1}\frac{n(n+1)}{2} = \frac{n}{2}. \quad ■$$

EXAMPLE 4.2 Let X have a binomial density with parameters n and p. Then

$$\begin{aligned} E[X] &= \sum_{x=0}^{n} x\binom{n}{x}p^x q^{n-x} \\ &= np\sum_{x=1}^{n}\frac{(n-1)!}{(x-1)!(n-x)!}p^{x-1}q^{(n-1)-(x-1)} \\ &= np\sum_{x=0}^{n-1} b(x; n-1, p) \\ &= np, \end{aligned}$$

the last equation holding because $\sum_{x=0}^{n-1} b(x; n-1, p)$ is the sum of all the probabilities making up a binomial density. ■

EXAMPLE 4.3 Let X have a geometric density with parameter p so that $f_X(x) = pq^{x-1}, x = 1, 2, \ldots$. Regard the series $\sum_{x=0}^{\infty} q^x$ as a power series in q with $(-1, 1)$ as its interval of convergence. Since

$$\sum_{x=0}^{\infty} q^x = \frac{1}{1-q},$$

within the interval of convergence

$$\frac{1}{(1-q)^2} = \frac{d}{dq}\frac{1}{1-q} = \frac{d}{dq}\sum_{x=0}^{\infty} q^x = \sum_{x=0}^{\infty}\frac{d}{dq}(q^x) = \sum_{x=0}^{\infty} xq^{x-1},$$

with the latter series also converging absolutely in the interval $(-1, 1)$. Returning to the geometric density,

$$E[X] = \sum_{x=1}^{\infty} xpq^{x-1} = p\sum_{x=0}^{\infty} xq^{x-1} = \frac{p}{(1-q)^2} = \frac{1}{p}. \quad ■$$

EXAMPLE 4.4 Let X have a Poisson density with parameter $\lambda > 0$ so that

$$E[X] = \sum_{x=0}^{\infty} x \cdot \frac{\lambda^x e^{-\lambda}}{x!},$$

provided that the series converges absolutely. Since the terms of the series are nonnegative, absolute convergence and convergence are the same thing and we need only verify the latter. Since

$$\sum_{x=0}^{\infty} x\frac{\lambda^x e^{-\lambda}}{x!} = \lambda e^{-\lambda}\sum_{x=1}^{\infty}\frac{\lambda^{x-1}}{(x-1)!} = \lambda e^{-\lambda}\sum_{x=0}^{\infty}\frac{\lambda^x}{x!}$$

and the latter series is the Maclaurin series expansion of e^{λ}, which is known to converge absolutely on $(-\infty, +\infty)$, $E[X]$ is defined and $E[X] = \lambda e^{-\lambda}e^{\lambda} = \lambda$. ■

If X is a random variable and ϕ is a real-valued function on R, then $Z = \phi(X)$ is also a random variable. According to the definition of expected value, to calculate $E[Z]$ we must first determine the density f_Z of the random variable Z. This need not be done according to the following theorem.

Theorem 4.2.1 *Let X be a random variable with range $\{x_1, x_2, \ldots\}$ and let ϕ be a real-valued function on R. Then $E[\phi(X)]$ is defined and*

$$E[\phi(X)] = \sum_j \phi(x_j) f_X(x_j)$$

provided the series converges absolutely.

In applying this result, the sum on the right is formed by replacing X in $\phi(X)$ by a typical value x_j, multiplying by the probability that X takes on that value, and then summing over j.

PROOF: Assume that the series converges absolutely. Any rearrangement of the series will not affect the convergence or sum of the series. Let $\{z_1, z_2, \ldots\}$ be the range of $Z = \phi(X)$. By rearranging the terms of the series,

$$\begin{aligned}\sum_j \phi(x_j) f_X(x_j) &= \sum_i \sum_{\{j:\phi(x_j)=z_i\}} z_i f_X(x_j) = \sum_i z_i P(\phi(X) = z_i) \\ &= \sum_i z_i f_Z(z_i) = E[Z].\end{aligned}$$

The same steps applied with $\phi(x_j)$ replaced by $|\phi(x_j)|$ show that the series $\sum_i z_i f_Z(z_i)$ converges absolutely. ■

EXAMPLE 4.5 Let X be a random variable with binomial density $b(\cdot; n, p)$ and let $\phi(x) = x^2, x \in R$. Then

$$\begin{aligned}E[\phi(X)] &= \sum_{k=0}^{n} k^2 \binom{n}{k} p^k q^{n-k} \\ &= \sum_{k=0}^{n} k(k-1)\binom{n}{k} p^k q^{n-k} + \sum_{k=0}^{n} k \binom{n}{k} p^k q^{n-k}.\end{aligned}$$

We have seen that the second sum on the right is equal to np. Since

$$\begin{aligned}\sum_{k=0}^{n} k(k-1)\binom{n}{k} p^k q^{n-k} &= \sum_{k=2}^{n} \frac{n!}{(k-2)!(n-k)!} p^k q^{n-k} \\ &= n(n-1)p^2 \sum_{k=2}^{n} \binom{n-2}{k-2} p^{k-2} q^{(n-2)-(k-2)} \\ &= n(n-1)p^2 \sum_{k=0}^{n-2} \binom{n-2}{k} p^k q^{(n-2)-k} \\ &= n(n-1)p^2 \sum_{k=0}^{n-2} b(k; n-2, p) \\ &= n(n-1)p^2,\end{aligned}$$

$E[X^2] = n(n-1)p^2 + np = n^2p^2 - np^2 + np.$ ■

It is not hard to construct examples of random variables X for which $E[X]$ is not defined as a real number.

EXAMPLE 4.6 Let X be a random variable with density $f_X(x) = \frac{1}{x(1+x)}$, $x = 1, 2, \ldots$ (see Exercise 1.5.10). Since the series

$$\sum_{x=1}^{\infty} x f_X(x) = \sum_{x=1}^{\infty} \frac{1}{1+x} = \frac{1}{2} + \frac{1}{3} + \frac{1}{4} + \cdots$$

is the divergent harmonic series except for a missing first term, the series $\sum_{x=1}^{\infty} x f_X(x)$ does not converge, and therefore $E[X]$ is not defined as a real number. ■

In some instances, as in the previous example, when the terms of the series $\sum_j x_j f_X(x_j)$ are nonnegative but the series does not converge, we say that the series diverges to $+\infty$ and we write $E[X] = +\infty$.

Theorem 4.2.2 *If X has finite expectation and c is any real number, then*

(i) If $P(X \geq 0) = 1$, then $E[X] \geq 0$.

(ii) If $P(X = c) = 1$, then $E[X] = c$.

(iii) $E[cX] = cE[X]$.

PROOF: If $P(X \geq 0) = 1$, then $f_X(x) = 0$ whenever $x < 0$, and so $E[X] = \sum_j x_j f_X(x_j) = \sum_{x_j \geq 0} x_j f_X(x_j) \geq 0$. If $P(X = c) = 1$, then $f_X(c) = 1$ and $f_X(x) = 0$ whenever $x \neq c$, so that $E[X] = \sum_j x_j f_X(x_j) = c f_X(c) = c$. By Theorem 4.3.1, $E[cX] = \sum_j c x_j f_X(x_j) = c \sum_j x_j f_X(x_j) = cE[X]$. ■

If the density function of a nonnegative integer-valued random variable X is not known but its generating function is, the expected value of X can be calculated indirectly using the generating function. The following notation is useful for this purpose. Let f be a real-valued function defined on an interval having a as its right endpoint. Then $f(a-)$ is defined to be $\lim_{x \to a-} f(x)$, even if infinite.

Theorem 4.2.3 (Abel) *Let $\{a_j\}_{j=0}^{\infty}$ be a sequence with $a_j \geq 0$ and generating function $A(t)$ on $(-1, 1)$, and let $A(1) = \sum_{j=0}^{\infty} a_j$. Then*

$$A(1-) = \lim_{t \to 1-} A(t) = A(1),$$

even if the series diverges to $+\infty$.

PROOF: Suppose first that $\sum_{j=0}^{\infty} a_j$ diverges to $+\infty$. Given any M, there is an $N \geq 1$ such that $\sum_{j=0}^{n} a_j > M$ for all $n \geq N$. Since $\lim_{t \to 1-} \sum_{j=0}^{n} a_j t^j = \sum_{j=1}^{n} a_j > M$ for all $n \geq N$,

$$A(1-) = \lim_{t \to 1-} \sum_{j=0}^{\infty} a_j t^j \geq \lim_{t \to 1-} \sum_{j=0}^{n} a_j t^j > M.$$

Since M is arbitrary, $A(1-) = +\infty$. Suppose now that $\sum_{j=0}^{\infty} a_j < +\infty$. Let $L = \sum_{j=0}^{\infty} a_j$. Given any $\varepsilon > 0$, there is an $N \geq 1$ such that $\sum_{j=0}^{n} a_j > L - \varepsilon$ for all $n \geq N$. Thus,

$$L \geq \lim_{t \to 1-} \sum_{j=0}^{\infty} a_j t^j \geq \lim_{t \to 1-} \sum_{j=0}^{n} a_j t^j > L - \varepsilon.$$

Since ε is arbitrary, $A(1-) = \lim_{t \to 1-} \sum_{j=0}^{\infty} a_j t^j = L = \sum_{j=0}^{\infty} a_j$. ■

The point of Abel's theorem is that if we formally put $t = 1$ in the equation $A(t) = \sum_{j=0}^{\infty} a_j t^j$, then $A(1-) = A(1)$, even if infinite. In working with power series on $(-1, 1)$ having nonnegative coefficients, we will put $A(1) = \sum_{j=0}^{\infty} a_j$, even if infinite.

Let X be a nonnegative integer-valued random variable with density function f_X and generating function $\hat{f}_X$. In this case, it is possible for $E[X] = \sum_{x=0}^{\infty} x f_X(x) = +\infty$. We will need a *tail probability function* defined by

$$g_X(x) = P(X > x), \qquad x = 0, 1, \ldots$$

and its generating function

$$\hat{g}_X(t) = \sum_{x=0}^{\infty} g_X(x) t^x, \qquad -1 < t \leq 1.$$

Note that g_X is not a density function. The functions $\hat{f}_X$ and $\hat{g}_X$ are related by the following equation:

$$\hat{g}_X(t) = \frac{1 - \hat{f}_X(t)}{1 - t}, \qquad -1 < t < 1. \tag{4.1}$$

To see this, write

$$\begin{aligned}(1 - t)\hat{g}_X(t) &= (1 - t) \sum_{x=0}^{\infty} g_X(x) t^x \\ &= \sum_{x=0}^{\infty} g_X(x) t^x - \sum_{x=0}^{\infty} g_X(x) t^{x+1}\end{aligned}$$

$$= g_X(0) + \sum_{x=1}^{\infty} g_X(x)t^x - \sum_{x=1}^{\infty} g_X(x-1)t^x$$

$$= 1 - f_X(0) - \sum_{x=1}^{\infty} (g_X(x-1) - g_X(x))t^x.$$

Noting that $g_X(x-1) - g_X(x) = P(X > x-1) - P(X > x) = P(X = x) = f_X(x)$ for $x \geq 1$,

$$(1-t)\hat{g}_X(t) = 1 - \hat{f}_X(t).$$

This establishes Equation 4.1.

Since the interval of convergence of the power series defining $\hat{f}_X$ contains $(-1, 1)$,

$$\hat{f}_X'(t) = \frac{d}{dt}\sum_{x=0}^{\infty} f_X(x)t^x = \sum_{x=0}^{\infty} x f_X(x)t^{x-1}, \qquad -1 < t < 1,$$

and therefore $E[X] = \sum_{x=0}^{\infty} x f_X(x) = \hat{f}_X'(1)$. Similarly, $E[X(X-1)] = \sum_{x=0}^{\infty} x(x-1) f_X(x) = \hat{f}_X''(1)$, even if infinite.

Theorem 4.2.4 *If X is a nonnegative integer-valued random variable, then*

$$E[X] = \hat{f}_X'(1) = \hat{g}_X(1)$$

whether finite or infinite.

PROOF: By Equation 4.1 and the mean value theorem for derivatives,

$$\hat{g}_X(t) = \frac{1 - \hat{f}_X(t)}{1 - t} = \hat{f}_X'(\xi)$$

where $t < \xi < 1$. By Abel's theorem,

$$\hat{g}_X(1) = \hat{g}_X(1-) = \lim_{t \to 1^-} \frac{1 - \hat{f}_X(t)}{1-t} = \hat{f}_X'(1-) = \hat{f}_X'(1) = E[X]. \blacksquare$$

EXAMPLE 4.7 Let X have a negative binomial density with parameters r and p so that $\hat{f}_X(t) = p^r(1-qt)^{-r}$. Then $\hat{f}_X'(t) = rp^r q(1-qt)^{-r-1}$, and so $E[X] = \hat{f}_X'(1) = r(q/p)$. ■

It was shown in Section 2.7 that a remotely operated garage door opener is anything but secure. Let $\{X_j\}_{j=1}^{\infty}$ be an infinite sequence of Bernoulli random variables with probability of success $p = 1/2$. We have seen that the word 1001 will occur infinitely often in an outcome with probability 1.

EXAMPLE 4.8 (Password Problem) Consider the sequence just described and define a random waiting time T by putting $T = n$ if the word 1001 appears for the first time at the end of the nth trial so that $T \geq 4$. Let $g_T(n) = P(T > n)$; i.e., $g_T(n)$ is the probability that the word 1001 does not occur in the first n trials. Note that $g_T(0) = g_T(1) = g_T(2) = g_T(3) = 1$. The only way for the word 1001 not to appear in the first n trials for $n \geq 4$ is for an outcome to begin with one of the following starting patterns:

$$0\cdots, 11\cdots, 101\cdots, 1000\cdots,$$

and the word 1001 does not subsequently appear. The event consisting of those outcomes with the starting pattern $0\cdots$ and the word 1001 does not subsequently appear in the remaining $n - 1$ trials has probability $(1/2)g_T(n-1)$. A similar argument applies to the other three starting patterns. Therefore, the $g_T(n)$ must satisfy the difference equation

$$\begin{aligned} g_T(n) &= \frac{1}{2}g_T(n-1) + \frac{1}{4}g_T(n-2) \\ &+ \frac{1}{8}g_T(n-3) + \frac{1}{16}g_T(n-4), \qquad n \geq 4 \end{aligned} \tag{4.2}$$

subject to the initial conditions

$$g_T(0) = g_T(1) = g_T(2) = g_T(3) = 1. \tag{4.3}$$

Multiplying both sides of Equation 4.2 by t^n and summing over $n \geq 4$,

$$\begin{aligned} &\hat{g}_T(t) - 1 - t - t^2 - t^3 \\ &= \frac{t}{2}(\hat{g}_T(t) - 1 - t - t^2) + \frac{t^2}{4}(\hat{g}_T(t) - 1 - t) \\ &+ \frac{t^3}{8}(\hat{g}_T(t) - 1) + \frac{t^4}{16}\hat{g}_T(t). \end{aligned}$$

Solving for $\hat{g}_T$,

$$\hat{g}_T(t) = \frac{16 + 8t + 4t^2 + 2t^3}{16 - 8t - 4t^2 - 2t^3 - t^4}.$$

Since $\hat{f}_T(t) = 1 - (1-t)\hat{g}_T(t)$, which is a rational function of t, we could in principle determine the density f_T by applying the method of partial fraction expansions to $\hat{f}_T(t)$; this requires, however, finding the roots of the polynomial in the denominator of $\hat{g}_T$. If all we are interested in is $E[T]$, then these problems can be avoided by using Theorem 4.2.4 to obtain $E[T] = \hat{g}_T(1) = 30$. On the average, it will take about 30 trials for the word 1001 to appear in an outcome. ■

EXERCISES 4.2

1. Let X have a geometric density with parameter p. Find $E[X^2]$.
2. If X has a Poisson density function $p(\cdot; \lambda)$, calculate $E[X^2]$.
3. A random sample of size 3 is drawn from a bowl containing 10 white and 5 red balls. If X is the number of white balls in the sample, find $E[X]$.
4. Let X be a random variable having a Poisson density with parameter $\lambda > 0$. Calculate $E[1/(1+X)]$.
5. Let X be a random variable having a negative binomial density with parameters $r \geq 2$ and p. Calculate $E[1/(X+1)]$.
6. If X is a nonnegative integer-valued random variable, show that $E[X] = \sum_{x=1}^{\infty} P(X \geq x)$.
7. Let $\{X_j\}_{j=1}^{\infty}$ be a sequence of independent nonnegative integer-valued random variables all having the same density function for which $E[X_j] = E[X_1]$ is defined as a real number, and let N be a positive integer-valued random variable such that $E[N]$ is defined as a real number. Assume that $N, X_1, X_2, \ldots$ are independent. If $S_N = X_1 + X_2 + \cdots + X_N$, show that $E[S_N] = E[N]E[X_1]$.
8. A remotely operated garage door opener has an electronic combination lock of 10 binary digits. If a random device transmits a signal that has the probability properties of an infinite sequence of Bernoulli trials with probability of success $p = 1/2$, what is the expected number of digits required to activate the opener?

The following problem requires mathematical software such as Mathematica or Maple V.

9. Consider the random variable T of Example 4.8. Calculate $P(T \leq 11)$.

4.3 PROPERTIES OF EXPECTATION

In Chapter 3, we defined functions of random variables such as $\phi(X, Y)$ and $\psi(X_1, \ldots, X_n)$. In the case of a single random variable X, it was shown that the expected value of $Z = \phi(X)$ could be calculated without going through the intermediate step of determining the density function of Z. A similar result applies to a function of several random variables.

Theorem 4.3.1 *If $X_1, \ldots, X_n$ are random variables and ψ is a real-valued function of n variables, then*

$$E[\psi(X_1, \ldots, X_n)] = \sum_{x_1,\ldots,x_n} \psi(x_1, \ldots, x_n) f_{X_1,\ldots,X_n}(x_1, \ldots, x_n) \quad (4.4)$$

provided the multiple series on the right converges absolutely.

Operationally, the sum on the right is obtained by replacing $X_1, \ldots, X_n$ by typical values $x_1, \ldots, x_n$, respectively, multiplying by the probability that the random variables will take on those values, and then summing in any order over all possible values of the random variables.

The proof of Theorem 4.3.1 amounts to a justification of the rearrangement of the terms of the series. The reader is referred to Theorem 12–42 in the book by Apostol listed at the end of the chapter.

Theorem 4.3.2 *If X and Y are random variables with finite expectations and $P(X \geq Y) = 1$, then $E[X] \geq E[Y]$.*

PROOF: Since $f_{X,Y}(x_i, y_j) = 0$ whenever $x_i < y_j$, by Theorem 4.3.1,

$$\begin{aligned} E[X] &= \sum_{x_i, y_j} x_i f_{X,Y}(x_i, y_j) \\ &= \sum_{x_i \geq y_j} x_i f_{X,Y}(x_i, y_j) \\ &\geq \sum_{x_i \geq y_j} y_j f_{X,Y}(x_i, y_j) \\ &= \sum_{x_i, y_j} y_j f_{X,Y}(x_i, y_j) \\ &= E[Y]. \ \blacksquare \end{aligned}$$

Theorem 4.3.3 *If $X_1, \ldots, X_n$ are any random variables with finite expectations and $c_1, \ldots, c_n$ are any real constants, then $\sum_{j=1}^{n} c_j X_j$ has finite expectation and*

$$E\left[\sum_{j=1}^{n} c_j X_j\right] = \sum_{j=1}^{n} c_j E[X_j].$$

PROOF: By Theorem 4.2.2, we can assume that the $c_j = 1, j = 1, \ldots, n$. Taking $\psi(x_1, \ldots, x_n) = x_1 + \cdots + x_n$ in Theorem 4.3.1, we must first show that the series therein is absolutely convergent. Since $|x_1 + \cdots + x_n| \leq$

$|x_1| + \cdots + |x_n|$,

$$\begin{aligned}
\sum_{x_1,\ldots,x_n} & |x_1 + \cdots + x_n| f_{X_1,\ldots,X_n}(x_1, \ldots, x_n) \\
&\le \sum_{x_1,\ldots,x_n} (|x_1| + \cdots + |x_n|) f_{X_1,\ldots,X_n}(x_1, \ldots, x_n) \\
&= \sum_{x_1,\ldots,x_n} |x_1| f_{X_1,\ldots,X_n}(x_1, \ldots, x_n) + \cdots + \sum_{x_1,\ldots,x_n} |x_n| f_{X_1,\ldots,X_n}(x_1, \ldots, x_n).
\end{aligned}$$

By Theorem 3.4.3, a suitable order of iterated summation can be chosen so that

$$\begin{aligned}
\sum_{x_1,\ldots,x_n} & |x_j| f_{X_1,\ldots,X_n}(x_1, \ldots, x_n) \\
&= \sum_{x_j} \left(\sum_{x_1,\ldots,x_{j-1},x_{j+1},\ldots,x_n} |x_j| f_{X_1,\ldots,X_n}(x_1, \ldots, x_n) \right) \\
&= \sum_{x_j} |x_j| \left(\sum_{x_1,\ldots,x_{j-1},x_{j+1},\ldots,x_n} f_{X_1,\ldots,X_n}(x_1, \ldots, x_n) \right) \\
&= \sum_{x_j} |x_j| f_{X_j}(x_j),
\end{aligned}$$

and so

$$\begin{aligned}
\sum_{x_1,\ldots,x_n} |x_1 + \cdots + x_n| f_{X_1,\ldots,X_n}(x_1, \ldots, x_n) &\le \sum_{x_1} |x_1| f_{X_1}(x_1) \\
&\quad + \cdots + \sum_{x_n} |x_n| f_{X_n}(x_n).
\end{aligned}$$

Since each X_j has finite expectation, each term on the right is finite and the multiple series converges absolutely. Therefore, $E[X_1 + \cdots + X_n]$ is defined, and

$$\begin{aligned}
E[X_1 + \cdots + X_n] &= \sum_{x_1,\ldots,x_n} (x_1 + \cdots + x_n) f_{X_1,\ldots,X_n}(x_1, \ldots, x_n) \\
&= \sum_{x_1,\ldots,x_n} x_1 f_{X_1,\ldots,X_n}(x_1, \ldots, x_n) \\
&\quad + \cdots + \sum_{x_1,\ldots,x_n} x_n f_{X_1,\ldots,X_n}(x_1, \ldots, x_n) \\
&= E[X_1] + \cdots + E[X_n]. \quad \blacksquare
\end{aligned}$$

EXAMPLE 4.9 Consider an infinite sequence of Bernoulli random variables $\{X_j\}_{j=1}^{\infty}$ with probability of success p and let $S_n = X_1 + \cdots + X_n$. Then

$E[X_j] = 1 \cdot p + 0 \cdot q = p, j = 1, \ldots, n$. By Theorem 4.3.3, $E[S_n] = np$, a result obtained previously using the fact that S_n has the binomial density $b(\cdot; n, p)$. ■

The introduction of auxiliary random variables as in the next example can simplify the computation of expected value.

EXAMPLE 4.10 Suppose a population of n objects consists of n_1 objects of Type 1, n_2 objects of Type 2, ..., n_s objects of Type s, where $n = n_1 + n_2 + \cdots + n_s$. A random sample of size $r \leq n$ is taken without replacement from the population. Let X_1 be the number of Type 1 objects, X_2 the number of Type 2 objects, ..., X_s the number of Type s objects in the sample. To calculate $E[X_j]$, we define auxiliary random variables $I_{j,1}, \ldots, I_{j,r}$ as follows. Let $I_{j,k} = 1$ or 0 according to whether the kth object in the sample is of Type j or not. The value of $I_{j,k}$ is determined by looking at the kth object chosen from the population and totally disregarding the other choices. This amounts to selecting just one object. Thus, $P(I_{j,k} = 1) = n_j/n, P(I_{j,k} = 0) = (n - n_j)/n$, and therefore $E[I_{j,k}] = 0 \cdot ((n - n_j)/n) + 1 \cdot (n_j/n) = n_j/n$. Since $X_j = I_{j,1} + \cdots + I_{j,r}$,

$$E[X_j] = \sum_{k=1}^{r} E[I_{j,k}] = r\frac{n_j}{n}$$

by Theorem 4.3.3. ■

Theorem 4.3.4 *If X and Y are independent random variables with finite expectations, then XY has finite expectation and*

$$E[XY] = E[X]E[Y].$$

PROOF: Let x_i and y_j be arbitrary elements of the range of X and Y, respectively. We must first show that

$$\sum_{i,j} |x_i y_j| f_{X,Y}(x_i, y_j) < +\infty.$$

By Theorem 3.4.3,

$$\sum_{i,j} |x_i y_j| f_{X,Y}(x_i, y_j) = \sum_{i,j} |x_i||y_j| f_X(x_i) f_Y(y_j)$$

$$= \sum_i |x_i| \left(\sum_j |y_j| f_Y(y_j) \right) f_X(x_i).$$

Since the sum within the parentheses is a constant, it can be taken outside the summation over i to obtain

$$\sum_{i,j} |x_i y_j| f_{X,Y}(x_i, y_j) = \left(\sum_j |y_j| f_Y(y_j)\right)\left(\sum_i |x_i| f_X(x_i)\right) < +\infty.$$

Therefore, XY has finite expectation and

$$\begin{aligned} E[XY] &= \sum_{i,j} x_i y_j f_{X,Y}(x_i, y_j) \\ &= \sum_i x_i \left(\sum_j y_j f_Y(y_j)\right) f_X(x_i) \\ &= E[Y] \sum_i x_i f_X(x_i) \\ &= E[Y]E[X]. \blacksquare \end{aligned}$$

It is important to remember that Theorem 4.3.4 applies only to independent random variables.

EXAMPLE 4.11 Suppose two dice, one red and one white, are rolled. Let X and Y be the number of pips on the red and white die, respectively. By Exercise 1.2.6,

$$E[X] = E[Y] = \sum_{j=1}^{6} j f_Y(j) = \sum_{j=1}^{6} \frac{j}{6} = \frac{7}{2}.$$

Since X and Y are independent random variables, $E[XY] = E[X]E[Y] = 49/4$. ■

Definition 4.2 *The random variable X has a* finite second moment *if $E[X^2]$ is finite.* ■

We will need the following fact: if x is any real number, then $|x| \leq x^2 + 1$. To see this, note that if $|x| \leq 1$, then $|x| \leq x^2 + 1$ whereas if $|x| \geq 1$, then $|x| \leq |x|^2 \leq x^2 + 1$.

Consider a random variable X with finite second moment and range $\{x_1, x_2, \ldots\}$. Since $\sum_i |x_i| f_X(x_i) \leq \sum_i (x_i^2 + 1) f_X(x_i) \leq E[X^2] + 1$, X has finite expectation. In this case, we can define a parameter μ_X by

$$\mu_X = E[X]$$

which is called the *mean* or *expected value* of X.

Consider the random variable $(X-\mu_X)^2 = X^2 - 2\mu_X X + \mu_X^2$. Since X^2 and X have finite expectation, $(X-\mu_X)^2$ has finite expectation by Theorems 4.2.2 and 4.3.3, and we can define a second parameter

$$\sigma_X^2 = E[(X-\mu_X)^2],$$

called the *variance* of X. The variance of X is also denoted by var X. $\sigma_X = \sqrt{\operatorname{var} X}$ is called the *standard deviation* of X. If the random variable X is clear from the context, the subscript X on μ_X and σ_X will be omitted. Since $(X-\mu_X)^2 = X^2 - 2\mu_X X + \mu_X^2$ and $E[\mu_X X] = \mu_X E[X] = (E[X])^2$, by Theorem 4.2.2

$$\operatorname{var} X = \sigma_X^2 = E[(X-\mu_X)^2] = E[X^2] - (E[X])^2.$$

It is easily checked that $\operatorname{var}(aX) = a^2 \operatorname{var} X$ and $\operatorname{var}(X+c) = \operatorname{var} X$.

EXAMPLE 4.12 Let X be a random variable having a uniform density on $\{0, 1, \ldots, n\}$. It was shown in the previous section that $E[X] = n/2$. By Exercise 1.2.6,

$$E[X^2] = \sum_{x=0}^{n} x^2 \frac{1}{n+1} = \frac{1}{n+1}[1^2 + 2^2 + \cdots + n^2] = \frac{n(2n+1)}{6},$$

$$\operatorname{var} X = \sigma^2 = \frac{n(2n+1)}{6} - \left(\frac{n}{2}\right)^2 = \frac{n(n+2)}{12}. \blacksquare$$

EXAMPLE 4.13 Let X be a random variable having a binomial density $b(\cdot; n, p)$. It was shown in the previous section that $E[X] = np$ and that $E[X^2] = n^2p^2 - np^2 + np$, so that

$$\operatorname{var} X = E[X^2] - (E[X])^2 = np(1-p). \blacksquare$$

EXAMPLE 4.14 Let $X_1, \ldots, X_n$ be independent random variables all having the same density function and finite second moments. Let $\mu = E[X_j]$ and $\sigma^2 = \operatorname{var} X_j$, $1 \le j \le n$, and let $S_n = X_1 + \cdots + X_n$. By Theorem 4.3.3, $E[S_n] = n\mu$. The variance of S_n can be calculated using the equation

$$\begin{aligned}(S_n - n\mu)^2 &= \left(\sum_{j=1}^{n}(X_j-\mu)\right)^2 \\ &= \sum_{j=1}^{n}(X_j-\mu)^2 + \sum_{i\neq j}(X_i-\mu)(X_j-\mu),\end{aligned}$$

provided the terms on the right have finite expectations. The terms of the first sum have finite expectations because the X_j have finite second moments. By independence and Theorem 4.3.4, the terms of the second sum have finite expectations, and $E[(X_i - \mu)(X_j - \mu)] = E[X_i - \mu]E[X_j - \mu] = 0$. Thus,

$$\operatorname{var} S_n = E[(S_n - n\mu)^2] = \sum_{j=1}^{n} E[(X_j - \mu)^2] = n\sigma^2. \quad \blacksquare$$

Let X be a nonnegative integer-valued random variable with finite second moment. We have seen that $E[X]$ can be calculated from the generating function $\hat{f}_X$ by the equation $E[X] = \hat{f}'_X(1)$. $\operatorname{var} X$ can also be calculated from the generating function. In fact,

$$\operatorname{var} X = \hat{f}''_X(1) + \hat{f}'_X(1) - [\hat{f}'_X(1)]^2. \tag{4.5}$$

To see this, recall that $\hat{f}_X(t) = \sum_{x=0}^{\infty} f_X(x)t^x$, so that

$$\hat{f}''_X(t) = \sum_{x=0}^{\infty} x(x-1)f_X(x)t^{x-2}$$

on the interval $(-1, 1)$. By Abel's theorem,

$$\hat{f}''_X(1) = \sum_{x=0}^{\infty} x(x-1)f_X(x) = E[X(X-1)] = E[X^2] - E[X],$$

and so $E[X^2] = \hat{f}''_X(1) + \hat{f}'_X(1)$ and $\operatorname{var} X = \hat{f}''_X(1) + \hat{f}'_X(1) - [\hat{f}'_X(1)]^2$.

EXAMPLE 4.15 Let X be a random variable having a Poisson density with parameter $\lambda > 0$. Then $\hat{f}_X(t) = e^{\lambda(t-1)}, \hat{f}'_X(t) = \lambda e^{\lambda(t-1)}, \hat{f}''_X(t) = \lambda^2 e^{\lambda(t-1)}$. Thus, $\operatorname{var} X = \hat{f}''_X(1) + \hat{f}'_X(1) - [\hat{f}'_X(1)]^2 = \lambda^2 + \lambda - \lambda^2 = \lambda$. ■

The mean and variance of a random variable X are just two parameters that summarize some of the information in its density function. Even though in most cases they do not determine the density, they can provide information about probabilities.

Lemma 4.3.5 (Markov's Inequality) *If X is any random variable with finite expectation and $t > 0$, then*

$$P(|X| \geq t) \leq \frac{E[|X|]}{t}.$$

PROOF: Let $\{x_1, x_2, \ldots\}$ be the range of X. Since the series defining $E[X]$ is absolutely convergent, $E[|X|] < +\infty$. By Theorem 4.2.1,

$$\begin{aligned} E[|X|] &= \sum_j |x_j| f_X(x_j) \\ &\geq \sum_{|x_j| \geq t} |x_j| f_X(x_j) \geq t \sum_{|x_j| \geq t} f_X(x_j) \\ &= tP(|X| \geq t). \quad \blacksquare \end{aligned}$$

The next inequality is an easy consequence of Markov's inequality.

Theorem 4.3.6 (Chebyshev's Inequality) *Let X be a random variable with mean μ and finite variance σ^2. Then*

$$P(|X - \mu| \geq \delta) \leq \frac{\sigma^2}{\delta^2}$$

for all $\delta > 0$.

PROOF: By Markov's inequality,

$$P(|X - \mu| \geq \delta) = P((X - \mu)^2 \geq \delta^2) \leq \frac{E[(X - \mu)^2]}{\delta^2} = \frac{\sigma^2}{\delta^2}. \quad \blacksquare$$

Consider an infinite sequence of Bernoulli trials $\{X_j\}_{j=1}^{\infty}$ with probability of success p and let $S_n = X_1 + \cdots + X_n$ be the number of successes in n trials. We know that S_n has a $b(\cdot; n, p)$ density, that $E[S_n] = np$, and that $\operatorname{var} S_n = np(1 - p)$. By Chebyshev's inequality,

$$P\left(\left|\frac{S_n}{n} - p\right| \geq \delta\right) = P(|S_n - np| \geq n\delta) \leq \frac{\operatorname{var} S_n}{n^2\delta^2} = \frac{p(1-p)}{n\delta^2}.$$

By maximizing the function $g(p) = p(1 - p), 0 \leq p \leq 1$, it is easily seen that the maximum value of g is $1/4$. Thus,

$$P\left(\left|\frac{S_n}{n} - p\right| \geq \delta\right) \leq \frac{1}{4n\delta^2}. \tag{4.6}$$

Since S_n represents the number of successes in n trials, S_n/n represents the relative frequency of successes in n trials. Taking the limit as $n \to \infty$,

$$\lim_{n \to \infty} P\left(\left|\frac{S_n}{n} - p\right| \geq \delta\right) = 0 \tag{4.7}$$

for all $\delta > 0$; i.e., given a prescribed error $\delta > 0$, the probability that the relative frequency S_n/n will differ from p by more than δ goes to zero as $n \to \infty$. This sounds suspiciously like the empirical law for relative frequencies, but it is, in fact, a mathematical theorem.

Inequality 4.6 can be used to determine how many repetitions of an experiment are required to pin down the probability of success p when it is unknown.

EXAMPLE 4.16 Consider an infinite sequence of Bernoulli trials with probability of success p. How many repetitions are required to be 97 percent confident that the relative frequency of success S_n/n will be within .05 of p? That is, how do we choose n so that

$$P\left(\left|\frac{S_n}{n} - p\right| \geq .05\right) \leq .03\,?$$

By Inequality 4.6, if we choose n so that

$$\frac{1}{4n(.05)^2} \leq .03,$$

then the above condition will be satisfied. Therefore,

$$n \geq \frac{1}{4(.05)^2(.03)} = 3333.3,$$

and so n can be taken to be 3334. ■

The number $n = 3334$ in this example is rather large, but it must be remembered that nothing has been assumed about p. Any preliminary information about p can reduce the number n by several factors; e.g., if it is known that p pertains to an event that is relatively uncommon, say $p \leq 1/10$, then $p(1-p) \leq 9/100$ and n can be reduced to 1200.

The fact that $\lim_{n\to\infty} P(|(S_n/n) - p| \geq \delta) = 0$ for all $\delta > 0$ was first proved by Jacob Bernoulli around 1713. It is a special case of a slightly more general result.

Theorem 4.3.7 (Weak Law of Large Numbers) *Let $\{X_j\}_{j=1}^{\infty}$ be a sequence of independent random variables all having the same density function and finite second moments. If $S_n = X_1 + \cdots + X_n$ and $\mu = E[X_j], j \geq 1$, then*

$$\lim_{n\to\infty} P\left(\left|\frac{S_n}{n} - \mu\right| \geq \delta\right) = 0$$

for all $\delta > 0$.

PROOF: Let $\sigma^2 = \operatorname{var} X_j, j \geq 1$. By Theorem 4.3.3 and Example 4.14, $E[S_n] = n\mu$ and $\operatorname{var} S_n = n\sigma^2$. Thus, for each $\delta > 0$,

$$P\left(\left|\frac{S_n}{n} - \mu\right| \geq \delta\right) = P(|S_n - n\mu| \geq n\delta) \leq \frac{\operatorname{var} S_n}{n^2\delta^2} = \frac{\sigma^2}{n\delta^2} \to 0$$

as $n \to \infty$. ■

There is a strong law of large numbers that reflects the empirical law more precisely than the weak law. The strong law is beyond the scope of this book.

EXERCISES 4.3

1. Suppose two dice are rolled, one red and one white. Let X be the number of pips on the red die, let Y be the number of pips on the white die, and let Z be the larger of the two numbers of pips. The joint density of $f_{X,Z}$ is given by Equation 3.6. Calculate $E[XZ]$.
2. Let X and Y be as in Problem 1 and let $U = \min(X, Y)$. Calculate $E[U]$.
3. Let X be a random variable having a geometric density with parameter p. Use the generating function $\hat{f}_X(t)$ to find $\operatorname{var} X$.
4. Let X be a random variable having generating function $\hat{f}(t) = e^{2(t^4-1)}$. Calculate $\operatorname{var} X$.
5. Let X be a random variable having a negative binomial density with parameters r and p. Calculate $\operatorname{var} X$.
6. A manufacturer produces items of which 3 percent are defective. The manufacturer contracts to sell 10,000 items to a buyer with the stipulation that if the number of defective items exceeds d units, then the buyer can claim a full refund. How should d be chosen so that the manufacturer does not have to give a refund to more than 5 percent of the buyers?
7. Consider an infinite sequence of Bernoulli trials with probability of success p for which it is known that $p \leq 1/4$. How many trials are required to be 90 percent confident that the relative frequency of successes S_n/n will be within .05 of p?

The next three problems pertain to a population of n objects of which n_1 are of Type 1, n_2 are of Type 2, . . . , n_s are of Type s.

8. If Type 1 objects have value V_1, Type 2 have value $V_2, \ldots$, Type s have value V_s, and V is the value of a random sample of size $r \leq n$ without replacement from the population, derive a formula for $E[V]$.

9. A commercial fisherman is allowed to net 50 game fish each month from a lake in which 30 percent of the fish are largemouth bass, 10 percent are smallmouth bass, 20 percent are white bass, and 40 percent are walleyes. If the largemouth bass average 2.5 pounds, the smallmouth bass 1.8 pounds, the white bass 1.2 pounds, and the walleye 2.4 pounds, what is the expected weight of his catch?
10. For $j = 1, \ldots, s$, let X_j be the number of Type j objects in a random sample of size $r \leq n$ without replacement from the population. Use the auxiliary random variables of Example 4.10 to calculate var X_j.
11. Let X be a random variable having a finite second moment. If $\mu = E[X]$ and $\sigma^2 = \text{var}\, X = 0$, show that $X = \mu$ with probability 1.
12. If X is a nonnegative integer-valued random variable, then $E[X] = \hat{f}'_X(1) = \hat{g}_X(1)$. Assuming that $E[X^2]$ is finite, express var X in terms of $\hat{g}_X$.
13. Calculate the standard deviation σ_T of the waiting time T of Example 4.8.

4.4 COVARIANCE AND CORRELATION

It was shown in the previous section that the expected value of a sum of random variables is the sum of the expected values. Is this true of variances? Generally speaking, it is not true.

A simple inequality will be needed for the next result. If a, b are any real numbers, then

$$(a + b)^2 \leq 2(a^2 + b^2).$$

This follows from the fact that $a^2 - 2ab + b^2 = (a - b)^2 \geq 0$, so that $2ab \leq a^2 + b^2$ and $(a + b)^2 = a^2 + 2ab + b^2 \leq 2(a^2 + b^2)$.

Lemma 4.4.1 *If $X_1, \ldots, X_n$ are random variables with finite second moments and $c_1, \ldots, c_n$ are any real numbers, then $\sum_{j=1}^{n} c_j X_j$ has a finite second moment.*

PROOF: If the random variable X with range $\{x_1, x_2, \ldots\}$ has finite second moment and $c \in R$, then cX has finite second moment since $\sum_j (cx_j)^2 f_X(x_j) = c^2 \sum_j x_j^2 f_X(X_j) < +\infty$. Thus, each $c_j X_j$ has finite second moment and it can be assumed that $c_j = 1, 1 \leq j \leq n$. We prove the result for the $n = 2$ case first. Since $(X_1 + X_2)^2 \leq 2(X_1^2 + X_2^2)$, by Theorem 4.3.2 $E[(X_1 + X_2)^2] \leq 2(E[X_1^2] + E[X_2^2]) < +\infty$, and $X_1 + X_2$ has finite second moment. The general case follows from a mathematical induction argument. ■

Consider two random variables X and Y with finite second moments. Since $\mu_{X+Y} = E[X + Y] = \mu_X + \mu_Y$,

$$\begin{aligned}\operatorname{var}(X+Y) &= E[((X+Y)-(\mu_X+\mu_Y))^2] \\ &= E[((X-\mu_X)+(Y-\mu_Y))^2] \\ &= E[(X-\mu_X)^2]+E[(Y-\mu_Y)^2]+2E[(X-\mu_X)(Y-\mu_Y)] \\ &= \operatorname{var} X+\operatorname{var} Y+2E[(X-\mu_X)(Y-\mu_Y)].\end{aligned}$$

The last term will be given a name of its own. But we must first establish that it is finite.

Theorem 4.4.2 (Schwarz's Inequality) *If X and Y have finite second moments, then*

$$(E[XY])^2 \leq E[X^2]E[Y^2]; \tag{4.8}$$

equality holds if and only if $P(X = 0) = 1$ or $P(Y = aX) = 1$ for some constant a.

PROOF: Either $P(X = 0) = 1$ or $P(X = 0) < 1$. In the first case, equality holds in Equation 4.8 because both sides are zero. We can therefore assume that $P(X = 0) < 1$, which means that X takes on some value $x_0 \neq 0$ with positive probability, so that $E[X^2] = \sum_j x_j^2 f_X(x_j) > 0$. Define a quadratic function by the equation

$$g(\lambda) = E[(Y-\lambda X)^2] = E[Y^2] - 2\lambda E[XY] + \lambda^2 E[X^2].$$

This function has a minimum value at

$$\lambda_0 = \frac{E[XY]}{E[X^2]}.$$

Thus, $0 \leq E[(Y-\lambda_0 X)^2] \leq E[(Y-\lambda X)^2]$ for all real numbers λ. Replacing λ_0 by $E[XY]/E[X^2]$,

$$\begin{aligned}E[(Y-\lambda_0 X)^2] &= E[Y^2] - 2\lambda_0 E[XY] + \lambda_0^2 E[X^2] \\ &= E[Y^2] - 2\frac{(E[XY])^2}{E[X^2]} + \frac{(E[XY])^2}{E[X^2]} \\ &= E[Y^2] - \frac{(E[XY])^2}{E[X^2]},\end{aligned}$$

and so

$$0 \leq E[(Y-\lambda_0 X)^2] = E[Y^2] - \frac{(E[XY])^2}{E[X^2]} \leq E[(Y-\lambda X)^2].$$

On the one hand, this implies that

$$(E[XY])^2 \leq E[X^2]E[Y^2];$$

on the other hand, if there is equality then $E[(Y - \lambda_0 X)^2] = 0$. If $Y - \lambda_0 X$ takes on some nonzero value with positive probability, we would have $E[(Y - \lambda_0 X)^2] > 0$, a contradiction. Thus, $P(Y - \lambda_0 X = 0) = 1$. ■

If X and Y have finite second moments, then we know that both $E[(X - \mu_X)^2]$ and $E[(Y - \mu_Y)^2]$ are finite. Applying Inequality 4.8 to the random variables $X - \mu_X$ and $Y - \mu_Y$,

$$(E[(X - \mu_X)(Y - \mu_Y)])^2 \leq E[(X - \mu_X)^2]E[(Y - \mu_Y)^2] < +\infty,$$

and therefore $E[(X - \mu_X)(Y - \mu_Y)]$ is defined.

Definition 4.3 *If X and Y have finite second moments, the* covariance *of X and Y, denoted by* $\operatorname{cov}(X, Y)$, *is defined by*

$$\operatorname{cov}(X, Y) = E[(X - \mu_X)(Y - \mu_Y)].$$

Alternatively,

$$\operatorname{cov}(X, Y) = E[XY] - E[X]E[Y]. \blacksquare$$

Note that $\operatorname{cov}(X, c) = E[(X - \mu_X)(c - c)] = 0$ whenever c is a constant, that $\operatorname{cov}(X, X) = E[X^2] - (E[X])^2 = \operatorname{var} X$, and also that $\operatorname{cov}(X, Y) = 0$ whenever X and Y are independent, by Theorem 4.3.4. We now return to the variance of a sum.

Theorem 4.4.3 *If $X_1, \ldots, X_n$ have finite second moments, then*

$$\operatorname{var}\left(\sum_{j=1}^{n} X_j\right) = \sum_{j=1}^{n} \operatorname{var} X_j + 2 \sum_{1 \leq i < j \leq n} \operatorname{cov}(X_i, X_j).$$

PROOF: Since $E[\sum_{j=1}^{n} X_j] = \sum_{j=1}^{n} \mu_{X_j}$,

$$\begin{aligned}
\operatorname{var}\left(\sum_{j=1}^{n} X_j\right) &= E\left[\left(\sum_{j=1}^{n} X_j - \sum_{j=1}^{n} \mu_{X_j}\right)^2\right] \\
&= E\left[\left(\sum_{j=1}^{n} (X_j - \mu_{X_j})\right)^2\right] \\
&= E\left[\sum_{i,j=1}^{n} (X_i - \mu_{X_i})(X_j - \mu_{X_j})\right]
\end{aligned}$$

$$= \sum_{j=1}^{n} E[(X_j - \mu_{X_j})^2] + \sum_{i,j=1, i \neq j}^{n} E[(X_i - \mu_{X_i})(X_j - \mu_{X_j})]$$

$$= \sum_{j=1}^{n} \operatorname{var} X_j + 2 \sum_{1 \leq i < j \leq n} \operatorname{cov}(X_i, X_j). \blacksquare$$

Corollary 4.4.4 *If $X_1, \ldots, X_n$ are independent random variables having finite second moments, then*

$$\operatorname{var}\left(\sum_{j=1}^{n} X_j\right) = \sum_{j=1}^{n} \operatorname{var} X_j.$$

PROOF: For $i \neq j, X_i$ and X_j are independent and $\operatorname{cov}(X_i, X_j) = 0$. ■

There is a more general version of Theorem 4.4.3.

Theorem 4.4.5 *Let $X_1, \ldots, X_m, Y_1, \ldots, Y_n$ have finite second moments and let $a_1, \ldots, a_m$, $b_1, \ldots, b_n$ be arbitrary real numbers. Then*

$$\operatorname{cov}\left(\sum_{i=1}^{m} a_i X_i, \sum_{j=1}^{n} b_j Y_j\right) = \sum_{i=1}^{m} \sum_{j=1}^{n} a_i b_j \operatorname{cov}(X_i, Y_j).$$

PROOF: By Theorem 4.3.3,

$$E\left[\sum_{i=1}^{m} a_i X_i\right] = \sum_{i=1}^{m} a_i E[X_i]$$

and

$$E\left[\sum_{j=1}^{n} b_j Y_j\right] = \sum_{j=1}^{n} b_j E[Y_j].$$

Since

$$E\left[\left(\sum_{i=1}^{m} a_i X_i\right)\left(\sum_{j=1}^{n} b_j Y_j\right)\right] = E\left[\sum_{i=1}^{m} \sum_{j=1}^{n} a_i b_j X_i X_j\right] = \sum_{i=1}^{m} \sum_{j=1}^{n} a_i b_j E[X_i Y_j],$$

$$\begin{aligned}\operatorname{cov}\left(\sum_{i=1}^{m} a_i X_i, \sum_{j=1}^{n} b_j Y_j\right) &= E\left[\left(\sum_{i=1}^{m} a_i X_i\right)\left(\sum_{j=1}^{n} b_j Y_j\right)\right] \\ &\quad - \left(\sum_{i=1}^{m} a_i E[X_i]\right)\left(\sum_{j=1}^{n} b_j E[Y_j]\right) \\ &= \sum_{i=1}^{m}\sum_{j=1}^{n} a_i b_j E[X_i Y_j] \\ &\quad - \left(\sum_{i=1}^{m} a_i E[X_i]\right)\left(\sum_{j=1}^{n} b_j E[Y_j]\right) \\ &= \sum_{i=1}^{m}\sum_{j=1}^{n} a_i b_j (E[X_i Y_j] - E[X_i]E[Y_j]) \\ &= \sum_{i=1}^{m}\sum_{j=1}^{n} a_i b_j \operatorname{cov}(X_i, Y_j). \quad \blacksquare\end{aligned}$$

EXAMPLE 4.17 Consider an experiment in which balls numbered 1, $2, \ldots, n$ are distributed at random in n boxes so that the total number of outcomes is $n!$. Let S_n be the number of matches; i.e., the number of balls in boxes having the same number. The range of S_n is $\{0, 1, \ldots, n\}$. Suppose we want to calculate $E[S_n]$ and $\operatorname{var} S_n$. For $j = 1, \ldots, n$, let $X_j = 1$ or 0 according to whether the jth ball is in the jth box or not. Then $S_n = X_1 + \cdots + X_n$. Since $P(X_j = 1) = (n-1)!/n! = 1/n, E[X_j] = 1/n$. Thus, $E[S_n] = 1$. Since $X_j^2 = X_j$,

$$\operatorname{var} X_j = E[X_j^2] - (E[X_j])^2 = E[X_j] - (E[X_j])^2 = \frac{1}{n} - \frac{1}{n^2} = \frac{n-1}{n^2}.$$

We now calculate $E[X_j X_k]$ for $j \neq k$. Now $X_j X_k$ is 1 or 0 according to whether the jth and kth balls are in the corresponding boxes or not. Thus, $P(X_j X_k = 1) = (n-2)!/n! = 1/(n(n-1))$, so that for $j \neq k$,

$$\operatorname{cov}(X_j X_k) = E[X_j X_k] - E[X_j]E[X_k] = \frac{1}{n^2(n-1)}.$$

By Theorem 4.4.3,

$$\operatorname{var} S_n = \sum_{j=1}^{n} \operatorname{var} X_j + 2 \sum_{1 \le i < j \le n} \operatorname{cov}(X_i, X_j).$$

Since all the terms in the second sum are equal to $1/(n^2(n-1))$ and the number of terms is the number of ways of selecting two distinct integers i and j from $\{1,\ldots,n\}$ without regard to order,

$$\operatorname{var} S_n = n \cdot \frac{n-1}{n^2} + 2\binom{n}{2}\frac{1}{n^2(n-1)} = 1.$$

Therefore, $E[S_n] = 1$ and $\operatorname{var} S_n = 1$. ■

There are good reasons for replacing the random variable X by a centered and normalized random variable $(X - \mu_X)/\sigma_X$.

Definition 4.4 *If X and Y are two random variables having finite second moments, the* correlation *between X and Y, denoted by $\rho(X,Y)$, is defined by*

$$\rho(X,Y) = E\left[\left(\frac{X-\mu_X}{\sigma_X}\right)\left(\frac{Y-\mu_Y}{\sigma_Y}\right)\right] = \frac{\operatorname{cov}(X,Y)}{\sigma_X\sigma_Y}. \quad ■$$

The following result is of interest in its own right and also tells us something about $\rho(X,Y)$.

If X and Y are independent random variables with finite second moments, then $\rho(X,Y) = 0$ since $\operatorname{cov}(X,Y) = 0$ in this case. The converse is not true in general. It is possible for $\rho(X,Y) = 0$ without X and Y being independent. Also, replacing X and Y in $\rho(X,Y)$ by certain linear functions of X and Y, respectively, does not change the correlation; i.e.,

$$\rho(aX+b, cY+d) = \rho(X,Y) \text{ whenever } a > 0, c > 0.$$

This follows from the fact that $\operatorname{var}(aX+b) = E[((aX+b)-(a\mu_X+b))^2] = E[a^2(X-\mu_X)^2] = a^2 \operatorname{var} X$, $\operatorname{var}(cY+d) = c^2 \operatorname{var} Y$, and

$$\begin{aligned}\operatorname{cov}(aX+b, cY+d) &= E[((aX+b)-(a\mu_X+b))((cY+d)-(c\mu_Y+d))] \\ &= E[ac(X-\mu_X)(Y-\mu_Y)] \\ &= ac \operatorname{cov}(X,Y),\end{aligned}$$

so that

$$\rho(aX+b, cY+d) = \frac{ac \operatorname{cov}(X,Y)}{a\sigma_X c\sigma_Y} = \rho(X,Y) \tag{4.9}$$

whenever $a > 0, c > 0$.

Theorem 4.4.6 *Let X and Y be random variables with finite second moments, $\sigma_X > 0$, $\sigma_Y > 0$. Then $|\rho(X,Y)| \le 1$ with equality if and only if there are constants a and b such that $P(Y = aX + b) = 1$.*

PROOF: Let $X^* = (X - \mu_X)/\sigma_X$, $Y^* = (Y - \mu_Y)/\sigma_Y$. By Equation 4.9, $\rho(X^*, Y^*) = \rho(X, Y)$. Since $E[X^{*2}] = E[((X - \mu_X)/\sigma_X)^2] = 1/\sigma_X^2 E[(X - \mu_X)^2] = 1$ and likewise $E[Y^{*2}] = 1$, by Inequality 4.8,

$$\rho(X, Y)^2 = \rho(X^*, Y^*)^2 \leq E[X^{*2}]E[Y^{*2}] = 1.$$

Thus, $|\rho(X, Y)| \leq 1$ with equality if and only if $P(Y^* = \alpha X^*) = 1$ for some $\alpha \in R$, in which case there are constants a and b such that $P(Y = aX + b) = 1$. ■

EXAMPLE 4.18 Let X_1, X_2, and X_3 be independent random variables with $\sigma_{X_1}^2 = 2$, $\sigma_{X_2}^2 = 4$, and $\sigma_{X_3}^2 = 3$, respectively, and consider the problem of calculating the correlation between the random variables $2X_1 - 3X_2 + 5X_3$ and $X_1 + 2X_2 - 4X_3$. By independence, $\text{cov}(X_i, X_j) = 0$ whenever $i \neq j$. By Theorem 4.4.5,

$$\begin{aligned}
&\text{cov}(2X_1 - 3X_2 + 5X_3, X_1 + 2X_2 - 4X_3) \\
&= (2)(1)\,\text{cov}(X_1, X_1) + (2)(2)\,\text{cov}(X_1, X_2) \\
&+ (2)(-4)\,\text{cov}(X_1, X_3) + (-3)(1)\,\text{cov}(X_2, X_1) \\
&+ (-3)(2)\,\text{cov}(X_2, X_2) + (-3)(-4)\,\text{cov}(X_2, X_3) \\
&+ (5)(1)\,\text{cov}(X_3, X_1) + (5)(2)\,\text{cov}(X_3, X_2) \\
&+ (5)(-4)\,\text{cov}(X_3, X_3) \\
&= 2\,\text{cov}(X_1, X_1) - 6\,\text{cov}(X_2, X_2) - 20\,\text{cov}(X_3, X_3) \\
&= 2\sigma_{X_1}^2 - 6\sigma_{X_2}^2 - 20\sigma_{X_3}^2 \\
&= -80
\end{aligned}$$

Since the random variables $2X_1$, $-3X_2$, and $5X_3$ are independent,

$$\begin{aligned}
\text{var}(2X_1 - 3X_2 + 5X_3) &= \text{var}(2X_1) + \text{var}(-3X_2) + \text{var}(5X_3) \\
&= 4\sigma_{X_1}^2 + 9\sigma_{X_2}^2 + 25\sigma_{X_3}^2 \\
&= 119
\end{aligned}$$

Similarly,

$$\text{var}(X_1 + 2X_2 - 4X_3) = 66.$$

Therefore,

$$\rho(2X_1 - 3X_2 + X_3, X_1 + 2X_2 - 4X_3) = \frac{-80}{\sqrt{119}\sqrt{66}} \approx -.903. \quad ■$$

The correlation between X and Y measures the *linear dependence* between X and Y. In the case $\rho(X, Y) = \pm 1$, there is a linear functional relationship between X and Y. It is possible for two random variables U and V to be related functionally in the same way as two random variables X and Y with wide disparities between the correlations $\rho(U, V)$ and $\rho(X, Y)$.

EXAMPLE 4.19 Let X be a random variable that takes on values $-1, 0, 1$ with probabilities $1/4$, $1/2$, $1/4$, respectively, and let $Y = X^2$. It is easy to calculate that $\rho(X, Y) = 0$. If we let $U = X + 1$ and $V = U^2$, then U and V have the same functional relationship as X and Y. It is also easy to calculate that $\rho(U, V) = 2\sqrt{2/3} \approx .94$ if use is made of the fact that $X^3 = X$ and $X^4 = X^2$; for example,

$$\begin{aligned}
\operatorname{var} V &= E[V^2] - (E[V])^2 \\
&= E[(X+1)^4] - (E[(X+1)^2])^2 \\
&= E[X^4] + 4E[X^3] + 6E[X^2] + 4E[X] + 1 \\
&\quad - (E[X^2] + 2E[X] + 1)^2 \\
&= E[X^2] + 4E[X] + 6E[X^2] + 4E[X] + 1 \\
&\quad - (E[X^2] + 2E[X] + 1)^2
\end{aligned}$$

Since $E[X] = 0$ and $E[X^2] = 1/2$, $\operatorname{var} V = 9/4$. Even though U and V are functionally related in the same way as X and Y, one pair has correlation zero and the other pair has correlation close to 1. ■

EXERCISES 4.4

1. The joint density function $f_{X,Y}(x, y)$ of the random variables X and Y is tabulated below. Calculate $\rho(X, Y)$.

	$y = 1$	$y = 2$	$y = 3$	$y = 4$	$y = 5$
$x = -1$	$\frac{1}{20}$	$\frac{2}{20}$	$\frac{1}{20}$	0	0
$x = 0$	0	$\frac{3}{20}$	$\frac{2}{20}$	$\frac{1}{20}$	0
$x = 1$	$\frac{1}{20}$	$\frac{2}{20}$	$\frac{3}{20}$	0	0
$x = 2$	0	$\frac{1}{20}$	$\frac{1}{20}$	$\frac{1}{20}$	$\frac{1}{20}$

2. Let X and Y be random variables with $\rho(X, Y) = 3/4$, $\operatorname{var} X = 2$, and $\operatorname{var} Y = 1$. Calculate $\operatorname{var}(X + 2Y)$.

3. A bowl contains r red balls and b black balls. An unordered random sample of size 2 is selected from the bowl. Let X be the number of red balls and Y the number of black balls in the sample. Calculate $\rho(X, Y)$ without using the joint density of X and Y.

4. Let X_1, X_2, and X_3 be independent random variables with $\sigma^2_{X_1} = 4$, $\sigma^2_{X_2} = 3$, and $\sigma^2_{X_3} = 1$. Calculate $\rho(X_1 + 2X_2 - X_3, 3X_1 - X_2 + X_3)$.
5. A bowl contains three balls numbered 1, 2, 3. Two balls are successively selected at random from the bowl without replacement. If X is the number on the first ball and Y the number on the second ball, calculate $\rho(X, Y)$.
6. Suppose n distinguishable balls are randomly distributed into r boxes. If S_r is the number of empty boxes, $S_r = X_1 + \cdots + X_r$ where X_i is 1 or 0 according to whether box i is empty or not, $1 \leq i \leq r$.
   (a) Calculate $E[X_i]$.
   (b) Calculate $E[X_i X_j]$, $i \neq j$.
   (c) Calculate $E[S_r]$.
   (d) Calculate var S_r.
7. Consider a basic experiment with r outcomes $1, 2, \ldots, r$ having probabilities $p_1, p_2, \ldots, p_r$, respectively, and consider n independent repetitions of this basic experiment. For $i = 1, 2, \ldots, r$, let Y_i be the number of trials resulting in the outcome i. Writing $Y_i = I_{i,1} + I_{i,2} + \cdots + I_{i,n}$ where $I_{i,j} = 1$ or 0 according to whether the jth trial results in i or not,
   (a) Calculate $E[I_{i,k} I_{j,k}]$ for $i \neq j, k = 1, \ldots, n$.
   (b) Calculate $E[I_{i,k} I_{j,\ell}]$ for $i \neq j, k \neq \ell, 1 \leq k, \ell \leq n$.
   (c) Calculate $E[Y_i]$ and $E[Y_i Y_j]$, $i \neq j$.
   (d) Calculate var Y_i.
   (e) Calculate $\rho(Y_i, Y_j)$, $i \neq j$.
8. Let X and Y be two random variables that take on only two values each. If cov $(X, Y) = 0$, show that X and Y are independent.
9. Let X and Y be random variables with finite second moments. The linear function $aX + b$ of X is called the *best mean square linear predictor* of Y if

$$E[(Y - aX - b)^2] \leq E[(Y - cX - d)^2]$$

for all real numbers $c, d \in R$. Calculate a and b.

4.5 CONDITIONAL EXPECTATION

We have seen in some instances that conditional probabilities can be used to simplify computations and, in fact, some probability models are defined in terms of conditional probabilities. We will look at this concept in the context of random variables.

Let X and Y be two random variables with ranges $\{x_1, x_2, \ldots\}$ and $\{y_1, y_2, \ldots\}$, respectively. If $P(X = x_j) > 0$, then $P(Y = y_k|X = x_j)$ is defined, and we will let $f_{Y|X}(y_k|x_j)$ denote this conditional probability. Thus,

$$f_{Y|X}(y_k|x_j) = P(Y = y_k|X = x_j) = \frac{f_{X,Y}(x_j, y_k)}{f_X(x_j)}.$$

When $P(X = x_j) = f_X(x_j) = 0$, the above quotient is undefined, and we define $f_{Y|X}(y_k|x_j) = 0$ whenever $f_X(x_j) = 0$. The function $f_{Y|X}(y_k|x_j)$ of the two variables x_j, y_k is called the *conditional density* of Y given $X = x_j$. The x_j variable is usually thought of as a parameter. It follows from the definition that

$$f_{X,Y}(x_j, y_k) = f_{Y|X}(y_k|x_j)f_X(x_j). \tag{4.10}$$

EXAMPLE 4.20 A bowl contains chips numbered from 1 to 10. A chip is selected at random from the bowl. If the chip selected is numbered x, $1 \leq x \leq 10$, then a second chip is selected at random from the chips numbered $1, 2, \ldots, x$. This is an experiment for which the probability model is defined in terms of conditional densities. Let X be the number on the first chip and Y the number on the second chip. Then

$$f_X(x) = \begin{cases} 1/10 & \text{for } x = 1, 2, \ldots, 10 \\ 0 & \text{otherwise.} \end{cases}$$

The remainder of the description of the experiment specifies $f_{Y|X}(y|x)$. For $x = 1, 2, \ldots, 10$,

$$f_{Y|X}(y|x) = \begin{cases} 1/x & \text{for } y = 1, 2, \ldots, x \\ 0 & \text{otherwise.} \end{cases}$$

Thus,

$$f_{X,Y}(x, y) = \begin{cases} 1/10x & \text{for } 1 \leq y \leq x,\ x = 1, 2, \ldots, 10 \\ 0 & \text{otherwise.} \end{cases} \quad \blacksquare$$

Conditional probabilities can also be defined for collections of random variables. In what follows, x_j will denote a typical value of X_j and y a typical value of Y.

Definition 4.5 *If $Y, X_1, X_2, \ldots, X_m$ are random variables, the* conditional density *of Y given $X_1, X_2, \ldots, X_m$ is the function*

$$f_{Y|X_1,\ldots,X_m}(y|x_1, \ldots, x_m) = \frac{f_{X_1,\ldots,X_m,Y}(x_1, \ldots, x_m, y)}{f_{X_1,\ldots,X_m}(x_1, \ldots, x_m)}$$

whenever the denominator is different from zero and is equal to zero otherwise. $\blacksquare$

The conditional density satisfies the following equation:

$$f_{X_1,\ldots,X_m,Y}(x_1,\ldots,x_m,y) = f_{Y|X_1,\ldots,X_m}(y|x_1,\ldots,x_m)f_{X_1,\ldots,X_m}(x_1,\ldots,x_m). \tag{4.11}$$

Theorem 4.5.1 *If the random variable Y is independent of the collection of random variables $\{X_1,\ldots,X_m\}$ (i.e., $f_{X_1,\ldots,X_m,Y}(x_1,\ldots,x_m,y) = f_{X_1,\ldots,X_m}(x_1,\ldots,x_m)f_Y(y)$), then*

$$f_{Y|X_1,\ldots,X_m}(y|x_1,\ldots,x_m) = f_Y(y)$$

whenever $f_{X_1,\ldots,X_m}(x_1,\ldots,x_m) > 0$.

PROOF: The result follows directly from the definition of the conditional density. ■

EXAMPLE 4.21 Consider an infinite sequence of Bernoulli random variables $\{X_j\}_{j=1}^{\infty}$ with probability of success p. Fixing $n \geq 1$,

$$f_{X_n|X_1,\ldots,X_{n-1}}(x_n|x_1,\ldots,x_{n-1}) = f_{X_n}(x_n)$$

whenever $f_{X_1,\ldots,X_{n-1}}(x_1,\ldots,x_{n-1}) > 0$. This follows from the fact that the $X_1,\ldots,X_n$ are independent random variables, so that

$$\begin{aligned} f_{X_1,\ldots,X_n}(x_1,\ldots,x_n) &= f_{X_1}(x_1) \times \cdots \times f_{X_n}(x_n) \\ &= f_{X_1,\ldots,X_{n-1}}(x_1,\ldots,x_{n-1})f_{X_n}(x_n), \end{aligned}$$

and therefore X_n is independent of the collection $\{X_1,\ldots,X_{n-1}\}$. ■

Let $\{y_1,y_2,\ldots\}$ be the range of the random variable Y. For any values $x_1,\ldots,x_n$ of $X_1,\ldots,X_n$, respectively, such that $f_{X_1,\ldots,X_n}(x_1,\ldots,x_n) > 0$, the conditional density $f_{Y|X_1,\ldots,X_n}(y|x_1,\ldots,x_n)$ is a density function as a function of y since

$$f_{Y|X_1,\ldots,X_n}(y_k|x_1,\ldots,x_n) = P(Y = y_k|X_1 = x_1,\ldots,X_n = x_n),$$

the conditional probabilities on the right are nonnegative, and the union of the disjoint events $(Y = y_k)$ is all of Ω, so that

$$\begin{aligned} \sum_{y_k} f_{Y|X_1,\ldots,X_n}(y_k|x_1,\ldots,x_n) &= \sum_{y_k} P(Y = y_k|X_1 = x_1,\ldots,X_n = x_n) \\ &= P\left(\bigcup_{y_k}(Y = y_k)|X_1 = x_1,\ldots,X_n = x_n\right) \\ &= P(\Omega|X_1 = x_1,\ldots,X_n = x_n) \\ &= 1 \end{aligned}$$

Definition 4.6 *Let $\{Y, X_1, \ldots, X_n\}$ be a collection of random variables with $E[Y]$ finite. The* conditional expectation *of Y given $X_1 = x_1, \ldots, X_n = x_n$ is defined by*

$$E[Y|X_1 = x_1, \ldots, X_n = x_n] = \sum_{y_j} y_j f_{Y|X_1,\ldots,X_n}(y_j|x_1, \ldots, x_n)$$

whenever $f_{X_1,\ldots,X_n}(x_1, \ldots, x_n) > 0$ and is defined arbitrarily when $f_{X_1,\ldots,X_n}(x_1, \ldots, x_n) = 0$. ■

The definition of $E[Y]$ required that the defining series be absolutely convergent, but no mention is made of absolute convergence of the series above defining $E[Y|X_1 = x_1, \ldots, X_n = x_n]$. The absolute convergence is inherent in the requirement that $E[Y]$ be finite; i.e., that $E[|Y|] < +\infty$. By Theorem 3.4.3,

$$\begin{aligned}
+\infty &> \sum_{y_j} |y_j| f_Y(y_j) \\
&= \sum_{y_j} |y_j| \sum_{x_1,\ldots,x_n} f_{X_1,\ldots,X_n,Y}(x_1, \ldots, x_n, y_j) \\
&= \sum_{y_j} \sum_{x_1,\ldots,x_n} |y_j| f_{X_1,\ldots,X_n,Y}(x_1, \ldots, x_n, y_j) \\
&= \sum_{y_j} \sum_{x_1,\ldots,x_n} |y_j| f_{Y|X_1,\ldots,X_n}(y|x_1, \ldots, x_n) f_{X_1,\ldots,X_n}(x_1, \ldots, x_n) \\
&= \sum_{x_1,\ldots,x_n} \left(\sum_{y_j} |y_j| f_{Y|X_1,\ldots,X_n}(y|x_1, \ldots, x_n) \right) f_{X_1,\ldots,X_n}(x_1, \ldots, x_n).
\end{aligned}$$

Thus, for any term with $f_{X_1,\ldots,X_n}(x_1, \ldots, x_n) > 0$, the series within the parentheses converges, and so the series defining $E[Y|X_1 = x_1, \ldots, X_n = x_n]$ converges absolutely.

EXAMPLE 4.22 Consider the random variables X and Y of Example 4.20. Suppose $x \in \{1, 2, \ldots, 10\}$. By Exercise 1.2.6,

$$E[Y|X = x] = \sum_{y=1}^{x} y f_{Y|X}(y|x) = \sum_{y=1}^{x} y \frac{1}{x} = \frac{1}{x} \frac{x(x+1)}{2} = \frac{x+1}{2}$$

for $x = 1, 2, \ldots, 10$. ■

We will now consider operational properties of the conditional expectation.

Theorem 4.5.2 *If $Y, X_1, \ldots, X_n$ are any random variables with $E[Y]$ finite, then*

$$E[Y] = \sum_{x_1,\ldots,x_n} E[Y|X_1 = x_1, \ldots, X_n = x_n] f_{X_1,\ldots,X_n}(x_1, \ldots, x_n).$$

PROOF: By definition of the conditional expectation,

$$\begin{aligned}
\sum_{x_1,\ldots,x_n} & E[Y|X_1 = x_1,\ldots,X_n = x_n] f_{X_1,\ldots,X_n}(x_1,\ldots,x_n) \\
&= \sum_{x_1,\ldots,x_n} \sum_{y_k} y_k f_{Y|X_1,\ldots,X_n}(y_k|x_1,\ldots,x_n) f_{X_1,\ldots,X_n}(x_1,\ldots,x_n) \\
&= \sum_{x_1,\ldots,x_n} \sum_{y_k} y_k f_{X_1,\ldots,X_n,Y}(x_1,\ldots,x_n,y_k) \\
&= \sum_{y_k} \sum_{x_1,\ldots,x_n} y_k f_{X_1,\ldots,X_n,Y}(x_1,\ldots,x_n,y_k) \\
&= \sum_{y_k} y_k f_Y(y_k) = E[Y].
\end{aligned}$$

The interchange of order of summation is justifiable by absolute convergence and Theorem 3.4.4. ■

EXAMPLE 4.23 Consider the random variables X and Y of Example 4.22. The expected value of Y is given by $E[Y] = \sum_{x=1}^{10} E[Y|X = x] f_X(x) = \sum_{x=1}^{10}((x+1)/2)(1/10) = 1/20 \sum_{x=1}^{10}(x+1) = 3.25$. ■

Theorem 4.5.3 *Let Y be a random variable with finite expectation that is independent of $X_1,\ldots,X_n$. Then*

$$E[Y|X_1 = x_1,\ldots,X_n = x_n] = E[Y]$$

whenever $f_{X_1,\ldots,X_n}(x_1,\ldots,x_n) > 0$.

PROOF: Suppose $f_{X_1,\ldots,X_n}(x_1,\ldots,x_n) > 0$. By Theorem 4.5.1,

$$\begin{aligned}
E[Y|X_1 = x_1,\ldots,X_n = x_n] &= \sum_{y_k} y_k f_{Y|X_1,\ldots,X_n}(y_k|x_1,\ldots,x_n) \\
&= \sum_{y_k} y_k f_Y(y_k) = E[Y]. \ ■
\end{aligned}$$

We mention in passing that if $\{Y_1,\ldots,Y_m\}$ and $\{X_1,\ldots,X_n\}$ are two collections of random variables and $\psi(Y_1,\ldots,Y_m)$ has finite expectation, then

$$\begin{aligned}
E[&\psi(Y_1,\ldots,Y_m)|X_1 = x_1,\ldots,X_n = x_n] \\
&= \sum_{y_1,\ldots,y_m} \psi(y_1,\ldots,y_m) f_{Y_1,\ldots,Y_m|X_1,\ldots,X_n}(y_1,\ldots,y_m|x_1,\ldots,x_n).
\end{aligned}$$

The proof of this result again involves rearranging the terms of an infinite series. Using this result, properties of conditional expected value analogous to those of expected value can be proved; e.g., the conditional expectation of a sum of random variables is equal to the sum of the conditional expectations.

In the remainder of this section, we will deal with the expected duration of play for the gambler's ruin problem. Consider an infinite sequence of Bernoulli random variables $\{X_j\}_{j=1}^{\infty}$ with probability of success p and the associated gambler's ruin problem. We will use the notation of Section 3.5 where we were able to calculate the probabilities of eventual ruin given in Equations 3.14 and 3.15. We will now consider how long the play will last. If the gambler's initial capital is x, let $T_x = n$ if play terminates on the nth play (i.e., either the gambler or his adversary is ruined on the nth play) and let $D_x = E[T_x]$ be the expected duration of play. Suppose $1 < x < a - 1$. If the gambler wins one unit on the first play (with probability p), then his capital becomes $x + 1$, and the subsequent expected duration of play is D_{x+1}; if he loses one unit on the first play (with probability q), then his capital becomes $x - 1$, and the subsequent expected duration of play is D_{x-1}. Since one play has already taken place and these two possibilities are mutually exclusive,

$$D_x = p(D_{x+1} + 1) + q(D_{x-1} + 1) \qquad \text{if } 1 < x < a - 1.$$

If $x = 1$ and the gambler loses on the first play, then his subsequent expected duration is zero; since one play has already taken place, the second term in this equation becomes just q. Similarly, the first term becomes just p if $x = a - 1$ and he wins on the first play. This means that this equation holds for $x = 1$ and $x = a - 1$ provided we put $D_0 = D_a = 0$. Therefore, the expected duration D_x satisfies the difference equation

$$D_x = pD_{x+1} + qD_{x-1} + 1 \qquad \text{if } 1 \leq x \leq a - 1. \tag{4.12}$$

subject to the boundary conditions

$$D_0 = 0, D_a = 0. \tag{4.13}$$

It should be emphasized that the derivation of the difference equation and boundary conditions is heuristic and not mathematical. Were it not for the constant term in Equation 4.12, we could solve this problem as in Section 3.5. The procedure for solving this problem is as follows. Note that if u_x satisfies the equation

$$u_x = pu_{x+1} + qu_{x-1} \qquad \text{for } 1 \leq x \leq a - 1 \tag{4.14}$$

and $v_x^{(p)}$ satisfies the equation

$$v_x^{(p)} = pv_{x+1}^{(p)} + qv_{x-1}^{(p)} + 1 \qquad \text{for } 1 \leq x \leq a - 1, \tag{4.15}$$

then $u_x + v_x^{(p)}$ satisfies Equation 4.12. Since Equation 4.14 is the same as Equation 3.12 in Section 3.5, we can use the results of Section 3.5 to solve Equation 4.14 depending upon whether $p \neq q$ or $p = q$. In the $p \neq q$ case,

$$u_x = A + B\left(\frac{q}{p}\right)^x,$$

where A and B are arbitrary constants, and in the $p = q$ case

$$u_x = A + Bx.$$

Since there are two arbitrary constants A and B in these solutions, it suffices to find some $v_x^{(p)}$, called a particular solution, of Equation 4.15. In the $p \neq q$ case, we can take

$$v_x^{(p)} = \frac{x}{q - p}$$

and in the $p = q$ case,

$$v_x^{(p)} = -x^2.$$

We can therefore find a solution to Equation 4.12 in the $p \neq q$ case of the form

$$D_x = \frac{x}{q - p} + A + B\left(\frac{q}{p}\right)^x, \qquad 1 \leq x \leq a - 1$$

and in the $p = q$ case of the form

$$D_x = -x^2 + A + Bx, \qquad 1 \leq x \leq a - 1.$$

Choosing A and B to satisfy the boundary conditions 4.13, in the $p \neq q$ case,

$$D_x = \frac{x}{q - p} - \frac{a}{q - p}\frac{1 - (q/p)^x}{1 - (q/p)^a}, \qquad 1 \leq x \leq a - 1 \quad (4.16)$$

and in the $p = q$ case,

$$D_x = x(a - x), \qquad 1 \leq x \leq a - 1. \quad (4.17)$$

Against an infinitely rich adversary in the unfair case $q > p$, $\lim_{a \to \infty} D_x = x/(q - p)$; in the fair case $q = p$, $\lim_{a \to \infty} D_x = +\infty$.

EXAMPLE 4.24 Suppose a gambler and his adversary each have \$100 and \$1 is wagered each time in a fair game. The expected duration is then $D_{100} = 100(200 - 100) = 10{,}000$. If one-half dollar is wagered each time, then this has the effect of doubling the units, and the expected duration is then $D_{200} = 200(400 - 200) = 40{,}000$. We saw in Section 3.5 that doubling the number of units has no effect on the probability of eventual ruin in the fair case; doubling the units by wagering one-half unit on each play only prolongs the agony. ■

The heuristic argument used to derive the difference equation 4.12 and the boundary conditions 4.13 should not be confused with a mathematical derivation. Although a proper mathematical argument can be made, the details are too tedious at this stage.

EXERCISES 4.5

1. An experiment consists of selecting an integer X at random from $\{1, 2, \ldots, 100\}$ and then selecting an integer Y at random from $\{1, 2, \ldots, X\}$. Calculate $E[Y]$ and var Y.
2. An experiment consists of selecting an integer X at random from $\{0, 1, \ldots, 100\}$ and then selecting an integer Y at random from $\{0, 1, \ldots, X\}$. Use the results of Example 4.1 and Example 4.12 to identify $E[Y|X = x]$ and $E[Y^2|X = x]$ for $x = 0, 1, \ldots, 100$ and then calculate $E[Y]$ and var Y.
3. Let X_1 and X_2 be independent random variables with Poisson densities $p(\cdot; \lambda_1)$ and $p(\cdot; \lambda_2)$, respectively. If n is a positive integer, show that the conditional density of X_1 given that $X_1 + X_2 = n$ is a binomial density with parameters n and $p = \lambda_1/(\lambda_1 + \lambda_2)$.
4. If X and Y are independent random variables with binomial densities $b(\cdot; m, p)$ and $b(\cdot; n, p)$, respectively, calculate $E[X|X + Y = z], z = 0, 1, \ldots, m + n$.
5. A number P is selected from the set $\{1/10, 2/10, \cdots, 9/10\}$ according to a uniform density. Given that $P = j/10$, a number X is selected from $\{1, 2, \ldots, 100\}$ according to a binomial density with parameters $n = 100$ and $p = j/10$. Calculate $E[X]$.
6. Let $\{X_j\}$ be a sequence of random variables having finite expectations and let N be a nonnegative integer-valued random variable that is independent of each X_j. Show that
$$E[X_N|N = n] = E[X_n]$$
whenever $f_N(n) > 0$.
7. Let $\{X_j\}$ be a sequence of independent random variables having the same mean μ and finite variance σ^2, let $S_0 = 0$, and let $S_n = X_1 + \cdots + X_n, n \geq 1$. Also let N be a nonnegative integer-valued random variable having finite mean and finite variance that is independent of the X_j. Use the result of the previous problem to show that
$$E[S_N] = \mu E[N] \quad \text{and} \quad \operatorname{var} S_N = \sigma^2 E[N] + \mu^2 \operatorname{var} N.$$

8. If c is a constant and X is any random variable, show that $E[c|X = x] = c$ whenever $f_X(x) > 0$.
9. Let X and Y be random variables with Y having a finite second moment, and let $\phi(X)$ be a real-valued function of X having finite second moment. Show that $E[\phi(X)Y|X = x] = \phi(x)E[Y|X = x]$ whenever $f_X(x) > 0$.
10. Calculate the expected duration of play D_x for the modified gambler's ruin problem described in Exercise 3.5.4.

4.6 ENTROPY

In 1948, in a fundamental paper on the transmission of information (see the Supplemental Reading List at the end of the chapter), C. E. Shannon proposed a measure to quantify the uncertainty of an event. The basic idea of his measure is that frequently occurring events convey less information than infrequently occurring events. For example, the frequently occurring letter E in an English message conveys less information than the infrequently occurring Q, X, or Z.

The two words *uncertainty* and *information* are used repeatedly in what follows, and it is necessary to have some understanding of the relationship between the two. For example, consider a random variable X that takes on the values $1, \ldots, 6$ with equal probabilities. Initially, there is uncertainty about the value of X. But if X is observed and we are told that the value of X is 3 or 4, then there is a decrease in uncertainty and an increase in information.

Consider an event A with probability p. A measure of the uncertainty of A should be some nonnegative monotone decreasing function $I(p)$ of p so that $I(p)$ is large when p is small. Moreover, if A_1 and A_2 are independent events with probabilities p_1 and p_2, respectively, then the uncertainty of $A_1 \cap A_2$ is $I(p_1 p_2)$. If it becomes known that A_2 has occurred, the uncertainty $I(p_1 p_2)$ should be decreased by $I(p_2)$, and we should be left with the uncertainty $I(p_1)$; i.e., $I(p_1 p_2) - I(p_2) = I(p_1)$ or

$$I(p_1 p_2) = I(p_1) + I(p_2). \tag{4.18}$$

This additive property of uncertainty for independent events is an assumption on our part. While we are at it, we might as well assume that $I(p)$ is a continuous function of p. The property expressed by Equation 4.18 is reminiscent of the log function, and it should not come as a surprise that $I(p)$ can be shown to be the log function except possibly for a multiplicative factor. The function

$$I(p) = -\log p = \log \frac{1}{p}, \qquad 0 < p \leq 1$$

satisfies all of the above requirements. But $I(p)$ has not been completely determined, because there is more than one choice for the base of the log function. In communication theory, the base 2 is used because an on-off relay records one unit of information, called a *bit*. In this section, it will be understood that the log function is to the base 2. If an event A has probability 1/2, then $I(1/2) = -\log 1/2 = 1$ bit.

Having defined the uncertainty of a single event, we now define the uncertainty associated with a random variable as the average uncertainty of the events $(X = x)$.

Definition 4.7 *Let X be a random variable with range $\{x_1, x_2, \ldots\}$. The* entropy *or* uncertainty *of X is the quantity*

$$H(X) = -\sum_i f_X(x_i) \log f_X(x_i) = \sum_i f_X(x_i) \log \frac{1}{f_X(x_i)}$$

where $f_X(x) \log f_X(x) = 0$ whenever $f_X(x) = 0$. ■

It must be emphasized that $H(X)$ is determined by the values of the density function and only indirectly by the random variable X.

EXAMPLE 4.25 Let X be a random variable taking on the values -1, 0, and 1 with probabilities 1/4, 1/2, and 1/4, respectively, and let Y be a random variable taking on the values 0, e, and π with probabilities 1/2, 1/4, and 1/4, respectively. Then

$$H(X) = -\frac{1}{4}\log\frac{1}{4} - \frac{1}{2}\log\frac{1}{2} - \frac{1}{4}\log\frac{1}{4} = \frac{3}{2}$$

and

$$H(Y) = -\frac{1}{2}\log\frac{1}{2} - \frac{1}{4}\log\frac{1}{4} - \frac{1}{4}\log\frac{1}{4} = \frac{3}{2}. \quad ■$$

It is apparent from this example that the entropy of a random variable is totally unrelated to the meaning of the random variable and is determined solely by the values of its density function. This situation would be better portrayed if the notation $H(f)$, where f is a density function, were used instead of $H(X)$. $H(X)$ is the expected value of a function of X in a rather complicated way; namely, $H(X) = E[-\log f_X(X)] = -\sum_i f_X(x_i) \log f_X(x_i)$.

Consider a typical term of the form $h(p) = -p \log p$ in the definition of $H(X)$. The function h is continuous on $(0, 1]$. If we define $h(p)$ to be zero when $p = 0$, then h is also continuous at 0 since $\lim_{p \to 0+}(-p \log p) = 0$ by l'Hôpital's rule. The graph of $h(p)$ is depicted in Figure 4.1.

Since the terms in the series defining $H(X)$ are nonnegative, $H(X)$ is defined even if the series diverges to $+\infty$. It is possible for $H(X)$ to be infinite.

EXAMPLE 4.26 Consider a random variable X having density function

$$f_X(n) = \frac{c}{n \log^2 n}, \qquad n = 2, 3, \ldots$$

where c is chosen so that $\sum_{n=2}^{\infty} f_X(n) = 1$. The integral test for infinite series can be used to show that the series $\sum_{n=2}^{\infty} 1/(n \log n)$ diverges to $+\infty$ and that the series $\sum_{n=2}^{\infty} 1/(n \log^2 n)$ converges. The entropy of X is then

$$\begin{aligned} H(X) &= -\sum_{n=2}^{\infty} \frac{c}{n \log^2 n} \log\left(\frac{c}{n \log^2 n}\right) \\ &= \sum_{n=2}^{\infty} \frac{c}{n \log^2 n} (-\log c + \log n + 2 \log \log n) \\ &= c \sum_{n=2}^{\infty} \left(-\frac{\log c}{n \log^2 n} + \frac{1}{n \log n} + \frac{2 \log \log n}{n \log^2 n}\right). \ \blacksquare \end{aligned}$$

Since the sum of the first terms converges, the sum of the second terms diverges to $+\infty$, and the sum of the third terms is nonnegative, the series defining $H(X)$ diverges to $+\infty$. Thus, $H(X) = +\infty$.

EXAMPLE 4.27 Let X be a random variable having a uniform density on $\{1, 2, \ldots, n\}$. Then

$$H(X) = -\sum_{k=1}^{n} \frac{1}{n} \log \frac{1}{n} = \log n \text{ bits.} \ \blacksquare$$

The following lemma will be needed to establish an important property of $H(X)$.

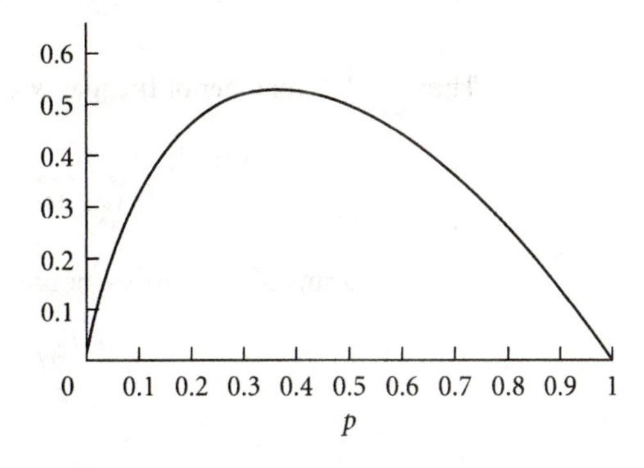

FIGURE 4.1 Graph of $-p \log p$.

Lemma 4.6.1 $\ln x \leq x - 1$ *for all* $x > 0$ *with equality holding if and only if* $x = 1$.

This result is proved by showing that the line $y = x - 1$ is tangent to the curve $y = \ln x$ when $x = 1$ and that the graph of the latter lies below the tangent line since $y = \ln x$ is concave downward.

The assumption that the random variables of the following theorem have the same range is not essential because their ranges can be replaced by the union of their ranges insofar as densities are concerned.

Theorem 4.6.2 (Gibbs' Inequality) *Let X and Y be discrete random variables having the same range such that* $f_X(z) = 0$ *if and only if* $f_Y(z) = 0$. *Then*

$$\sum_j f_X(z_j) \log \frac{f_Y(z_j)}{f_X(z_j)} \leq 0 \tag{4.19}$$

with equality holding if and only if $f_X(z_j) = f_Y(z_j)$ *for all j.*

PROOF: We need only consider those z_j for which $f_X(z_j) > 0$. By Lemma 4.6.1,

$$\begin{aligned}\sum_j f_X(z_j) \ln \frac{f_Y(z_j)}{f_X(z_j)} &\leq \sum_j f_X(z_j) \left(\frac{f_Y(z_j)}{f_X(z_j)} - 1 \right) \\ &= \sum_j f_X(z_j) - \sum_j f_Y(z_j) \\ &= 1 - 1 = 0 \end{aligned} \tag{4.20}$$

Since $\log a = (\ln a)/(\ln 2)$, multiplying both sides of this inequality by $1/\ln 2$ we obtain Inequality 4.19. If $f_X(z_j) = f_Y(z_j)$ for all j, then the left side of Inequality 4.19 is zero and there is equality therein. Assume now that there is equality in Inequality 4.19. Multiplying both sides by ln 2,

$$\sum_j f_X(z_j) \ln \frac{f_Y(z_j)}{f_X(z_j)} = 0.$$

Thus, the left member of Inequality 4.20 is zero, and therefore

$$\sum_j f_X(z_j) \left(\left(\frac{f_Y(z_j)}{f_X(z_j)} - 1 \right) - \ln \frac{f_Y(z_j)}{f_X(z_j)} \right) = 0 \tag{4.21}$$

Since the terms of the sum are nonnegative, they must all be zero; i.e.,

$$\ln \frac{f_Y(z_j)}{f_X(z_j)} = \frac{f_Y(z_j)}{f_X(z_j)} - 1,$$

and $f_X(z_j) = f_Y(z_j)$ by Lemma 4.6.1. ■

EXAMPLE 4.28 Consider all random variables X that take on exactly n values $x_1, \ldots, x_n$ with positive probabilities. Let Y be a random variable such that $P(Y = x_i) = 1/n, i = 1, \ldots, n$. Then $H(X) \leq H(Y) = \log n$ for all such X; i.e., the entropy is a maximum when the density is uniform. This follows from Gibbs' inequality, since

$$\sum_{i=1}^{n} f_X(x_i) \log \frac{1/n}{f_X(x_i)} \leq 0,$$

so that

$$H(X) = \sum_{i=1}^{n} f_X(x_i) \log \frac{1}{f_X(x_i)} \leq -\sum_{i=1}^{n} f_X(x_i) \log \frac{1}{n} = \log n = H(Y)$$

by Example 4.27. ■

If a random variable X takes on values including -1 and 1, then X^2 has the effect of lumping -1 and 1 together. Such lumping reduces entropy. Upon observing X^2, there is a loss of information gained as compared to observing X.

Theorem 4.6.3 *If X is a discrete random variable and ϕ is any real-valued function on the range of X, then $H(\phi(X)) \leq H(X)$.*

PROOF: Suppose first that $\{x_{i_1}, x_{i_2}, \ldots\}$ is a subset of the range of X, finite or infinite. For each $j \geq 1$, let $p_j = f_X(x_{i_j})$. Consider first the finite case $\{x_{i_1}, \ldots, x_{i_k}\}$. Since $p_i \leq \sum_{j=1}^{k} p_j$,

$$p_i \log p_i \leq p_i \log\left(\sum_{j=1}^{k} p_j\right) \qquad i = 1, \ldots, k.$$

Adding corresponding members of these k inequalities,

$$\sum_{i=1}^{k} p_i \log p_i \leq \left(\sum_{i=1}^{k} p_i\right) \log\left(\sum_{i=1}^{k} p_i\right)$$

If the sequence $\{x_{i_j}\}$ is infinite, since $p \log p$ is continuous on $[0, 1]$ we can let $k \to \infty$ in this inequality to obtain

$$\sum_{i} p_i \log p_i \leq \left(\sum_{i} p_i\right) \log\left(\sum_{i} p_i\right)$$

in the finite or infinite case. Therefore,

$$-\left(\sum_i p_i\right)\log\left(\sum_i p_i\right) \le -\sum_i p_i \log p_i.$$

Now let $\{z_1, z_2, \ldots\}$ be the range of $Z = \phi(X)$. Then

$$H(Z) = H(\phi(X)) = -\sum_j f_Z(z_j)\log f_Z(z_j).$$

Consider a fixed z_j and let $\{x_{j,1}, x_{j,2}, \ldots\}$ be the set of values of X such that $z_j = \phi(x_{j,k})$. Since the probabilities of the $x_{j,1}, x_{j,2}, \ldots$ are lumped together to produce $f_Z(z_j)$, by the above result

$$-f_Z(z_j)\log f_Z(z_j) \le -\sum_k f_X(x_{j,k})\log f_X(x_{j,k}).$$

Summing over j,

$$H(Z) \le \sum_j\sum_k\{-f_X(x_{j,k})\log f_X(x_{j,k})\}.$$

Since the terms in the iterated sums on the right are nonnegative, Theorem 3.4.3 can be applied to obtain

$$H(Z) \le -\sum_i f_X(x_i)\log f_X(x_i) = H(X). \quad \blacksquare$$

Since the definition of uncertainty applies to any discrete density, it makes sense to discuss the joint uncertainty of two random variables.

Definition 4.8 *If X and Y are discrete random variables, the* joint uncertainty *or* joint entropy *is defined by*

$$H(X, Y) = -\sum_{i,j} f_{X,Y}(x_i, y_j)\log f_{X,Y}(x_i, y_j). \quad \blacksquare$$

Theorem 4.6.4 *If X and Y are discrete random variables, then $H(X, Y) \le H(X) + H(Y)$ with equality holding if and only if X and Y are independent.*

PROOF: Since

$$H(X) = -\sum_i f_X(x_i)\log f_X(x_i) = -\sum_i\sum_j f_{X,Y}(x_i, y_j)\log f_X(x_i)$$

and

$$H(Y) = -\sum_j f_Y(y_j)\log f_Y(y_j) = -\sum_j\sum_i f_{X,Y}(x_i,y_j)\log f_Y(y_j),$$

by Theorem 3.4.3,

$$\begin{aligned} H(X)+H(Y) &= -\sum_i\sum_j f_{X,Y}(x_i,y_j)(\log f_X(x_i)+\log f_Y(y_j)) \\ &= -\sum_i\sum_j f_{X,Y}(x_i,y_j)\log f_X(x_i)f_Y(y_j). \end{aligned}$$

By Gibbs' inequality,

$$\sum_i\sum_j f_{X,Y}(x_i,y_j)\log\frac{f_X(x_i)f_Y(y_j)}{f_{X,Y}(x_i,y_j)} \le 0,$$

from which it follows that

$$\begin{aligned} H(X,Y) &= -\sum_i\sum_j f_{X,Y}(x_i,y_j)\log f_{X,Y}(x_i,y_j) \\ &\le -\sum_i\sum_j f_{X,Y}(x_i,y_j)\log f_X(x_i)f_Y(y_j) \\ &= H(X)+H(Y). \end{aligned}$$

There is equality in this application of Gibbs' inequality if and only if $f_{X,Y}(x_i,y_j) = f_X(x_i)f_Y(y_j)$ for all i and j; i.e., if and only if X and Y are independent.

Since the concept of uncertainty applies to any discrete density function, we can define conditional uncertainty.

Definition 4.9 1. *Let X and Y be discrete random variables. The* conditional uncertainty *or* conditional entropy *of Y given that $X = x$ is defined by*

$$H(Y|X=x) = -\sum_i f_{Y|X}(y_j|x)\log f_{Y|X}(y_j|x).$$

2. *The* conditional uncertainty *or* conditional entropy *of Y given X is defined by*

$$H(Y|X) = \sum_i f_X(x_i)H(Y|X=x_i);$$

i.e., $H(Y|X)$ is the weighted average of the $H(Y|X=x_i)$. ∎

Note that

$$H(Y|X) = -\sum_i f_X(x_i) \sum_j f_{Y|X}(y_j|x_i) \log f_{Y|X}(y_j|x_i)$$
$$= -\sum_{i,j} f_{X,Y}(x_i, y_j) \log f_{Y|X}(y_j|x_i).$$

Theorem 4.6.5 *If X and Y are discrete random variables, then*

$$H(X,Y) = H(X) + H(Y|X) = H(Y) + H(X|Y).$$

PROOF:

$$\begin{aligned} H(X,Y) &= -\sum_{i,j} f_{X,Y}(x_i, y_j) \log f_{X,Y}(x_i, y_j) \\ &= -\sum_{i,j} f_{Y|X}(y_j|x_i) f_X(x_i) \log f_{Y|X}(y_j|x_i) f_X(x_i) \\ &= -\sum_{i,j} f_{Y|X}(y_j|x_i) f_X(x_i) \log f_{Y|X}(y_j|x_i) \\ &\quad -\sum_{i,j} f_{Y|X}(y_j|x_i) f_X(x_i) \log f_X(x_i) \\ &= -\sum_i f_X(x_i) H(Y|X = x_i) - \sum_i f_X(x_i) \log f_X(x_i) \\ &= H(Y|X) + H(X). \quad \blacksquare \end{aligned}$$

We will now examine a procedure for selecting a density function called the *maximum entropy principle.* Consider a random variable X that takes on values $1, 2, \ldots, 6$ with unknown probabilities. What is known is that $E[X] = 9/2$ rather than the $7/2$ it would be if X took on the six values with equal probabilities. For $i = 1, \ldots, 6$, let $p_i = P(X = i)$. Can we choose $p_1, \ldots, p_6$ so that $E[X] = \sum_{i=1}^{6} i p_i = 9/2$, and how do we choose them? We might see if we can choose the $p_1, \ldots, p_6$ to maximize the entropy

$$H(X) = -\sum_{i=1}^{6} p_i \log p_i,$$

subject to the conditions that

$$\sum_{i=1}^{6} p_i = 1 \tag{4.22}$$

and

$$\sum_{i=1}^{6} i p_i = \frac{9}{2}. \tag{4.23}$$

Recall from the calculus that this is a maximization problem subject to two constraints, which can be dealt with by the method of Lagrange multipliers. Let

$$L(p_1, \ldots, p_6) = H(X) - \lambda\left(\sum_{i=1}^{6} p_i - 1\right) - \mu\left(\sum_{i=1}^{6} i p_i - \frac{9}{2}\right).$$

Setting $(\partial/\partial p_i)L = 0, i = 1, \ldots, 6$,

$$\frac{\partial}{\partial p_i}(p_i \log \frac{1}{p_i}) - \lambda - \mu i = 0, \qquad i = 1, \ldots, 6$$

or

$$-1 - \log p_i - \lambda - \mu i = 0, \qquad i = 1, \ldots 6.$$

Thus, $\log p_i = -(1 + \lambda + \mu i)$ or

$$p_i = e^{-(1+\lambda+\mu i)}.$$

Let $x = e^{-\mu}$ and $y = e^{(1+\lambda)}$. Then $p_i = x^i/y$ and Equations 4.22 and 4.23 become

$$\sum_{i=1}^{6} x^i = y$$

and

$$\sum_{i=1}^{6} i x^i = \frac{9}{2} y.$$

It follows from the first of these two equations that x cannot be zero. Therefore,

$$\sum_{i=1}^{6} i x^i = \frac{9}{2} \sum_{i=1}^{6} x^i.$$

Dividing by x,

$$\sum_{i=1}^{6} \left(\frac{9}{2} - i\right) x^{i-1} = 0.$$

Writing out the terms of this equation and clearing of fractions,

$$3x^5 + x^4 - x^3 - 3x^2 - 5x - 7 = 0.$$

This equation has only one positive root $x \approx 1.449254$. It follows that $y \approx 26.663653$, and using the equation $p_i = x^i/y$,

$$\begin{aligned} p_1 &\approx .05435 \\ p_2 &\approx .07877 \\ p_3 &\approx .11416 \\ p_4 &\approx .16544 \\ p_5 &\approx .23977 \\ p_6 &\approx .34749. \end{aligned}$$

EXERCISES 4.6

1. Calculate the entropy of a random variable X having density $f(1) = 1/2$, $f(2) = 1/4, f(3) = 1/8, f(4) = 1/16, f(5) = 1/16$.
2. Let X be a random variable having a geometric density $f_X(x) = (1/2)^x, x = 1, 2, \ldots$. Calculate the entropy $H(X)$. How much information is gained upon observing that $X = 3$?
3. Consider a random variable X that has a uniform density on $\{1, 2, \ldots, 2n\}$. If successive pairs of integers are lumped together (i.e., 1 and 2, 3 and 4, etc., are lumped together), by how much is the uncertainty decreased?
4. Let X be selected at random from the set of integers $\{1, 2, \ldots, n\}$. Given that $X = x$, Y is then selected at random from the set of integers $\{1, 2, \ldots, x\}$. Calculate $H(X, Y)$ without using the joint density of X and Y.
5. If X is the score on tossing two dice, calculate $H(X)$.
6. If a card is drawn at random from a deck of 52 cards and a king of diamonds is observed, how much information has been gained?
7. If a card is drawn at random from a deck of 52 cards and you are told that a king has been observed, how much information has been gained?
8. If X and Y are random variables, use Lemma 4.6.1 to show that

$$H(X|Y) \leq H(X)$$

by calculating $H(X|Y) - H(X)$.

9. Consider a random variable X that takes on the values $1, 2, \ldots, 6$. Given that $E[X] = 4$, determine the density of X that maximizes $H(X)$.

SUPPLEMENTAL READING LIST

1. T. M. Apostol (1957). *Mathematical Analysis.* Reading, Mass.: Addison-Wesley.
2. C. E. Shannon (1948). *A Mathematical Theory of Communication.* Monograph B-1598. Bell System Technical Journal.

CHAPTER 5

STOCHASTIC PROCESSES

5.1 INTRODUCTION

The topics discussed in this chapter have been selected not only to illustrate the concepts introduced in the previous chapters but also to expose the reader to the breadth and depth of applications of probability theory. In this elementary treatment, we will only scratch the surface of these topics. Topics within this chapter are independent, and subsequent chapters are independent of the topics of this chapter.

We first take up a model for randomly evolving processes having the property that probability statements about future developments given the past history depend only upon the immediate past and not the remote past. The section on random walks was chosen primarily because the topic involves more applications of generating functions and difference equations. After random walks comes a section on branching processes, which were developed as a model for survival of family names and nuclear chain reactions. Because some of the great successes of probability theory have to do with prediction theory and communication theory in general, the chapter concludes with an application to prediction theory.

Each of these topics could be expanded to book length and has been. Having learned some of the techniques for dealing with such topics, the reader can pursue them in greater depth in the book by Karlin and Taylor listed in the Supplemental Readings at the end of the chapter. More substantial applications to engineering can be found in the book by Helstrom. The section on prediction theory just barely scratches the surface of this subject. An excellent additional source is the book by Kendall and Ord.

5.2 MARKOV CHAINS

Several of the examples discussed in the previous chapters share a common structure that will be elaborated upon in this section.

Consider a countable set $S = \{s_1, s_2, \ldots\}$, finite or infinite, called a *state space* and consisting of objects called *states.* Since we can encode the states by giving s_j the label j, we can assume that $S = \{1, 2, \ldots, N\}$ for some N or $S = \{1, 2, \ldots\}$.

Definition 5.1 *A sequence of random variables $\{X_n\}_{n=0}^{\infty}$ is called a* Markov chain *if for all $n \geq 1$ and $j_0, j_1, \ldots, j_n \in S$,*

$$P(X_n = j_n \mid X_0 = j_0, \ldots, X_{n-1} = j_{n-1}) \\ = P(X_n = j_n \mid X_{n-1} = j_{n-1}). \quad \blacksquare \tag{5.1}$$

The significance of a Markov chain lies in the fact that if $(X_n = j_n)$ is a future event, then the conditional probability of this event given the past history $(X_0 = j_0, \ldots, X_{n-1} = j_{n-1})$ depends only upon the immediate past $(X_{n-1} = j_{n-1})$ and not upon the remote past $(X_0 = j_0, \ldots, X_{n-2} = j_{n-2})$.

Let $\{X_n\}_{n=0}^{\infty}$ be a Markov chain. If $X_n = j$, we say that the chain is in the state j at time n. The probabilities

$$p_{i,j}^{n-1,n} = P(X_n = j \mid X_{n-1} = i), \qquad n \geq 1, i, j \in S$$

are called *one-step transition probabilities* and depend upon the time that a transition from i to j takes place. If $P(X_n = j \mid X_{n-1} = i)$ is defined and is independent of n, the probabilities

$$p_{i,j} = P(X_n = j \mid X_{n-1} = i), \qquad n \geq 1, i, j \in S$$

are called *stationary transition probabilities*; if the conditional probability is not defined, we put $p_{i,j} = 0$. The numbers $p_{i,j}$ can be displayed in matrix form:

$$P = \begin{bmatrix} p_{1,1} & p_{1,2} & \cdots & \cdots \\ p_{2,1} & p_{2,2} & \cdots & \cdots \\ \vdots & \vdots & & \\ p_{i,1} & p_{i,2} & \cdots & \cdots \\ \vdots & \vdots & \ddots & \end{bmatrix}.$$

If $|S| = N$, this is an $N \times N$ matrix; if S is infinite, there are an infinite number of rows and columns. The matrix is customarily symbolized by $P = [p_{i,j}]$. The ith row of P is the conditional density of X_n given that $X_{n-1} = i$. Clearly, each $p_{i,j} \geq 0$. Since the union of the disjoint events $(X_n = j), j = 1, 2, \ldots,$ is Ω,

$$\sum_j p_{i,j} = \sum_j P(X_n = j \mid X_{n-1} = i) = P(\Omega \mid X_n = i) = 1.$$

A matrix $P = [p_{i,j}]$ with the last two properties is called a *stochastic matrix*. The density of X_0 is denoted by π_0—i.e., $\pi_0(j) = f_{X_0}(j), j \in S$—and is called the *initial density*. Suppose $n \geq 1$ and $j_0, j_1, \ldots, j_n \in S$. If the Markov chain $\{X_n\}_{n=0}^{\infty}$ has stationary transition probabilities, then

$$\begin{aligned}
P\ (X_0 &= j_0, \ldots, X_n = j_n) \\
&= P(X_n = j_n \mid X_0 = j_0, \ldots, X_{n-1} = j_{n-1})P(X_0 = j_0, \ldots, X_{n-1} = j_{n-1}) \\
&= P(X_n = j_n \mid X_{n-1} = j_{n-1})P(X_0 = j_0, \ldots, X_{n-1} = j_{n-1}) \\
&= p_{j_{n-1},j_n} P(X_{n-1} = j_{n-1} \mid X_0 = j_0, \ldots, X_{n-2} = j_{n-2}) \\
&\quad \times P(X_0 = j_0, \ldots, X_{n-2} = j_{n-2}) \\
&\vdots \\
&= p_{j_{n-1},j_n} p_{j_{n-2},j_{n-1}} \times \cdots \times p_{j_0,j_1} P(X_0 = j_0) \\
&= \pi_0(j_0) p_{j_0,j_1} \times \cdots \times p_{j_{n-1},j_n}. \qquad (5.2)
\end{aligned}$$

It should be noted that if at some stage the given event has zero probability, then the final result is still true since both sides are then zero.

The last equation provides the means for constructing Markov chains on a state space $S = \{1, 2, \ldots\}$. Given a stochastic matrix $P = [p_{i,j}]$ and a density function π_0 on S, a probability space $(\Omega, \mathcal{F}, \mathcal{P})$ and random variables $\{X_n\}_{n=0}^{\infty}$ can be constructed so that the probabilities $P(X_0 = j_0, \ldots, X_n = j_n)$ are defined by Equation 5.2.

Equation 5.2 can be used to reformulate the definition of a Markov chain. If $1 \leq m < n$ and $j_0, j_1, \ldots, j_n \in S$, then

$$\begin{aligned}
P\ (X_{m+1} &= j_{m+1}, \ldots, X_n = j_n \mid X_0 = j_0, \ldots, X_m = j_m) \\
&= P(X_{m+1} = j_{m+1}, \ldots, X_n = j_n \mid X_m = j_m); \qquad (5.3)
\end{aligned}$$

i.e., the probability of any future event given the past depends only upon the immediate past $(X_m = j_m)$ and not upon the remote past $(X_0 = j_0, \ldots, X_{m-1} = j_{m-1})$. To see this, first consider the left side of the equation. By Equation 5.2,

$$\begin{aligned}
P\ (X_{m+1} &= j_{m+1}, \ldots, X_n = j_n \mid X_0 = j_0, \ldots, X_m = j_m) \\
&= \frac{\pi_0(j_0) p_{j_0,j_1} \times \cdots \times p_{j_m,j_{m+1}} \times \cdots \times p_{j_{n-1},j_n}}{\pi_0(j_0) p_{j_0,j_1} \times \cdots \times p_{j_{m-1},j_m}} \\
&= p_{j_m,j_{m+1}} \times \cdots \times p_{j_{n-1},j_n}.
\end{aligned}$$

Since

$$P(X_m = j_m, \ldots, X_n = j_n) = \left(\sum_{j_0, \ldots, j_{m-1}} \pi_0(j_0) p_{j_1, j_2} \times \cdots \times p_{j_{m-1}, j_m} \right) p_{j_m, j_{m+1}} \times \cdots \times p_{j_{n-1}, j_n}$$

and

$$P(X_m = j_m) = \sum_{j_0, \ldots, j_{m-1}} \pi_0(j_0) p_{j_1, j_2} \times \cdots \times p_{j_{m-1}, j_m},$$

it follows that

$$P(X_m = j_m, \ldots, X_n = j_n) = P(X_m = j_m) p_{j_m, j_{m+1}} \times \cdots \times p_{j_{n-1}, j_n}.$$

Dividing by $P(X_m = j_m)$,

$$P(X_{m+1} = j_{m+1}, \ldots, X_n = j_n \mid X_m = j_m) = p_{j_m, j_{m+1}} \times \cdots p_{j_{n-1}, j_n}.$$

Thus, both sides of Equation 5.3 are equal to the product $p_{j_m, j_{m+1}} \times \cdots \times p_{j_{n-1}, j_n}$. This establishes Equation 5.3.

EXAMPLE 5.1 (Binary Information Source) A Markov information source is a sequential mechanism for which the chance that a certain symbol will be produced may depend upon the preceding symbol. Suppose the possible symbols are 0 and 1. If at some stage a 0 is produced, then at the next stage a 1 will be produced with probability p and a 0 will be produced with probability $1 - p$; if a 1 is produced, at the next stage a 0 will be produced with probability q and a 1 will be produced with probability $1 - q$. Since the p and q do not depend upon the number of times a symbol has been produced, this experiment can be described by the stationary transition matrix

$$P = \begin{bmatrix} 1-p & p \\ q & 1-q \end{bmatrix}. \quad \blacksquare$$

EXAMPLE 5.2 (Ehrenfest Diffusion Model) In 1907, P. and T. Ehrenfest described a conceptual experiment for the movement of N molecules between two containers A and B. The state of the system at any given time is the number of molecules in A so that $S = \{0, 1, \ldots, N\}$. At any given time, a molecule is chosen at random from among the N and moved to the other container. This chance mechanism is repeated indefinitely. Since the mechanism does not depend upon how many changes have occurred, the process has stationary transition probabilities given by $p_{i,i-1} = i/N$, $p_{i,i+1} = 1 - (i/N)$ and $p_{i,j} = 0$ otherwise. In this case,

$$P = \begin{bmatrix} 0 & 1 & 0 & 0 & \dots & 0 \\ 1/N & 0 & 1-(1/N) & 0 & \dots & 0 \\ \vdots & \vdots & \vdots & \vdots & \ddots & \vdots \\ 0 & 0 & 0 & \dots & 0 & 1/N \\ 0 & 0 & 0 & \dots & 1 & 0 \end{bmatrix}. \blacksquare$$

EXAMPLE 5.3 (Random Walk on the Integers) Consider the space $S = \{\dots, -2, -1, 0, 1, 2, \dots\}$. Let $\{Y_j\}_{j=0}^{\infty}$ be a sequence of independent random variables such that Y_0 has a specified density π_0 and $P(Y_j = 1) = p, P(Y_j = -1) = q$ where $j \geq 1, p, q > 0, p + q = 1$. For $n \geq 1$, let $X_n = \sum_{j=0}^{n} Y_j$. Then the sequence $\{X_n\}_{n=0}^{\infty}$ is a Markov chain with stationary transition probabilities

$$p_{i,j} = P(X_n = j \mid X_{n-1} = i) = \begin{cases} p & \text{if } j = i+1 \\ q & \text{if } j = i-1 \\ 0 & \text{otherwise.} \end{cases}$$

To show that $\{X_n\}_{n=0}^{\infty}$ is a Markov chain, consider

$$P(X_n = j \mid X_0 = j_0, \dots, X_{n-1} = j_{n-1}).$$

Note that $X_0 = j_0, X_1 = j_1, \dots, X_{n-1} = j_{n-1}$ if and only if $Y_0 = j_0, Y_1 = j_1 - j_0, Y_{n-1} = j_{n-1} - j_{n-2}$. By independence,

$$\begin{aligned} &P\ (X_n = j \mid X_0 = j_0, X_1 = j_1, \dots, X_{n-1} = j_{n-1}) \\ &= P\left(\sum_{i=0}^{n} Y_i = j \,\middle|\, Y_0 = j_0, Y_1 = j_1 - j_0, \dots, Y_{n-1} = j_{n-1} - j_{n-2}\right) \\ &= \frac{P(Y_0 = j_0, \dots, Y_{n-1} = j_{n-1} - j_{n-2}, Y_n = j - j_{n-1})}{P(Y_0 = j_0, \dots, Y_{n-1} = j_{n-1} - j_{n-2})} \\ &= P(Y_n = j - j_{n-1}). \end{aligned}$$

Also by independence and Lemma 3.3.3,

$$\begin{aligned} P(X_n = j \mid X_{n-1} = j_{n-1}) &= P\left(\sum_{i=0}^{n} Y_i = j \,\middle|\, \sum_{i=0}^{n-1} Y_i = j_{n-1}\right) \\ &= \frac{P(\sum_{i=0}^{n-1} Y_i = j_{n-1}, Y_n = j - j_{n-1})}{P(\sum_{i=0}^{n-1} Y_i = j_{n-1})} \\ &= P(Y_n = j - j_{n-1}). \end{aligned}$$

Therefore,

$$P(X_n = j \,|\, X_0 = j_0, \ldots, X_{n-1} = j_{n-1}) = P(X_n = j \,|\, X_{n-1} = j_{n-1}),$$

and the sequence $\{X_n\}_{n=0}^{\infty}$ is a Markov chain. Since

$$p_{i,j} = P(X_n = j \,|\, X_{n-1} = i) = P(Y_n = j - i) = \begin{cases} p & \text{if } j = i + 1 \\ q & \text{if } j = i - 1 \\ 0 & \text{otherwise,} \end{cases}$$

the chain has stationary transition probabilities. The chain $\{X_n\}_{n=0}^{\infty}$ is interpreted as follows. A particle starts off at an initial position j_0 in accordance with the initial density π_0. The particle will then jump to $j_0 + 1$ with probability p or jump to $j_0 - 1$ with probability q; in general, if after n jumps it is at i, it will then jump to $i + 1$ with probability p or to $i - 1$ with probability q, independently of n. ■

Let $\{X_n\}_{n=0}^{\infty}$ be a Markov chain with stationary transition probabilities. The conditional probabilities

$$p_{i,j}(n) = P(X_{m+n} = j \,|\, X_m = i)$$

are independent of m (see Exercise 5.2.11) and are called *n-step transition probabilities.* More generally, if $m, n \geq 1$ and $j_1, \ldots, j_n \in S$, the conditional probabilities

$$P(X_{m+1} = j_1, \ldots, X_{m+n} = j_n \,|\, X_m = j_0)$$

are independent of m. This property of a Markov chain with stationary transition probabilities is called the *stationarity property.* The property simply means that an integer $m \geq 1$ can be subtracted from all the indices appearing in a conditional probability; e.g.,

$$\begin{aligned} &P\,(X_4 = j_4, X_5 = j_5, X_6 = j_6 \,|\, X_3 = j_3) \\ &\quad = P(X_1 = j_4, X_2 = j_5, X_3 = j_6 \,|\, X_0 = j_3). \end{aligned}$$

Letting $P(n) = [p_{i,j}(n)]$, $P(n)$ is called the *n-step transition matrix.* We define

$$p_{i,j}(0) = \begin{cases} 1 & \text{if } i = j \\ 0 & \text{if } i \neq j. \end{cases}$$

If $n = 1$, then $p_{i,j}(1) = P(X_{m+1} = j \,|\, X_m = i) = p_{i,j}$, and therefore $P(1) = P$.

Theorem 5.2.1 (Chapman-Kolmogorov Equation) *For all $m, n \geq 1$ and $i, j \in S$,*

$$p_{i,j}(m+n) = \sum_k p_{i,k}(m) p_{k,j}(n). \tag{5.4}$$

PROOF: We can assume that $P(X_0 = i) > 0$, because otherwise $p_{i,j}(m+n) = P(X_{m+n} = j | X_0 = i) = 0$, $p_{i,k}(m) = P(X_m = k | X_0 = i) = 0$, and both sides are zero. Suppose $1 \leq m < n$, $j_0, \ldots, j_n \in S$, and $P(X_0 = j_0) > 0$. Then it follows from Equation 5.3 that

$$\begin{aligned} P&(X_0 = j_0, \ldots, X_m = j_m, \ldots, X_n = j_n) \\ &= P(X_{m+1} = j_{m+1}, \ldots, X_n = j_n | X_0 = j_0, \ldots, X_m = j_m) \\ &\times P(X_0 = j_0, \ldots, X_m = j_m) \\ &= P(X_{m+1} = j_{m+1}, \ldots, X_n = j_n | X_m = j_m) \\ &\times P(X_1 = j_1, \ldots, X_m = j_m | X_0 = j_0) P(X_0 = j_0). \end{aligned}$$

Summing over $j_1, \ldots, j_{m-1}, j_{m+1}, \ldots, j_{n-1}$,

$$\begin{aligned} P&(X_0 = j_0, X_m = j_m, X_n = j_n) \\ &= P(X_n = j_n | X_m = j_m) P(X_m = j_m | X_0 = j_0) P(X_0 = j_0). \end{aligned} \tag{5.5}$$

Replacing n by $m+n$, j_0 by i, j_m by k, and j_n by j in Equation 5.5,

$$\begin{aligned} p_{i,j}(m+n) &= P(X_{m+n} = j | X_0 = i) \\ &= \sum_k P(X_{m+n} = j, X_m = k | X_0 = i) \\ &= \sum_k \frac{P(X_{m+n} = j, X_m = k, X_0 = i)}{P(X_0 = i)} \\ &= \sum_k P(X_{m+n} = j | X_m = k) P(X_m = k | X_0 = i) \\ &= \sum_k p_{k,j}(n) p_{i,k}(m). \quad \blacksquare \end{aligned}$$

Equation 5.4 can be interpreted in terms of matrix multiplication. Noting that the first factors $p_{i,1}(m), p_{i,2}(m), \ldots$ constitute the ith row of $P(m)$ and the second factors $p_{1,j}(n), p_{2,j}(n), \ldots$ constitute the jth column of $P(n)$, the sum on the right of Equation 5.4 is the product of the elements of the ith row of $P(m)$ and the corresponding elements of the jth column of $P(n)$; but this is just the definition of the element in the ith row and jth column of the product of the two matrices $P(m)$ and $P(n)$. In terms of matrix multiplication, the Chapman-Kolmogorov equation simply says that $P(m+n) = P(m)P(n)$. Since $P(1) = P$, $P(n+1) = P(n)P$ for all $n \geq 1$; iterating this result, we see

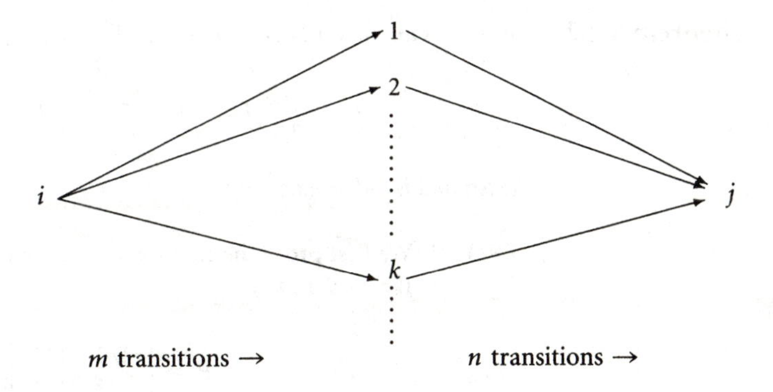

FIGURE 5.1 Stopping and restarting a chain.

that $P(n) = P^n$ and that $P(n)$ is simply the nth power of the transition matrix P.

Note also that $P(n)$ is a stochastic matrix for each $n \geq 1$. Since $P(1) = P$, the statement is true for $n = 1$. Assuming the statement is true for $n - 1$, it follows from Equation 5.4 and Theorem 3.4.3 that

$$\begin{aligned}\sum_j p_{i,j}(n) &= \sum_j \sum_k p_{i,k}(n-1)p_{k,j}(1) \\ &= \sum_k p_{i,k}(n-1) \sum_j p_{k,j}(1) \\ &= \sum_k p_{i,k}(n-1) = 1\end{aligned}$$

and that $P(n)$ has nonnegative entries. It follows from the principle of mathematical induction that $P(n)$ is a stochastic matrix for every $n \geq 1$.

Figure 5.1 is a graphical illustration of the Chapman-Kolmogorov equation. If at some time the Markov chain is in the state i, the probability of going from i to j in $m + n$ steps can be obtained by stopping the chain after m steps in state k, restarting the chain with initial state k, ending up in state j after n additional steps, and summing over k.

In the case of the Ehrenfest diffusion model, Example 5.2, it is reasonable to ask how the molecules will be distributed between the two containers after much time has lapsed; i.e., what happens to $p_{i,j}(n)$, the probability that starting from state i the chain will be in state j after n transitions, as $n \to \infty$. In general, determining the limiting behavior of the $p_{i,j}(n)$ can be difficult. To keep things as simple as possible in this chapter, we will limit the remaining discussion to Markov chains having a finite state space $S = \{1, 2, \ldots, r\}$ and $r \times r$ transition matrix $P = [p_{i,j}]$. We assume that $r \geq 2$ to avoid the trivial case of a chain with just one state.

Theorem 5.2.2 *If there is an integer N such that $p_{i,j}(N) > 0$ for $1 \le i,j \le r$, then*

$$\lim_{n\to\infty} p_{i,j}(n) = \pi_j, \qquad j = 1,\ldots,r$$

exists and is independent of i.

PROOF: We first prove the result assuming that $N = 1$; i.e., $p_{i,j} > 0$, $1 \le i,j \le r$. If $r = 2$ and

$$P = \begin{bmatrix} 1/2 & 1/2 \\ 1/2 & 1/2 \end{bmatrix},$$

then $P^n = P$ for all $n \ge 1$ and the assertion is true with $\pi_1 = \pi_2 = 1/2$. We can therefore assume that

$$\delta = \min_{1\le i,j\le r} p_{i,j} < \frac{1}{2}.$$

Consider a fixed j and define

$$M_n = \max_{1\le i\le r} p_{i,j}(n) \qquad m_n = \min_{1\le i\le r} p_{i,j}(n).$$

If we can show that the sequence $\{m_n\}$ increases, the sequence $\{M_n\}$ decreases as depicted below,

$$m_n \rightarrow \qquad \pi_j \qquad \leftarrow M_n$$

and $\lim_{n\to\infty}(M_n - m_n) = 0$, then there would be a number π_j such that

$$\lim_{n\to\infty} M_n = \lim_{n\to\infty} m_n = \pi_j,$$

and since $m_n \le p_{i,j}(n) \le M_n$, it would follow that

$$\lim_{n\to\infty} p_{i,j}(n) = \pi_j, \qquad j = 1,\ldots,r.$$

Since m_n is the minimum of a finite collection of numbers, it must be one of them, say

$$m_n = p_{k_n,j}(n).$$

By Equation 5.4,

$$\begin{aligned}
M_{n+1} &= \max_{1 \le i \le r} p_{i,j}(n+1) = \max_{1 \le i \le r} \sum_{k=1}^{r} p_{i,k} p_{k,j}(n) \\
&= \max_{1 \le i \le r} \left(p_{i,k_n} p_{k_n,j}(n) + \sum_{k \ne k_n} p_{i,k} p_{k,j}(n) \right) \\
&\le \max_{1 \le i \le r} \left(p_{i,k_n} m_n + M_n \sum_{k \ne k_n} p_{i,k} \right) \\
&= \max_{1 \le i \le r} \left(p_{i,k_n} m_n + M_n (1 - p_{i,k_n}) \right) \\
&= \max_{1 \le i \le r} \left(M_n - (M_n - m_n) p_{i,k_n} \right) \\
&= M_n - (M_n - m_n) \min_{1 \le i \le r} p_{i,k_n} \\
&\le M_n - (M_n - m_n)\delta.
\end{aligned}$$

Therefore,

$$M_{n+1} \le M_n - (M_n - m_n)\delta \le M_n,$$

and the sequence $\{M_n\}$ is decreasing. A similar argument shows that the sequence $\{m_n\}$ is increasing, and

$$m_{n+1} \ge m_n + (M_n - m_n)\delta \ge m_n.$$

By combining the last two inequalities,

$$M_{n+1} - m_{n+1} \le (M_n - m_n) - 2\delta(M_n - m_n) = (1 - 2\delta)(M_n - m_n).$$

Since $M_1 - m_1 < 1$,

$$0 \le M_n - m_n \le (1 - 2\delta)^{n-1}, \qquad n \ge 1,$$

and therefore $M_n - m_n \to 0$ as $n \to \infty$. Thus, $\lim_{n\to\infty} p_{i,j}(n) = \pi_j$ exists. Since $\lim_{n\to\infty} p_{i,j}(n) = \lim_{n\to\infty} M_n$ and the latter does not depend on i, π_j is independent of i. Suppose now that N is a positive integer for which $p_{i,j}(N) > 0$ for $1 \le i,j \le r$, let $\tilde{P} = [\tilde{p}_{i,j}] = P^N = [p_{i,j}(N)]$, and let $\tilde{P}(n) = \tilde{P}^n = P^{nN} = [p_{i,j}(nN)]$. Since $\tilde{P}(1) = \tilde{P}$ is a stochastic matrix with $\tilde{p}(1) > 0$, $1 \le i,j \le r$, the first part of the proof implies that

$$\lim_{n\to\infty} \tilde{p}_{i,j}(n) = \lim_{n\to\infty} p_{i,j}(nN) = \pi_j$$

exists for $1 \leq i,j \leq r$, independently of i. This means that given $\varepsilon > 0$, for each pair i,j there is an $N_{i,j} \geq 1$ such that

$$|p_{i,j}(nN) - \pi_j| < \varepsilon$$

whenever $n \geq N_{i,j}$. Letting $N_\varepsilon = \max_{1 \leq i,j \leq r} N_{i,j}$,

$$|p_{i,j}(nN) - \pi_j| < \varepsilon$$

simultaneously for all i,j with $1 \leq i,j \leq r$ whenever $n \geq N_\varepsilon$. Any positive integer n can be written $n = k(n)N + \ell(n)$ where $\lim_{n\to\infty} k(n) = +\infty$ and $0 \leq \ell(n) < r$. By Equation 5.4,

$$p_{i,j}(n) = \sum_k p_{i,k}(\ell(n))(p_{k,j}(k(n)N).$$

Since $\sum_k p_{i,k}(\ell(n)) = 1$,

$$p_{i,j}(n) - \pi_j = \sum_k p_{i,k}(\ell(n))(p_{k,j}(k(n)N) - \pi_j).$$

Since $\lim_{n\to\infty} k(n) = +\infty$, there is an $M \geq 1$ such that $k(n) \geq N_\varepsilon$ whenever $n \geq M$. Thus, for $n \geq M, k(n) \geq N_\varepsilon$ and

$$\begin{aligned} |p_{i,j}(n) - \pi_j| &\leq \sum_k p_{i,k}(\ell(n))|p_{k,j}(k(n)N) - \pi_j| \\ &< \varepsilon \sum_k p_{i,k}(\ell(n)) = \varepsilon. \end{aligned}$$

Therefore,

$$\lim_{n\to\infty} p_{i,j}(n) = \pi_j, \qquad 1 \leq j \leq r,$$

independently of i. ■

Definition 5.2 *(1) The state j can be reached from the state i if there is a positive integer n such that $p_{i,j}(n) > 0$; (2) the transition matrix P or chain $\{X_n\}_{n=0}^\infty$ is irreducible if each state can be reached from every other state.* ■

Consider a Markov chain with irreducible transition matrix $P = [p_{i,j}]$. There is then a positive integer N such that $p_{i,j}(N) > 0$ for all $i,j \in S$, and $\nu_j = \lim_{n\to\infty} p_{i,j}(n)$ is defined for all $i,j \in S$, independently of i, according to Theorem 5.2.2. Taking $n = 1$ and letting $m \to \infty$ in the Chapman-Kolmogorov equation,

$$\nu_j = \sum_{k=1}^r \nu_k p_{k,j}, \qquad j = 1,\ldots,r. \tag{5.6}$$

The ν_j are clearly nonnegative. Since $\sum_{j=1}^{r} p_{i,j}(n) = 1$ and

$$\sum_{j=1}^{r} \nu_j = \sum_{j=1}^{r} \lim_{n\to\infty} p_{i,j}(n) = \lim_{n\to\infty} \sum_{j=1}^{r} p_{i,j}(n) = 1,$$

$\{\nu_j\}_{j=1}^{r}$ is a probability density called the *asymptotic distribution* or *limiting distribution*. It is left as an exercise to show that the density $\{\nu_j\}_{j=1}^{r}$ satisfying Equation 5.6 is unique. According to Equation 5.6, the determination of the ν_j amounts to solving a system of linear equations.

EXAMPLE 5.4 Consider a Markov chain with state space $S = \{1, 2, 3\}$ and stationary transition matrix

$$P = \begin{bmatrix} 0 & 1/2 & 1/2 \\ 1 & 0 & 0 \\ 1/2 & 1/2 & 0 \end{bmatrix}.$$

In this case, Equation 5.6 becomes

$$\begin{aligned} \nu_1 &= \nu_2 + \frac{1}{2}\nu_3 \\ \nu_2 &= \frac{1}{2}\nu_1 + \frac{1}{2}\nu_3 \\ \nu_3 &= \frac{1}{2}\nu_1. \end{aligned}$$

These equations are not linearly independent, and one of them must be discarded and replaced by the equation

$$\nu_1 + \nu_2 + \nu_3 = 1.$$

If this is done and the resulting equations are solved, we find

$$\nu_1 = \frac{4}{9}, \quad \nu_2 = \frac{1}{3}, \quad \nu_3 = \frac{2}{9}. \quad \blacksquare$$

EXAMPLE 5.5 Consider the Ehrenfest diffusion model with $S = \{0, 1, \ldots, N\}$, $p_{i,i-1} = i/N$, $p_{i,i+1} = 1 - (i/N)$, and $p_{i,j} = 0$ otherwise. In this case, Equation 5.6 reads

$$\nu_j = \sum_{k=0}^{N} \nu_k p_{k,j}, \qquad j = 0, \ldots, N.$$

The equation corresponding to $j = 0$ is

$$\nu_0 = \frac{\nu_1}{N},$$

and the equation corresponding to $j = N$ is

$$\nu_N = \frac{\nu_{N-1}}{N}.$$

For $1 \leq j \leq N - 1$,

$$\begin{aligned}\nu_j &= \left(1 - \frac{j-1}{N}\right)\nu_{j-1} + \left(\frac{j+1}{N}\right)\nu_{j+1} \\ &= \left(\frac{N-j+1}{N}\right)\nu_{j-1} + \left(\frac{j+1}{N}\right)\nu_{j+1}. \qquad (5.7)\end{aligned}$$

Disregarding the N in the denominators on the right side, the equations suggest that ν_j has a form that allows the $j+1$ and $N-j+1$ coefficients to be cancelled. This suggests that the ν_j have the form $\binom{N}{j}$; but since the sum of all the ν_j is equal to 1 and $\sum_{j=0}^{N} \binom{N}{j} = 2^N$, the ν_j must have the form

$$\nu_j = \binom{N}{j}\frac{1}{2^N}, \qquad j = 0, \ldots, N.$$

It is easily verified that the ν_j given by this equation do in fact satisfy the above equations and represent the asymptotic distribution. The interpretation is that whatever the initial number of molecules in container A, ultimately each of the N molecules is assigned to one of the two containers with equal probabilities. ■

EXERCISES 5.2

1. A gambler and his adversary have a combined capital of N units. In successive wagers, the gambler can win one unit with probability p or lose one unit with probability $q = 1 - p$. Describe the transition matrix.
2. Consider a Markov chain with state space $S = \{1, 2, 3, 4\}$ and transition matrix

$$P = \begin{bmatrix} 0 & 0 & 1/4 & 3/4 \\ 1/2 & 0 & 0 & 1/2 \\ 0 & 1/3 & 0 & 2/3 \\ 0 & 0 & 1 & 0 \end{bmatrix}.$$

   Find the smallest integer n such that $p_{i,j}(n) > 0, i, j = 1, 2, 3, 4$ and determine the asymptotic distribution of the chain.

3. Determine the asymptotic distribution of the binary information source of Example 5.1.

4. Consider a Markov chain with state space $S = \{1, 2, 3\}$ and transition matrix

$$P = \begin{bmatrix} 0 & 1/2 & 1/2 \\ 1/2 & 0 & 1/2 \\ 1/2 & 1/2 & 0 \end{bmatrix}.$$

Find the smallest integer n for which $p_{i,j}(n) > 0$ for $i, j = 1, 2, 3$ and find the asymptotic distribution of the chain.

5. Let $P = [p_{i,j}]$ be an $N \times N$ transition matrix and suppose that $\mu_j = \sum_{k=1}^{N} \mu_k p_{k,j}, j = 1, \ldots, N$. Show that for each $n \geq 1$,

$$\mu_j = \sum_{k=1}^{N} \mu_k p_{k,j}(n), \qquad j = 1, \ldots, N.$$

6. An $N \times N$ transition matrix $P = [p_{i,j}]$ is *doubly stochastic* if $\sum_{i=1}^{N} p_{i,j} = 1$ for $j = 1, \ldots, N$. Assuming that the limits exist, show that

$$\nu_j = \lim_{n \to \infty} p_{i,j}(n) = \frac{1}{N}, \qquad j = 1, \ldots, N.$$

7. Consider a Markov chain with state space $S = \{1, 2, 3, 4, 5\}$ and transition matrix

$$P = \begin{bmatrix} 1/3 & 1/6 & 1/6 & 1/6 & 1/6 \\ 1/6 & 1/3 & 1/6 & 1/6 & 1/6 \\ 1/6 & 1/6 & 1/3 & 1/6 & 1/6 \\ 1/6 & 1/6 & 1/6 & 1/3 & 1/6 \\ 1/6 & 1/6 & 1/6 & 1/6 & 1/3 \end{bmatrix}.$$

Find the asymptotic distribution of the chain.

8. Let $P = [p_{i,j}]$ be an $N \times N$ transition matrix for which there is an $n \geq 1$ such that $p_{i,j}(n) > 0$ for all $i, j = 1, \ldots, N$. Let $\{\mu_j\}_{j=1}^{N}$ be a probability density that satisfies the equation

$$\mu_j = \sum_{k=1}^{N} \mu_k p_{k,j}, \qquad j = 1, \ldots, N.$$

Show that $\{\mu_j\}_{j=1}^{N}$ is unique.

9. Consider a Markov chain with state space $S = \{1, 2, 3\}$ and transition matrix

$$P = \begin{bmatrix} 0 & 1/2 & 1/2 \\ 1 & 0 & 0 \\ 1 & 0 & 0 \end{bmatrix}.$$

Calculate $P(2n)$ and $P(2n - 1)$ for all $n \geq 1$ and draw conclusions about the asymptotic distribution of the chain.

10. N red balls and N white balls are placed into two containers A and B so that each contains N balls. The number of red balls in A is the state of a system. At each step of a continuing process, a ball is selected at random from each container and transferred to the other container. Determine the transition matrix P and find the asymptotic distribution.

11. Let $\{X_n\}_{n=0}^{\infty}$ be a Markov chain with stationary transition probabilities. If $m, n \geq 1, i, j \in S$, show that $P(X_{m+n} = j \mid X_m = i)$ is independent of m.

The next two problems require mathematical software such as Mathematica or Maple V.

12. Consider a Markov chain with state space $S = \{1, 2, 3, 4, 5\}$ and transition matrix

$$P = \begin{bmatrix} .1 & .2 & .1 & .5 & .1 \\ .2 & .2 & .3 & .1 & .2 \\ .2 & .2 & .2 & .2 & .2 \\ .3 & .3 & .2 & .2 & 0 \\ .3 & .3 & .3 & 0 & .1 \end{bmatrix}.$$

Find the asymptotic distribution of the chain.

13. Consider a Markov chain with state space $S = \{1, 2, 3, 4, 5, 6\}$ and transition matrix

$$P = \begin{bmatrix} 0 & .12 & .38 & 0 & .40 & .10 \\ .11 & 0 & .29 & .20 & 0 & .40 \\ .10 & .10 & 0 & .15 & .25 & .40 \\ .40 & 0 & 0 & 0 & .30 & .30 \\ .05 & .05 & .30 & .40 & 0 & .20 \\ 0 & 0 & 0 & .30 & .30 & .40 \end{bmatrix}.$$

Find the asymptotic distribution of the chain.

5.3 RANDOM WALKS

The probability model for the gambler's ruin problem consists of an infinite sequence of independent random variables $\{Y_j\}_{j=1}^{\infty}$ with $P(Y_j = 1) = p$ and $P(Y_j = -1) = q = 1 - p$ where $0 < p < 1$. Fixing integers x and a with $1 \le x \le a - 1$, $S_n = x + Y_1 + \cdots + Y_n$ is the gambler's capital as of the nth play of a game, with x representing the gambler's initial capital. The S_n can also serve as model for a particle taking a random walk on the integer points of the line. We interpret x as the initial position of the particle, and starting at x the particle will jump one unit to the right with probability p and one unit to the left with probability q; i.e., its position after the first jump will be $S_1 = x + Y_1$. Starting from this new position, the particle will jump one unit to the right with probability p and one unit to the left with probability q, so that its position after the second jump will be $S_2 = x + Y_1 + Y_2$. After the nth jump, its position will be $S_n = x + Y_1 + \cdots + Y_n$. Let q_x be the probability that the particle will reach 0 before reaching a. Since only the interpretation of the S_n has changed and not the probability model, the q_x are given by Equations 3.14 and 3.15. In the language of random walks, 0 and a are boundary points for the interval of integers $\{0, 1, \ldots, a\}$, and q_x is the probability of absorption at 0.

Let $\{Y_j\}$ be an infinite sequence of independent random variables as described above and let $S_n = Y_1 + \cdots + Y_n$. Then S_n can be interpreted as the position of a particle after n moves with the particle starting at 0 and successively jumping one unit to the right with probability p and one unit to the left with probability $q = 1 - p$. After the first jump, the particle is no longer at 0 and may or may not eventually return to 0. We will endeavor to calculate the probability that the particle will eventually return to 0. The notation introduced above will be used throughout this section. Recall that Z is the set of integers $\{\ldots, -2, -1, 0, 1, 2 \ldots\}$.

Definition 5.3 *If $p = q$, the sequence $\{S_n\}_{n=1}^{\infty}$ is called a* symmetric random walk *on Z; if $p \neq q$, the sequence $\{S_n\}_{n=1}^{\infty}$ is called a* random walk *on Z with* drift *to the left if $q > p$ and to the right if $p > q$.* ■

We will introduce two sequences of numbers $\{u_j\}$ and $\{f_j\}$ by defining

$$\begin{aligned} u_j &= P(S_j = 0), & j &\ge 1 \\ f_j &= P(S_1 \neq 0, \ldots, S_{j-1} \neq 0, S_j = 0), & j &\ge 1. \end{aligned}$$

Since the numbers u_0 and f_0 have no meaning, we are free to define them however we choose and put $u_0 = 1, f_0 = 0$. Since $|u_j| \le 1$ and $|f_j| \le 1$ for all $j \ge 0$, the generating functions

$$U(s) = \sum_{j=0}^{\infty} u_j s^j,$$

$$F(s) = \sum_{j=0}^{\infty} f_j s^j$$

converge absolutely in the interval $(-1, 1)$. The probability that the particle will eventually return to the origin is $P(S_n = 0 \text{ for some } n \geq 1)$. Since this event can be stratified according to the first time the particle reaches the origin,

$$\begin{aligned} P(S_n = 0 \text{ for some } n \geq 1) &= \sum_{j=1}^{\infty} P(S_1 \neq 0, \ldots, S_{j-1} \neq 0, S_j = 0) \\ &= \sum_{j=1}^{\infty} f_j \leq 1. \end{aligned}$$

Letting $f = \sum_{j=1}^{\infty} f_j$, $1 - f$ is the probability that the particle will never return to the origin. The $\{u_j\}$ and $\{f_j\}$ sequences are related by the equation

$$u_j = f_0 u_j + f_1 u_{j-1} + f_2 u_{j-2} + \cdots + f_j u_0, \qquad j \geq 1. \tag{5.8}$$

Note that the first term on the right is zero since $f_0 = 0$. This equation follows from the fact that

$$(S_j = 0) = \bigcup_{k=1}^{j} (S_1 \neq 0, \ldots, S_{k-1} \neq 0, S_k = 0, S_j = 0)$$

and the following argument. The event $(S_1 \neq 0, \ldots, S_{k-1} \neq 0, S_k = 0)$ depends only upon $X_1, \ldots, X_k$, and since $S_k = X_1 + \cdots + X_k = 0$, the condition $S_j = 0$ is the same as the condition $X_{k+1} + \cdots + X_j = 0$, which depends only upon $X_{k+1}, \ldots, X_j$. By independence and the fact that the joint density of $X_{k+1}, \ldots, X_j$ is the same as the joint density of $X_1, \ldots, X_{j-k}$,

$$\begin{aligned} u_j &= P(S_j = 0) \\ &= \sum_{k=1}^{j} P(S_1 \neq 0, \ldots, S_{k-1} \neq 0, S_k = 0) P(X_{k+1} + \cdots + X_j = 0) \\ &= \sum_{k=1}^{j} P(S_1 \neq 0, \ldots, S_{k-1} \neq 0, S_k = 0) P(X_1 + \cdots + X_{j-k} = 0) \\ &= \sum_{k=1}^{j} f_k u_{j-k} = \sum_{k=0}^{j} f_k u_{j-k}. \end{aligned}$$

This establishes Equation 5.8. Multiplying both sides of Equation 5.8 by s^j, summing over $j \geq 1$, and using the fact that $f_0 = 0$,

$$\sum_{j=1}^{\infty} u_j s^j = \sum_{j=0}^{\infty} (f_0 u_j + f_1 u_{j-1} + \cdots + f_j u_0) s^j.$$

Since $u_0 = 1$, the left side of this equation is $U(s) - 1$, and according to the discussion preceding Definition 3.10, the right side is equal to

$$\Big(\sum_{j=0}^{\infty} f_j s^j\Big)\Big(\sum_{j=0}^{\infty} u_j s^j\Big) = U(s)F(s).$$

Thus, $U(s) - 1 = F(s)U(s)$ and the generating functions are related as follows:

$$U(s) = \frac{1}{1 - F(s)}, \qquad -1 < s < 1. \tag{5.9}$$

Since the u_j and f_j are nonnegative, by Abel's theorem (Theorem 4.2.3) both limits $\lim_{s \to 1-} U(s) = \sum_{j=0}^{\infty} u_j$ and $\lim_{s \to 1-} F(s) = \sum_{j=0}^{\infty} f_j = f \leq 1$ exist, even if the first is infinite.

Theorem 5.3.1 *$f < 1$ if and only if $\sum_{j=0}^{\infty} u_j < +\infty$.*

PROOF: Note that $0 \leq f < 1$ or $f = 1$. When $f < 1$,

$$\sum_{j=0}^{\infty} u_j = \lim_{s \to 1-} U(s) = \frac{1}{1 - \lim_{s \to 1-} F(s)} = \frac{1}{1 - f} < \infty;$$

when $f = 1$, $\lim_{s \to 1-} F(s) = 1$ and $\lim_{s \to 1-} U(s) = \sum_{j=0}^{\infty} u_j = +\infty$. In the latter case, $\sum_{j=0}^{\infty} u_j < +\infty$ implies that $f < 1$. ■

This theorem gives us a workable criterion for deciding if the particle will eventually return to the origin with probability 1. Before applying the criterion, it is necessary to take up approximations to factorials.

A sequence $\{a_j\}_{j=1}^{\infty}$ is said to be *asymptotically equivalent* to the sequence $\{b_j\}_{j=1}^{\infty}$, written $a_j \sim b_j$, if $\lim_{j \to \infty} a_j / b_j = 1$. It is easy to see that if $a_j \sim b_j$ and $c_j \sim d_j$, then $a_j / c_j \sim b_j / d_j$. The following relationship is known as *Stirling's formula:*

$$n! \sim \sqrt{2\pi} n^{n+(1/2)} e^{-n}. \tag{5.10}$$

An elementary proof of this result can be found in the book by R. Ash listed at the end of this chapter.

Returning to the series $\sum_{j=0}^{\infty} u_j$, note that $u_j = 0$ whenever j is odd because a return to the origin can occur only in an even number of jumps (i.e., the number of jumps to the right must be equal to the number of jumps to the left). Consider u_{2n} for $n \geq 1$. Since the number of jumps to the left and to the right must be equal,

$$u_{2n} = \binom{2n}{n} p^n q^n, \qquad n \geq 1.$$

By Stirling's formula,

$$\binom{2n}{n} p^n q^n = \frac{(2n)!}{n!n!} p^n q^n \sim \frac{(4pq)^n}{\sqrt{n\pi}}.$$

Since

$$\lim_{n \to \infty} \frac{u_{2n}}{(4pq)^n / \sqrt{n\pi}} = 1,$$

there is an $N \geq 1$ such that

$$\frac{u_{2n}}{(4pq)^n / \sqrt{n\pi}} < 2 \qquad \text{for all } n \geq N,$$

and since $\sqrt{n\pi} \geq 1$ for all positive integers n,

$$u_{2n} \leq 2\frac{(4pq)^n}{\sqrt{\pi n}} \leq 2(4pq)^n \qquad \text{for all } n \geq N.$$

We now take up the $p \neq q$ cases and $p = q$ cases separately. Suppose first that $p \neq q$. In this case, $4pq = 4p(1 - p) < 1$ since the maximum value $1/4$ of $p(1 - p)$ is attained only when $p = q = 1/2$. By the comparison test for positive series, the series $\sum_{n=0}^{\infty} u_{2n}$ converges since the geometric series $\sum_{n=0}^{\infty} (4pq)^n$ converges. Thus, $p \neq q$ implies that $\sum_{j=0}^{\infty} u_j$ converges. By Theorem 5.3.1, $f < 1$. This means that in the $p \neq q$ case, there is a positive probability that the particle will never return to the origin. Consider now the $p = q = 1/2$ case. Then $4pq = 1$ and $u_{2n} \sim 1/\sqrt{\pi n}$. That is,

$$\lim_{n \to \infty} \frac{u_{2n}}{1/\sqrt{n\pi}} = 1,$$

and there is an $N \geq 1$ such that

$$\frac{u_{2n}}{1/\sqrt{n\pi}} > \frac{1}{2} \qquad \text{for all } n \geq N$$

or

$$u_{2n} > \frac{1}{2}\frac{1}{\sqrt{\pi n}} \qquad \text{for all } n \geq N.$$

By comparison with the divergent p-series $\sum_{n=1}^{\infty} 1/n^{1/2}$, the series $\sum_{n=0}^{\infty} u_{2n}$ diverges. Thus, $p = q = 1/2$ implies that $\sum_{j=0}^{\infty} u_j$ diverges. By Theorem 5.3.1, $f = 1$ and the particle will return to the origin with probability 1. In summary, we have the following theorem.

Theorem 5.3.2 *If $p \neq q$, there is a positive probability that the random walk $\{S_n\}_{n=1}^{\infty}$ will never return to the origin; if $p = q$, the random walk $\{S_n\}_{n=1}^{\infty}$ will return to the origin with probability 1.*

Can the probability of eventually returning to the origin be determined in the $p \neq q$ case? With a little more work we can answer this question since the u_j are known. In fact,

$$U(s) = \sum_{j=0}^{\infty} \binom{2j}{j} p^j q^j s^{2j}.$$

Using the easily verified fact that

$$\binom{2n}{n} = (-4)^n \binom{-1/2}{n},$$

$$\begin{aligned} U(s) &= \sum_{j=0}^{\infty} (-4)^j \binom{-1/2}{j} p^j q^j s^{2j} \\ &= \sum_{j=0}^{\infty} \binom{-1/2}{j} (-4pqs^2)^j \\ &= (1 - 4pqs^2)^{-1/2}. \end{aligned}$$

By Equation 5.9,

$$F(s) = 1 - (1 - 4pqs^2)^{1/2}.$$

Since $F(1) = 1 - (1 - 4pq)^{1/2}$ and also $F(1) = \sum_{j=0}^{\infty} f_j = f, f = 1 - (1 - 4pq)^{1/2}$. Noting that $1 - 4pq = 1 - 4p(1-p) = (1 - 2p)^2 = (q - p)^2, f = 1 - |q - p|$.

Theorem 5.3.3 *The random walk $\{S_n\}_{n=1}^{\infty}$ will return to the origin with probability $f = 1 - |q - p|$.*

We can also use the previous result to determine the expected number of jumps to return to the origin. Define a waiting time random variable T by putting $T = n$ on $(S_1 \neq 0, \ldots, S_{n-1} \neq 0, S_n = 0)$ for $n \geq 1$ and $T = +\infty$ otherwise. Then $P(T = n) = P(S_1 \neq 0, \ldots, S_{n-1} \neq 0, S_n = 0) = f_n$. Consider the $p \neq q$ case. Since $f = \sum_{n=0}^{\infty} f_n < 1$, $P(T = +\infty) = 1 - P(T < \infty) = 1 - f > 0$ and $E[T] = +\infty$. Now consider the $p = q = 1/2$ case. This time $P(T < +\infty) = f = 1$. Note that $\hat{f}_T(s) = \sum_{n=0}^{\infty} P(T = n)s^n = \sum_{n=0}^{\infty} f_n s^n = F(s)$. Therefore,

$$E[T] = \hat{f}_T'(s) = F'(1) = \lim_{s \to 1-} \frac{s}{(1 - s^2)^{1/2}} = +\infty.$$

In summary, we have the following theorem.

Theorem 5.3.4 *The expected number of jumps for return to the origin is $+\infty$ for the one-dimensional random walk $\{S_n\}_{n=1}^{\infty}$.*

In the symmetric case, the random walk will return to the origin with probability 1, but the expected time for doing so is infinite.

A two-dimensional random walk on the points in the plane with integer coordinates can be described as follows. If at a given time a particle is at a point (x, y) with integer coordinates, then it will jump to one of the four neighboring points $(x + 1, y), (x - 1, y), (x, y + 1), (x, y - 1)$ with specified probabilities independently of what has taken place previously. A three-dimensional random walk on the points in 3-space with integer coordinates can be described similarly, except that jumps to six neighboring points will be allowed. To simplify the discussion, we will consider only symmetric two- and three-dimensional random walks.

For each $j \geq 1$, let (X_j, Y_j) be an ordered pair of random variables with joint density function

$$f_{X_j,Y_j}(x_j, y_j) = \begin{cases} 1/4 & \text{if } (x_j, y_j) = (\pm 1, 0) \text{ or } (0, \pm 1) \\ 0 & \text{otherwise.} \end{cases}$$

We will assume that a probability space $(\Omega, \mathcal{F}, P)$ can be constructed so that the pairs $(X_1, Y_1), (X_2, Y_2), \ldots$ are independent; i.e., for every $n \geq 1$,

$$f_{X_1,Y_1,\ldots,X_n,Y_n}(x_1, y_1, \ldots, x_n, y_n) = f_{X_1,Y_1}(x_1, y_1) \times \cdots \times f_{X_n,Y_n}(x_n, y_n).$$

Note that for each $j \geq 1$, the random variables X_j and Y_j are not independent. For each $n \geq 1$, let $S_n = X_1 + \cdots + X_n$ and $T_n = Y_1 + \cdots + Y_n$. Then the sequence of pairs $\{(S_n, T_n)\}_{n=1}^{\infty}$ describes a two-dimensional random walk starting at the origin on the points in the plane with integer coordinates. A particle taking such a random walk can be at the origin as of the nth jump if and only if both $S_n = 0$ and $T_n = 0$. As before, we can define

$$u_n = P(S_n = 0, T_n = 0), \qquad n \geq 1$$

with $u_0 = 1$ and also define

$$f_n = P(|S_1| + |T_1| \neq 0, \ldots, |S_{n-1}| + |T_{n-1}| \neq 0, S_n = 0, T_n = 0), \quad n \geq 1$$

with $f_0 = 0$. The probability f_n is the probability that the particle will return to the origin for the first time on the nth jump, and $f = \sum_{j=1}^{\infty} f_j$ is the probability that the particle will eventually return to the origin. The generating functions $U(s)$ and $F(s)$ are related as in Equation 5.9, so that again $f < 1$ if and only if $\sum_{j=0}^{\infty} u_j < +\infty$. As in the one-dimensional random walk, a return to the origin can occur only in an even number of jumps. For $S_{2n} = 0$ and $T_{2n} = 0$, the number k of jumps to the right must be equal to the number k of jumps to the left, and the number $n - k$ of jumps up must be equal to the number $n - k$ of jumps down where $k = 0, \ldots, n$. By the multinomial density and Equation 1.11,

$$\begin{aligned} u_{2n} &= \sum_{k=0}^{n} \frac{(2n)!}{k!k!(n-k)!(n-k)!}\left(\frac{1}{4}\right)^{2n} \\ &= \frac{(2n)!}{(n!)^2 4^{2n}} \sum_{k=0}^{n} \binom{n}{k}\binom{n}{n-k} \\ &= \frac{(2n)!}{(n!)^2 4^{2n}} \binom{2n}{n} \\ &= 4^{-2n}\binom{2n}{n}^2. \end{aligned}$$

By Stirling's formula,

$$u_{2n} \sim \frac{1}{n\pi}.$$

Thus, there is a positive integer $N \geq 1$ such that

$$u_{2n} \geq \frac{1}{2}\frac{1}{n\pi} \qquad \text{for all } n \geq N.$$

Since the series $\sum_{n=1}^{\infty} 1/n$ diverges, the series $\sum_{n=0}^{\infty} u_n$ diverges and therefore $f = 1$. Thus, a symmetric random walk in the plane will return to the origin with probability 1.

The situation changes, however, in higher dimensions. In the three-dimensional case, the random walk starting at the origin takes place on the points in 3-space with integer coordinates, and the particle will jump to any one of its nearest neighbors with probability 1/6. As in the previous cases, a return to the origin can occur only in an even number of jumps. For this to happen, the number of jumps in the positive x-direction must be equal to the number of jumps in the negative x-direction, and the same for the y-direction and z-direction. In this case,

$$\begin{aligned} u_{2n} &= \sum_{k=0}^{n} \sum_{j=0}^{k} \frac{(2n)!}{j!j!(k-j)!(k-j)!(n-k)!(n-k)!} \left(\frac{1}{6}\right)^{2n} \\ &= \frac{1}{2^{2n}} \binom{2n}{n} \sum_{k=0}^{n} \sum_{j=0}^{k} \left(\frac{n!}{3^n j!(k-j)!(n-k)!} \right)^2. \end{aligned}$$

Since the quantity in the parentheses is the general term of a multinomial density,

$$\begin{aligned} u_{2n} &\le \frac{1}{2^{2n}} \binom{2n}{n} \max_{j,k} \left(\frac{n!}{3^n j!(k-j)!(n-k)!} \right) \\ &\quad \times \sum_{k=0}^{n} \sum_{j=0}^{k} \left(\frac{n!}{3^n j!(k-j)!(n-k)!} \right) \\ &= \frac{1}{2^{2n}} \binom{2n}{n} \max_{j,k} \left(\frac{n!}{3^n j!(k-j)!(n-k)!} \right). \end{aligned}$$

The indicated maximum will be achieved when the three factorials in the denominator are equal and, since their sum is n, when each is equal to $(n/3)!$, assuming that $n/3$ is an integer. Putting aside such technical details,

$$u_{2n} \le \frac{1}{2^{2n}} \binom{2n}{n} \frac{n!}{3^n \left((n/3)!\right)^3}.$$

Applying Stirling's formula, Equation 5.10, to the factorials:

$$u_{2n} \le \frac{1}{2^{2n}} \binom{2n}{n} \frac{n!}{3^n \left((n/3)!\right)^3} \sim \frac{3}{2\pi} \sqrt{\frac{3}{\pi}} \frac{1}{n^{3/2}}.$$

By the comparison test, the series $\sum_{n=0}^{\infty} u_n$ can be compared with the convergent p-series $\sum_{n=1}^{\infty} 1/n^{3/2}$ with $p = 3/2 > 1$, and therefore the series $\sum_{n=0}^{\infty} u_n$ converges. In this case, $f < 1$ and there is a positive probability that a return to the origin will never occur.

The technical details glossed over previously can be taken care of by using the fact that for $0 \le j \le k \le n$,

$$j!(k-j)!(n-k)! \ge \left(\Gamma\left(\frac{n}{3}+1\right)\right)^3,$$

where Γ is the gamma function (see Section 6.5), and making use of known estimates of the gamma function for large values of the argument.

EXERCISES 5.3

The following terminology will be used in connection with a particle taking a random walk on the integers $\{0, 1, \ldots, a\}$, $a \geq 2$. The boundary point 0 is an *elastic barrier* for the walk if there is a number δ with $0 < \delta < 1$ such that the particle upon reaching 1 will jump to 2 with probability p or remain at 1 with probability δq or jump to 0 with probability $(1 - \delta)q$.

1. Verify that

$$\binom{2n}{n} = (-4)^n \binom{-1/2}{n}$$

for every positive integer n.

2. Consider a random walk on Z that jumps two units to the right with probability p and one unit to the left with probability $q, 0 < p < 1, p + q = 1$. If a particle starts at 0, for what values of p is return to 0 certain?

3. Let q_x be the probability that a random walk on $\{0, 1, \ldots, a\}$ with elastic barrier at 0 as described above will hit 0 before hitting a. Find a difference equation for the q_x, find boundary conditions, and determine q_x.

4. Let T_x, $1 \leq x \leq a - 1$, be the waiting time for the random walk of the previous problem to hit either 0 or a. Calculate $D_x = E[T_x]$, $1 \leq x \leq a - 1$, in the $p \neq q$ case.

5.4 BRANCHING PROCESSES

If a neutron collides with the nucleus of an atom, the nucleus may split and give rise to new neutrons, which in turn may collide with other nuclei and give rise to more neutrons, and so forth. This is an example of a branching process. Another commonly cited example involves the survival of family names, assumed to be passed on to male offspring. Starting with one individual, k offspring may be produced with probability $p_k, k = 0, 1, \ldots$. The number of offspring is a random variable X_1 that describes the size of the first generation. Each of the X_1 offspring can then produce k offspring with probability $p_k, k = 0, 1, 2, \ldots$, independently of X_1 and independently of the number of offspring of individuals of the same generation. The total number of the offspring of the X_1 individuals is then a random variable X_2 that describes the size of the second generation, and so forth. Continuing in this way, there is a sequence of random variables $X_0, X_1, \ldots$ where X_0 is the size of the initial generation and X_j describes the size of the jth generation.

A careful construction of a branching process in terms of random variables requires an infinite sequence of independent nonnegative integer-valued random variables all having the same density $p(k) = p_k, k = 0, 1, \ldots$. We will assume that there is such a sequence of random variables.

We commence with the density function p just described and assume throughout that $X_0 = 1$. Let X_1 be a random variable having density function p and let $Y_1^{(1)}, Y_2^{(1)}, \ldots$ be a sequence of independent random variables that all have the same density p and that are also independent of X_1. We then let

$$X_2 = Y_1^{(1)} + Y_2^{(1)} + \cdots + Y_{X_1}^{(1)};$$

i.e., X_2 is the sum of a random number of random variables. Letting $\hat{p}$ denote the generating function of the density p, by Theorem 3.4.5,

$$\hat{f}_{X_2}(t) = \hat{f}_{X_1}(\hat{p}(t)).$$

Since $\hat{f}_{X_1} = \hat{p}$,

$$\hat{f}_{X_2}(t) = \hat{p}(\hat{p}(t)).$$

Now let $Y_1^{(2)}, Y_2^{(2)}, \ldots$ be a sequence of independent random variables all having the same density and independent of all previously mentioned random variables, and let

$$X_3 = Y_1^{(2)} + Y_2^{(2)} + \cdots + Y_{X_2}^{(2)}.$$

Again by Theorem 3.4.5,

$$\hat{f}_{X_3}(t) = \hat{f}_{X_2}(\hat{p}(t)) = \hat{p}(\hat{p}(\hat{p}(t))).$$

Continuing in this manner, a sequence $X_1, X_2 \ldots$ is obtained whose generating functions satisfy

$$\hat{f}_{X_{j+1}}(t) = \hat{f}_{X_j}(\hat{p}(t) \qquad \text{for all } j \geq 1. \tag{5.11}$$

We will now show using mathematical induction that

$$\hat{f}_{X_{j+1}}(t) = \hat{p}(\hat{f}_{X_j}(t)) \qquad \text{for all } j \geq 1. \tag{5.12}$$

Since $\hat{f}_{X_2} = \hat{f}_{X_1}(\hat{p}(t))$ and $\hat{f}_{X_1} = \hat{p}, \hat{f}_{X_2}(t) = \hat{p}(\hat{f}_{X_1}(t))$ and Equation 5.12 is true for $j = 1$. Suppose Equation 5.12 is true for $j - 1$. By Equation 5.11,

$$\hat{f}_{X_{j+1}}(t) = \hat{f}_{X_j}(\hat{p}(t)) = \hat{p}(\hat{f}_{X_{j-1}}(\hat{p}(t)) = \hat{p}(\hat{f}_{X_j}(t)),$$

and the assertion is true for j. It follows from the principle of mathematical induction that Equation 5.12 is true for all $j \geq 1$.

The parameters $\mu = E[X_1] = \hat{p}'(1)$ and $\sigma^2 = \text{var}\, X_1 = \hat{p}''(1) + \hat{p}'(1) - (\hat{p}'(1))^2$, assumed to be finite, are useful in describing qualitative properties of the branching process. Since $E[X_{j+1}] = \hat{f}'_{X_{j+1}}(1)$, by Equation 5.12,

$$\hat{f}'_{X_{j+1}}(t) = \hat{p}'(\hat{f}_{X_j}(t))\hat{f}'_{X_j}(t)$$

and

$$E[X_{j+1}] = \hat{f}'_{X_{j+1}}(1) = \hat{p}'(\hat{f}_{X_j}(1))\hat{f}'_{X_j}(1) = \hat{p}'(1)\hat{f}'_{X_j}(1) = \mu E[X_j].$$

Iterating this result, $E[X_{j+1}] = \mu^{j+1}$. Therefore,

$$E[X_j] = \mu^j \qquad \text{for all } j \geq 1,$$

and the expected size of the *j*th generation increases or decreases geometrically according to whether $\mu > 1$ or $\mu < 1$.

Consider now the probability that the branching process will eventually terminate; i.e., $P(X_n = 0 \text{ for some } n \geq 1)$. We will want to exclude from consideration some special cases. Suppose first that $p_0 = 0$; i.e., the probability that an individual will have zero offspring is zero; in this case, extinction will never occur and we can henceforth assume that $p_0 > 0$. Suppose now that $p_0 = 1$. Then extinction will occur with the first generation, and the $p_0 = 1$ case will be excluded. Henceforth, we will assume that $0 < p_0 < 1$.

Consider the probability q_j that the size of the *j*th generation will be zero; i.e.,

$$q_j = P(X_j = 0) = \hat{f}_{X_j}(0).$$

By Equation 5.12, $q_{j+1} = \hat{f}_{X_{j+1}}(0) = \hat{p}(\hat{f}_{X_j}(0)) = \hat{p}(q_j)$. Thus, the q_j's are related by the equation

$$q_{j+1} = \hat{p}(q_j) \qquad \text{for all } j \geq 1. \tag{5.13}$$

Since $\hat{p}$ is supposedly given as part of the data describing the branching process and $q_1 = P(X_1 = 0) = p_0$, in principle Equation 5.13 can be used to determine the sequence $\{q_j\}_{j=1}^{\infty}$ by iteration. In general, however, $\hat{p}(s)$ will have nonlinear terms s^j, $j \geq 2$, which makes it difficult to find a formula for the q_j. As an alternative, we might examine the long-range behavior of the q_j by considering $\lim_{j\to\infty} q_j$, if it exists. Assuming that the limit exists and using the fact that $\hat{p}(s)$ is continuous on $[0, 1]$,

$$q = \lim_{j\to\infty} q_{j+1} = \lim_{j\to\infty} \hat{p}(q_j) = \hat{p}(q);$$

i.e., q is a solution of the equation $s = \hat{p}(s)$. Note that $s = 1$ solves this equation since $\hat{p}(1) = 1$, but there may be other solutions as well.

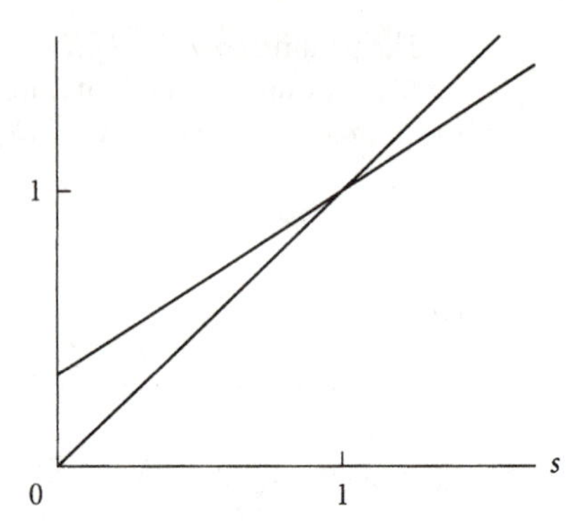

FIGURE 5.2 Graphs of $t = s$ and $t = \hat{p}(s), p_0 + p_1 = 1$.

To determine other roots of the equation $s = \hat{p}(s)$, we will examine the function $\hat{p}(\cdot)$ in greater detail. Since $0 < p_0 < 1, p_j > 0$ for some $j \geq 1$, and since $\hat{p}'(s) = \sum_{j=1}^{\infty} jp_j s^{j-1}, \hat{p}'(s) > 0$ on $(0, 1)$ and $\hat{p}(s)$ is strictly increasing on $[0, 1]$. Since $q_1 = p_0 > 0$, $q_2 = \hat{p}(q_1) > \hat{p}(0) = p_0 = q_1$. Assume that $q_j > q_{j-1}$. Then $q_{j+1} = \hat{p}(q_j) > \hat{p}(q_{j-1}) = q_j$. By mathematical induction, $q_j < q_{j+1}$ for all $j \geq 1$. Thus, $\{q_j\}_{j=1}^{\infty}$ is a monotone increasing sequence that is bounded above by 1, and therefore $q = \lim_{j\to\infty} q_j$ exists with $0 \leq q \leq 1$.

An alternative approach to solutions of the equation $s = \hat{p}(s)$ is to look at points of intersection of the graphs of the equations $t = s$ and $t = \hat{p}(s)$, $0 \leq s \leq 1$, since the s-coordinate of a point of intersection is a solution of the equation $s = \hat{p}(s)$. It will be necessary to consider two cases. Suppose first that $p_0 + p_1 = 1$. Thus, $\hat{p}(s) = p_0 + p_1 s$ with $0 < p_1 < 1$, and the two graphs are as depicted in Figure 5.2. In this case, it is clear that there is only one solution to the equation $s = \hat{p}(s)$; namely, $s = 1$, so that $q = 1$. This means that for large j, the probability that extinction will occur with the jth generation is very close to 1. Note that $\mu = \hat{p}'(1) = p_1 < 1$ in the $p_0 + p_1 = 1$ case. Suppose now that $p_0 + p_1 < 1$ so that $p_j > 0$ for some $j \geq 2$. In this case, $\hat{p}''(s) = \sum_{j=2}^{\infty} j(j-1)p_j s^{j-2} > 0$ on $(0, 1)$ and the function $\hat{p}(s)$ is convex and strictly increasing as depicted in Figure 5.3. In this case, it is clear that there are at most two solutions of the equation $s = \hat{p}(s)$. We will now show that q is the smallest solution of this equation. Let $r > 0$ be any solution. Then $q_1 = p_0 = \hat{p}(0) < \hat{p}(r) = r$. Assume that $q_{j-1} < r$. Then $q_j = \hat{p}(q_{j-1}) < \hat{p}(r) = r$. Thus, the sequence $\{q_j\}_{j=1}^{\infty}$ is bounded above by r, and therefore $q = \lim_{j\to\infty} q_j \leq r$; i.e., q is the smallest solution of the equation $s = \hat{p}(s)$.

EXAMPLE 5.6 According to a statistical study by A. J. Lotka, the number of male offspring of an American male is given by the modified geometric density $p_0 = .4823$ and $p_k = (.2126)(.5893)^{k-1}, k \geq 1$. The generating function $\hat{p}$ is then

$$\hat{p}(t) = .4823 + \frac{.2126t}{1 - .5893t}.$$

Using software such as Mathematica or Maple V to approximate the solution of the equation

$$.4823 + \frac{.2126t}{1 - .5893t} = t,$$

the probability of extinction is .8183. ■

We will now relate the probability of ultimate extinction to the expected number of offspring of a single individual in the $p_0 + p_1 < 1$ case. If $q < 1$, then there is a point s_0 in $(q, 1)$ such that $\hat{p}'(s_0) = 1$ by the mean value theorem; since $\hat{p}'(s)$ is strictly increasing on $[0, 1]$ and continuous from the left at 1 by Abel's theorem (Theorem 4.2.3), $\mu = \hat{p}'(1) > 1$. Thus, if $\mu = \hat{p}'(1) \leq 1$, then $q = 1$. Suppose now that $q = 1$ so that the graph of $t = \hat{p}(s)$ intersects the graph of $t = s$ in only one point. It follows that $\mu = \hat{p}'(1) \leq 1$. Thus, $q = 1$ if and only if $\mu \leq 1$. In the $p_0 + p_1 = 1$ case, $q = 1$ and $\mu \leq 1$. We thus have the following theorem.

Theorem 5.4.1 $q = \lim_{j \to \infty} q_j = 1$ *if and only if* $\mu \leq 1$.

EXAMPLE 5.7 Suppose $p_0 = 1/8$, $p_1 = 1/4$, and $p_2 = 5/8$. Then $\hat{p}(s) = 1/8 + (1/4)s + (5/8)s^2$ and the equation $s = \hat{p}(s)$ has the two solutions 1/5, 1. Therefore, the probability of ultimate extinction is 1/5 with $\mu = 3/2$. ■

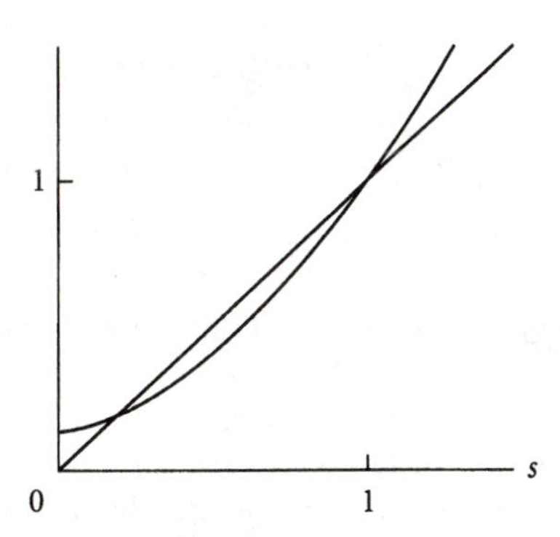

FIGURE 5.3 Graphs of $t = s$ and $t = \hat{p}(s), p_0 + p_1 < 1$.

EXERCISES 5.4

1. What is the probability of ultimate extinction q for a branching process with

$$\hat{p}(s) = \frac{1}{8}s^3 + \frac{3}{8}s^2 + \frac{3}{8}s + \frac{1}{8}?$$

2. Consider the branching process with $p_0 = 1/8$, $p_1 = 3/8$, $p_2 = 3/8$, $p_3 = 1/8$, and $p_n = 0$ for all $n \geq 4$. Calculate the probability q_3 that extinction will occur with the third generation.
3. Find a formula for the probability of ultimate extinction q for a branching process with $\hat{p}(s) = \alpha + \beta s^2$ where $0 < \alpha < 1$ and $\alpha + \beta = 1$.
4. If $\mu = E[X_1]$ and $\sigma^2 = \text{var } X_1$, show that

$$\text{var } X_{j+1} = \mu^2 \text{ var } X_j + \mu^j \sigma^2 \qquad \text{for } j \geq 1.$$

5. Show that

$$\text{var } X_j = \sigma^2(\mu^{2j-2} + \mu^{2j-3} + \cdots + \mu^{j-1}) \qquad \text{for } j \geq 1.$$

The following problems require mathematical software such as Mathematica or Maple V.

6. Consider the branching process with $p_0 = 1/4$, $p_1 = 1/2$, $p_2 = 1/8$, $p_3 = 3/32$, $p_4 = 1/32$, and $p_n = 0$ for all $n \geq 5$. Approximate the probability of ultimate extinction q.
7. Consider a branching process for which the number of offspring of an individual has a Poisson density with parameter $\lambda = 2$. Approximate the probability of ultimate extinction q.
8. Consider a branching process for which the number of offspring of an individual has a binomial density with parameters $n = 5$ and $p = .25$. Approximate the probability of ultimate extinction q.
9. Consider the branching process with $p_0 = 1/8$, $p_1 = 3/8$, $p_2 = 3/8$, $p_3 = 1/8$, and $p_n = 0$ for all $n \geq 4$. Calculate q_1 through q_{10}.

5.5 PREDICTION THEORY

Consider a sequence $\{X_j\}$ of random variables with finite second moments where j is allowed to range from $-\infty$ to $+\infty$. The sequence may correspond to a random process that has been going for some time. Suppose the index n corresponds to the present and the random variables $\ldots, X_{n-2}, X_{n-1}$ correspond to observations in the past. How can the past observations $\ldots, X_{n-2}, X_{n-1}$ be used to predict X_n? That is, is there some function $\psi(\ldots, X_{n-2}, X_{n-1})$ of the past that predicts X_n? Because prediction entails some probability of error, there must be some criterion for choosing a predictor. There also must be some internal coherence in the sequence $\{X_j\}$. For example, if X_n is independent of the past, then the past is of no use for predicting X_n.

Throughout this section, $\{X_j\}$ will denote a two-sided sequence of random variables $\{X_j\}_{j=-\infty}^{+\infty}$ having finite second moments. The construction of such sequences is similar to the construction of infinite sequences of Bernoulli random variables.

Definition 5.4 *The sequence $\{X_j\}$ is a* stationary sequence *if for each finite sequence of integers $j_1 < j_2 < \ldots < j_k$ and integers n,*

$$f_{X_{j_1+n},\ldots,X_{j_k+n}}(x_{j_1},\ldots,x_{j_k}) = f_{X_{j_1},\ldots,X_{j_k}}(x_{j_1},\ldots,x_{j_k}). \blacksquare$$

We have seen that a (one-sided) sequence of Bernoulli random variables has this property for positive integers n. Stationarity is stronger than what is required for this section.

Definition 5.5 *The sequence of random variables $\{X_j\}$ is* weakly stationary *if $E[X_j] = \mu$ independently of j and the covariance $E[(X_j - \mu)(X_k - \mu)]$ depends only upon $|j-k|$, $-\infty < j, k < +\infty$.* $\blacksquare$

Since $E[(X_j - \mu)(X_j - \mu)]$ is independent of j, $\sigma^2 = \text{var } X_0 = E[(X_0 - \mu)^2] = E[(X_j - \mu)^2] = \text{var } X_j$, $-\infty < j < +\infty$. We will assume throughout that $\sigma > 0$.

If $\{X_j\}$ is a weakly stationary process, the function

$$R(n) = E[(X_j - \mu)(X_{j+n} - \mu)], \qquad -\infty < n < +\infty$$

is independent of j and is called the *covariance function* of the sequence. Note that $R(0) = \sigma^2$ and that

$$R(-n) = E[(X_j - \mu)(X_{j-n} - \mu)] = E[(X_{j-n} - \mu)(X_j - \mu)] = R(n),$$

and so $R(n) = R(-n) = R(|n|)$. The function

$$\rho(n) = \frac{R(n)}{\sigma^2}, \qquad -\infty < n < +\infty$$

is called the *correlation function* of the sequence $\{X_j\}$. Note that $\rho(0) = 1$.

EXAMPLE 5.8 Let $\{X_j\}$ be a two-sided sequence of independent random variables having the same density function and let σ^2 be the common variance. Then $\text{cov}(X_j, X_k) = 0$ whenever $j \neq k$ and $\text{cov}(X_j, X_j) = \text{var } X_j = \sigma^2$. Thus,

$$R(n) = \begin{cases} \sigma^2 & \text{if } n = 0 \\ 0 & \text{if } n \neq 0. \end{cases}$$

The sequence $\{X_j\}$ is both stationary and weakly stationary. ■

The sequence of the previous example can be used to construct other weakly stationary sequences.

EXAMPLE 5.9 (Moving Average Process) Let $\{Y_j\}$ be a two-sided sequence of independent random variables with finite second moments having the same density function, and let $\mu = E[Y_0], \sigma^2 = \operatorname{var} Y_0$. Now let $a_0, \ldots, a_{m-1}$ be a finite sequence of real numbers and define

$$X_j = a_0 Y_j + a_1 Y_{j-1} + \cdots + a_{m-1} Y_{j-m+1}, \qquad -\infty < j < +\infty.$$

Each X_j has finite second moments by Lemma 4.4.1, and

$$E[X_j] = E\left[\sum_{k=0}^{m-1} a_k Y_{j-k}\right] = \mu(a_0 + \cdots + a_{m-1}).$$

This shows that $E[X_j]$ is independent of j. We will now show that $\operatorname{cov}(X_i, X_{i+n})$ is independent of i. Define $a_j = 0$ for $j \notin \{0, 1, \ldots, m-1\}$ and assume for the time being that $n \geq 0$. Since the Y_j are independent random variables and $E[(Y_{i-j} - \mu)(Y_{i+n-k} - \mu)] = 0$ except when $j = k - n$,

$$\begin{aligned}
\operatorname{cov}(X_i, X_{i+n}) &= E\left[\left(X_i - \mu \sum_{j=0}^{m-1} a_j\right)\left(X_{i+n} - \mu \sum_{k=0}^{m-1} a_k\right)\right] \\
&= E\left[\left(\sum_{j=0}^{m-1} a_j (Y_{i-j} - \mu)\right)\left(\sum_{k=0}^{m-1} a_k (Y_{i+n-k} - \mu)\right)\right] \\
&= \sum_{j,k=0}^{m-1} a_j a_k E[(Y_{i-j} - \mu)(Y_{i+n-k} - \mu)] \\
&= \sum_{k=0}^{m-1} a_{k-n} a_k \operatorname{var} Y_{i+n-k} \\
&= \sum_{k=n}^{m-1} a_{k-n} a_k \operatorname{var} Y_{i+n-k}.
\end{aligned}$$

Therefore,

$$\operatorname{cov}(X_i, X_{i+n}) = \begin{cases} \sigma^2(a_0a_n + \cdots + a_{m-1-n}a_{m-1}) & \text{if } n \le m-1 \\ 0 & \text{if } n > m-1. \end{cases}$$

Clearly, $\operatorname{cov}(X_i, X_{i+n})$ is independent of i. This is also true if n is replaced by $-n$, because then $\operatorname{cov}(X_i, X_{i-n}) = \operatorname{cov}(X_{i-n}, X_i)$ which is independent of i. In the particular case that $a_k = 1/\sqrt{m}, k = 0, \ldots, m-1$,

$$R(n) = \begin{cases} \sigma^2(1 - (|n|/m)) & \text{if } |n| \le m-1 \\ 0 & \text{if } |n| \ge m. \end{cases} \qquad \blacksquare$$

If we want to predict X_n using past observations $\ldots, X_{n-2}, X_{n-1}$, there are many ways to choose a predictor $\hat{X}_n = \psi(\ldots, X_{n-2}, X_{n-1})$, and we must formulate some criterion for deciding which is the best. One possible criterion for choosing a best predictor $\hat{X}_n$ is to choose $\hat{X}_n$ so that $E[(X_n - \hat{X}_n)^2]$ is a minimum, where $E[(X_n - \hat{X}_n)^2]$ is a measure of the distance between X_n and $\hat{X}_n$, called the *mean square error.*

Consider a finite collection of random variables $Y, Y_1, \ldots, Y_p$ having finite second moments and zero means and let $\mathcal{L}$ be the collection of all linear combinations of the $Y_1, \ldots, Y_p$; i.e.,

$$\mathcal{L} = \left\{ \sum_{i=1}^{p} a_i Y_i; a_1, \ldots, a_p \in R \right\}.$$

An element of $\mathcal{L}$ will be denoted by $\hat{Y}$ and called a *linear predictor* of Y. It is easy to see that if $\hat{Y}_1$ and $\hat{Y}_2$ are in $\mathcal{L}$ and a, b are any two real numbers, then $a\hat{Y}_1 + b\hat{Y}_2$ is in $\mathcal{L}$.

A proof of the following theorem would take us too far astray from probability theory. Proofs can be found in books on measure theory or Hilbert space theory.

Theorem 5.5.1 *There is a $\hat{Y}^* \in \mathcal{L}$ such that*

$$E[(Y - \hat{Y}^*)^2] \le E[(Y - \hat{Y})^2] \qquad \text{for all } \hat{Y} \in \mathcal{L}. \tag{5.14}$$

The $\hat{Y}^*$ of this theorem is called a *minimum mean square linear predictor* of Y. The quantity $E[(Y - \hat{Y}^*)^2]$ is called the *minimum mean square error.* It is possible to prove this result using calculus by writing

$$E[(Y - \hat{Y})^2] = E[Y^2] - 2\sum_{i=1}^{p} a_i E[Y Y_i] + \sum_{i,j=1}^{p} a_i a_j E[Y_i Y_j]$$

and minimizing the expression on the right as a quadratic function of the variables $a_1, \ldots, a_p$.

Theorem 5.5.2 *A predictor $\hat{Y}^*$ has minimum mean square error if and only if $E[(Y-\hat{Y}^*)\hat{Y}] = 0$ for every linear predictor $\hat{Y}$ in $\mathcal{L}$. Moreover, if $\hat{Y}_1^*$ and $\hat{Y}_2^*$ are any two linear predictors with minimum mean square error, then $\hat{Y}_1^* = \hat{Y}_2^*$ with probability 1.*

PROOF: Suppose first that $E[(Y-\hat{Y}^*)\hat{Y}] = 0$ for all $\hat{Y} \in \mathcal{L}$. Then

$$\begin{aligned} E[(Y-\hat{Y})^2] &= E[(Y-\hat{Y}^*+\hat{Y}^*-\hat{Y})^2] \\ &= E[(Y-\hat{Y}^*)^2] + 2E[(Y-\hat{Y}^*)(\hat{Y}^*-\hat{Y})] + E[(\hat{Y}^*-\hat{Y})^2]. \end{aligned}$$

Since $\hat{Y}^*-\hat{Y} \in \mathcal{L}$, $E[(Y-\hat{Y}^*)(\hat{Y}^*-\hat{Y})] = 0$ by hypothesis. Therefore,

$$E[(Y-\hat{Y})^2] \geq E[(Y-\hat{Y}^*)^2] \text{ for all } \hat{Y} \in \mathcal{L},$$

and $\hat{Y}^*$ has minimum mean square error. Now let $\hat{Y}^*$ be the predictor of Theorem 5.5.1 with minimum mean square error. Note that if $\hat{Y} \in \mathcal{L}$ and $E[\hat{Y}^2] = \operatorname{var}\hat{Y} = 0$, then $\hat{Y} = 0$ with probability 1, and therefore $E[(Y-\hat{Y}^*)\hat{Y}] = 0$. We can therefore assume that $E[\hat{Y}^2] \neq 0$. Suppose that

$$E[(Y-\hat{Y}^*)\hat{Y}] = \lambda \neq 0 \qquad \text{for some } \hat{Y} \in \mathcal{L}.$$

Consider

$$\hat{Z} = \hat{Y}^* + \frac{\lambda}{E[\hat{Y}^2]}\hat{Y},$$

which is in $\mathcal{L}$ since $\hat{Y}$ and $\hat{Y}^*$ are in $\mathcal{L}$. Writing

$$\hat{Z} - \hat{Y}^* = \frac{\lambda}{E[\hat{Y}^2]}\hat{Y},$$

$$\begin{aligned} E[(Y-\hat{Z})^2] &= E[((Y-\hat{Y}^*)+(\hat{Y}^*-\hat{Z}))^2] \\ &= E[(Y-\hat{Y}^*)^2] + 2E[(Y-\hat{Y}^*)(\hat{Y}^*-\hat{Z})] \\ &\quad + E[(\hat{Y}^*-\hat{Z})^2] \\ &= E[(Y-\hat{Y}^*)^2] - 2E\left[(Y-\hat{Y}^*)(\frac{\lambda}{E[\hat{Y}^2]}\hat{Y})\right] \\ &\quad + E\left[\frac{\lambda^2}{E[\hat{Y}^2]^2}\hat{Y}^2\right] \\ &= E[(Y-\hat{Y}^*)^2] - 2\frac{\lambda}{E[\hat{Y}^2]}E[(Y-\hat{Y}^*)\hat{Y}] + \frac{\lambda^2}{E[\hat{Y}^2]^2}E[\hat{Y}^2] \\ &= E[(Y-\hat{Y}^*)^2] - \frac{2\lambda^2}{E[\hat{Y}^2]} + \frac{\lambda^2}{E[\hat{Y}^2]} \end{aligned}$$

$$= E[(Y - \hat{Y}^*)^2] - \frac{\lambda^2}{E[\hat{Y}^2]}$$

$$< E[(Y - \hat{Y}^*)^2].$$

But this contradicts the fact that $\hat{Y}^*$ minimizes the mean square error. The assumption that $E[(Y - \hat{Y}^*)\hat{Y}] = \lambda \neq 0$ for some $\hat{Y} \in \mathcal{L}$ leads to a contradiction, and therefore $E[Y - \hat{Y}^*)\hat{Y}] = 0$ for all $\hat{Y} \in \mathcal{L}$. Finally, suppose that $\hat{Y}_1^*$ and $\hat{Y}_2^*$ are both minimum mean square linear predictors of Y. Then

$$0 = E[(Y - \hat{Y}_i^*)\hat{Y}] = E[Y\hat{Y}] - E[\hat{Y}_i^*\hat{Y}] \qquad \text{for } i = 1, 2, \hat{Y} \in \mathcal{L}.$$

Therefore, $E[(\hat{Y}_1^* - \hat{Y}_2^*)\hat{Y}] = 0$ for all $\hat{Y} \in \mathcal{L}$. Since the first factor is in $\mathcal{L}$, we can replace $\hat{Y}$ by $\hat{Y}_1^* - \hat{Y}_2^*$ to obtain $E[(\hat{Y}_1^* - \hat{Y}_2^*)^2] = 0$, and therefore $\hat{Y}_1^* = \hat{Y}_2^*$ with probability 1 (see Exercise 4.3.11). ■

We now return to the two-sided weakly stationary sequence $\{X_j\}$ and the problem of predicting X_n using the past $\ldots, X_{n-2}, X_{n-1}$. Computationally, we cannot expect to use all of the past $\ldots, X_{n-2}, X_{n-1}$ to predict X_n and must decide upon how many observations in the immediate past we will use. Suppose it has been decided to use just p observations $X_{n-p}, \ldots, X_{n-1}$. The most general predictor, not necessarily linear, will then have the form $\hat{X}_n = \psi(X_{n-p}, \ldots, X_{n-1})$. It is true, but cannot be proved here, that if we put

$$\psi(x_{n-p}, \ldots, x_{n-1}) = E[X_n \mid X_{n-p} = x_{n-p}, \ldots, X_{n-1} = x_{n-1}],$$

then $\hat{X}_n = \psi(X_{n-p}, \ldots, X_{n-1})$ is the minimum mean square predictor of X_n. This is not the same as the minimum mean square *linear* predictor of X_n. In practice, the computation of $E[X_n \mid X_{n-p} = x_{n-p}, \ldots, X_{n-1} = x_{n-1}]$ requires complete knowledge of the joint density $f_{X_{n-p},\ldots,X_{n-1},X_n}$; even if the joint density were known, the calculation of the conditional density might be intractable. The prediction problem is easier to handle if we limit ourselves to linear prediction.

To apply the above theorems, we must assume that the random variables X_j have been centered; i.e., that $E[X_j] = 0, -\infty < j + \infty$. Let p and n be fixed positive integers and let $\mathcal{L}$ be the collection of all linear combinations of $X_{n-p}, \ldots, X_{n-1}$. A typical element of $\mathcal{L}$ will be denoted by $\hat{X}_n$. Let $\hat{X}_n^* = \sum_{j=1}^p a_j X_{n-j}$ be a minimum mean square linear predictor of X_n. Then

$$E[(X_n - \hat{X}_n^*)\hat{X}_n] = 0 \qquad \text{for all } \hat{X}_n \in \mathcal{L}.$$

These equations hold if and only if

$$E[(X_n - \hat{X}_n^*)X_{n-j}] = 0 \qquad \text{for } j = 1, \ldots, p$$

or

$$E[(X_n - a_1X_{n-1} - \cdots - a_pX_{n-p})X_{n-j}] = 0 \qquad \text{for } j = 1, \ldots, p$$

or

$$\begin{aligned} &E\,[X_nX_{n-j}] \\ &\quad = a_1E[X_{n-1}X_{n-j}] + \cdots + a_pE[X_{n-p}X_{n-j}] \qquad \text{for } j = 1, \ldots, p. \end{aligned}$$

Thus, the above condition is equivalent to

$$R(j) = a_1R(1-j) + \cdots + a_pR(p-j) \qquad \text{for } j = 1, \ldots, p; \quad (5.15)$$

i.e., the $a_1, \ldots, a_p$ must satisfy the linear equations

$$\begin{aligned} R(1) &= a_1R(0) + \cdots + a_pR(p-1) \\ R(2) &= a_1R(-1) + \cdots + a_pR(p-2) \\ &\vdots \\ R(p) &= a_1R(1-p) + \cdots + a_pR(0). \end{aligned}$$

Since there is at least one minimum mean square linear predictor $\hat{X}_n^*$, there is at least one solution $a_1, \ldots, a_p$ of this system of equations.

We can also calculate the minimum mean square error σ_p^2 using $\hat{X}_n^*$ as follows:

$$\begin{aligned} \sigma_p^2 &= E[(X_n - \hat{X}_n^*)^2] \\ &= E[(X_n - \hat{X}_n^*)(X_n - \hat{X}_n^*)] \\ &= E[(X_n - \hat{X}_n^*)X_n] - E[(X_n - \hat{X}_n^*)\hat{X}_n^*]. \end{aligned}$$

The second term on the right is zero since $\hat{X}_n^* \in \mathcal{L}$. Thus,

$$\sigma_p^2 = E[X_nX_n] - E\left[\left(\sum_{j=1}^{p} a_jX_{n-j}\right)X_n\right] = R(0) - \sum_{j=1}^{p} a_jR(j).$$

EXAMPLE 5.10 Let $\{X_j\}$ be a two-sided weakly stationary process with covariance function

$$R(n) = \begin{cases} 1 - (|n|/3) & \text{for } n = 0, \pm 1, \pm 2 \\ 0 & \text{otherwise.} \end{cases}$$

Then $R(0) = 1, R(1) = 2/3$, and $R(2) = 1/3$. Suppose we take $p = 2$ so that the minimum mean square linear predictor $\hat{X}_n^* = a_1 X_{n-1} + a_2 X_{n-2}$ will be used to predict X_n. The coefficients a_1, a_2 must then satisfy the equations

$$\begin{cases} \frac{2}{3} = a_1 + \frac{2}{3}a_2 \\ \frac{1}{3} = \frac{2}{3}a_1 + a_2. \end{cases}$$

Solving for a_1 *and* a_2,

$$\hat{X}_n^* = \frac{4}{5}X_{n-1} - \frac{1}{5}X_{n-2},$$

and the minimum mean square error is $\sigma_p^2 = 8/15$. ■

EXERCISES 5.5

1. Let $\{Y_j\}$ be a sequence of independent random variables having the same density function with $E[Y_j] = 0$ and var $Y_j = 1$. For each $j \geq 1$, let $X_j = 1/4Y_j + 1/2Y_{j-1} + 1/4Y_{j-2}$. Find the covariance function for the $\{X_j\}$ sequence, a minimum mean square linear predictor $\hat{X}_n^*$ of X_n based on the last three observations $X_{n-1}, X_{n-2}, X_{n-3}$, and the minimum mean square error σ_3^2.
2. Would there be any improvement in the minimum mean square error in Problem 1 if the last four observations $X_{n-1}, X_{n-2}, X_{n-3}, X_{n-4}$ were used to predict X_n? Verify your answer by determining $\hat{X}_n^*$ and σ_4^2.
3. Consider a stationary sequence $\{X_n\}_{n=-\infty}^{+\infty}$ that satisfies the equation

$$X_n = \alpha X_{n-1} + \varepsilon_n, \qquad -\infty < n < +\infty,$$

where $\{\varepsilon_n\}_{n=-\infty}^{+\infty}$ is a stationary process with $\sigma_\varepsilon^2 = \text{var}\,\varepsilon_n > 0$ for which $E[\varepsilon_n X_m] = 0$ for all integers m and n. Show that $|\alpha| < 1$.
4. Consider a stationary process $\{X_n\}_{n=-\infty}^{+\infty}$ that satisfies the equation

$$X_{n+1} = a_1 X_n + a_2 X_{n-1} + \varepsilon_{n+1},$$

where $\{\varepsilon_n\}_{n=-\infty}^{+\infty}$ is a stationary process with $E[\varepsilon_n] = 0$, $\sigma_\varepsilon^2 = \text{var}\,\varepsilon_n > 0$ and $E[\varepsilon_n X_m] = 0$ for all integers m and n. If ρ is the correlation function of the X_n process, show that

$$\rho(1) = a_1 + a_2\rho(1)$$
$$\rho(2) = a_1\rho(2) + a_2$$

and determine a_1 and a_2 in terms of $\rho(1)$ and $\rho(2)$.

Solving the following problem without the benefit of mathematical software such as Mathematica or Maple V would be extremely tedious.

5. As a result of a statistical study of a stationary process, the values $R(0), R(1), R(2), R(3)$, and $R(4)$ of the covariance function $R(n)$ have been estimated to be 2, 1.68, 1.46, 1.22, and 1.08, respectively. If the last four observed values of the process are, in the order observed, $-2.25, -1.25, .25$, and 3.75, what is the minimum mean square linear predictor of the next value?

SUPPLEMENTAL READING LIST

R. B. Ash (1970). *Basic Probability Theory.* New York: Wiley.

C. W. Helstrom (1991). *Probability and Stochastic Processes for Engineers,* 2nd ed. New York: Macmillan.

S. Karlin and H. M. Taylor (1975). *A First Course in Stochastic Processes,* 2nd ed. New York: Academic Press.

M. Kendall and J. K. Ord (1990). *Time Series,* 3rd ed. New York: Oxford University Press.

CHAPTER 6

CONTINUOUS RANDOM VARIABLES

6.1 INTRODUCTION

At one time, a chance variable or random variable X was an undefined entity with an associated function $F(x)$, called the distribution function of X, that specified the probability that $X \leq x$. Probability theory at that time dealt with properties of distribution functions. The concept of probability space did not enter into the picture. Rapidly expanding applications of probability theory eventually necessitated a renewed look at the foundations. Most of this chapter discusses random variables as they were dealt with before the development of the probability space model.

Familiarity with the evaluation of double integrals by means of iterated integrals will be taken for granted. There will be situations in which it is necessary to interchange the order of integration of iterated integrals. The following statement justifies this procedure. Let $f : R^2 \to R$ be a nonnegative real-valued function that is Riemann integrable on each finite rectangle, and let (a, b), (c, d) be two intervals of real numbers, finite or infinite. Then

$$\int_a^b \left(\int_c^d f(x,y)dy \right) dx = \int_c^d \left(\int_a^b f(x,y)dx \right) dy.$$

Proofs of this result can be found in most calculus books.

6.2 RANDOM VARIABLES

The random variables considered in the previous chapters are customarily called discrete random variables, meaning that their ranges are countable sets. But because there are meaningful numerical attributes of outcomes of experiments not having this property, we must look at the concept of random variables anew.

Let $(\Omega, \mathcal{F}, P)$ be a probability space. In previous chapters, a mapping $X : \Omega \to R$ was called a random variable if its range is a countable set $\{x_1, x_2, \ldots\}$ and $(X = x) \in \mathcal{F}$ for all x in the range of X. Since $(X \leq x) = \cup_{x_i \leq x}(X = x_i) \in \mathcal{F}$, events of the type $(X \leq x), x \in R$, also belong to $\mathcal{F}$. We will take the latter property as the definition of a random variable.

Definition 6.1 *A mapping $X : \Omega \to R$ is called a* random variable *if $(X \leq x) = \{\omega : X(\omega) \leq x\} \in \mathcal{F}$ for all $x \in R$.* ■

In some instances, we allow X to take on the value $+\infty$, which is not in R, particularly for waiting times; in this case, X is called an *extended real-valued random variable.* The criterion is exactly the same; i.e., $(X \leq x) \in \mathcal{F}$ for all $x \in R$.

The fact that $(X \leq x) \in \mathcal{F}$ for $x \in R$ means that $P(X \leq x)$ is defined for all $x \in R$ and defines a function on R.

Definition 6.2 *If X is a random variable, the function $F_X : R \to R$ defined by*

$$F_X(x) = P(X \leq x), \qquad x \in R,$$

is called the distribution function *of the random variable X.* ■

Consider $a, b \in R$ with $a \leq b$. Since $P(a < X \leq b) = P((X \leq b) \cap (X \leq a)^c) = P(X \leq b) - P(X \leq a)$, probabilities of the type $P(a < X \leq b)$ can be calculated using F_X by the equation

$$P(a < X \leq b) = F_X(b) - F_X(a). \tag{6.1}$$

To illustrate these concepts, we need to enlarge our collection of probability spaces. In many experimental situations, the outcome of the experiment is a real number, and it is natural to take $\Omega = R$. It should be permissible to speak of the outcome being in some interval of real numbers. This means that $\mathcal{F}$ should at least include all intervals of the form (a, b), $[a, b)$, $(a, b]$, $[a, b]$, $(a, +\infty)$, $[a, +\infty)$, and so forth. It is a fact, but cannot be proved here, that there is a smallest σ-algebra of subsets of R that contains all intervals of the type just described. $\mathcal{F}$, however, does not contain all subsets of R. The reader may take comfort in the fact that any subset of R encountered at this level will be in $\mathcal{F}$. As usual, subsets of R in $\mathcal{F}$ are called *events.* Now that we have settled on Ω and $\mathcal{F}$, what do we do for a probability function P?

EXAMPLE 6.1 Consider a conceptual experiment in which a number is selected at random from the interval $(0, 1)$. What does this mean? It should mean that the probability that the number selected will be in the interval $(1/8, 1/4)$ should be the same as the probability that it will be in the interval $(7/8, 1)$ and also that it is twice as likely to be in the interval $(1/8, 3/8)$. This suggests that P should be determined by the length of the interval, provided the interval is a subinterval of $(0, 1)$; i.e., $P((a, b)) = b - a$ whenever $0 \leq a \leq b \leq 1$. Since no probability should be assigned to points outside $(0, 1)$, if (a, b) is any interval, $P((a, b))$ should be equal to $P((a, b) \cap (0, 1))$; e.g., $P((-3, 1/2)) = P((0, 1/2)) = 1/2$. It can be shown that there is a probability function P defined on $\mathcal{F}$ with these properties. ■

Example 6.1 can be modified by replacing the interval $(0, 1)$ by an interval (a, b) with $-\infty < a < b < +\infty$, as indicated below:

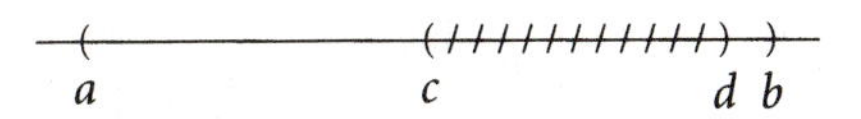

and defining

$$P((c, d)) = \frac{d - c}{b - a} \tag{6.2}$$

whenever $a \leq c \leq d \leq b$, $P((-\infty, a)) = 0$, and $P((b, +\infty)) = 0$. P is then called a *uniform probability measure* on (a, b). For this example and Example 6.1, it should be noted that $P(\{x\}) = 0$ for all $x \in R$. For example, if $x \in (a, b)$, then $\{x\} \subset (x - (1/n), x + (1/n)) \subset (a, b)$ for large n, and so

$$0 \leq P(\{x\}) \leq P\left(\left(x - \frac{1}{n}, x + \frac{1}{n}\right)\right) = \frac{2/n}{b - a} \to 0 \text{ as } n \to \infty.$$

Since $P(\{x\})$ does not depend upon n, $P(\{x\}) = 0$. Similar arguments can be used to show the same when $x \leq a$ or $x \geq b$. Since single points are assigned zero probability, $P((c, d]) = P([c, d)) = P([c, d]) = P((c, d))$.

Another way to modify Example 6.1, in addition to replacing $(0, 1)$ by (a, b), is to take $\Omega = (a, b)$ and define P only for intervals $(c, d) \subset (a, b)$ by Equation 6.2.

Every reader is familiar with experiments for which the above model is appropriate. The familiar pointer mounted on a circular disk is an example of an experiment in which a number between 0 and 2π is selected at random, although the interpretation of the outcome usually involves digitizing the outcome by assigning digits to equal sectors of the disk.

With such examples in mind, we can now exhibit nondiscrete random variables.

EXAMPLE 6.2 Let $(\Omega, \mathcal{F}, P)$ be a probability measure space where $\Omega = R$ and P is the uniform probability measure on (a, b), $-\infty < a < b < +\infty$. For each $\omega \in \Omega$, let $X(\omega) = \omega$. If $x \in R$, then $(X \leq x) = \{\omega : X(\omega) \leq x\} = \{\omega : \omega \leq x\} = (-\infty, x] \in \mathcal{F}$ and X is a random variable. X is not discrete since it can take on every value in R. The distribution function F_X of X can be calculated as follows:

(i) If $x \leq a$, then $F_X(x) = P(X \leq x) = P((-\infty, x]) = P((-\infty, x] \cap (a, b)) = P(\emptyset) = 0$ since $(-\infty, x] \cap (a, b) = \emptyset$.

(ii) If $a < x < b$, then $F_X(x) = P(X \leq x) = P((-\infty, x] \cap (a, b)) = P((a, x]) = P((a, x)) = (x - a)/(b - a)$.

(iii) If $x \geq b$, then $F_X(x) = P(X \leq x) = P((-\infty, x] \cap (a, b)) = P((a, b)) = 1$.

Thus,

$$F_X(x) = \begin{cases} 0 & \text{if } x \leq a \\ (x-a)/(b-a) & \text{if } a < x < b \\ 1 & \text{if } x \geq b. \end{cases} \quad \blacksquare$$

The random variable of this example is called a *continuous random variable.* The choice of "continuous" as a modifier is a traditional but poor one in that continuous in this context is analogous to a continuous distribution of mass as opposed to a discrete distribution and is in no way related to the concept of continuity of a function as studied in the calculus.

Other examples of continuous random variables can be constructed as follows. Let $f : R \to R$ be a real-valued nonnegative function that is Riemann integrable on every subinterval of R, and the improper integral $\int_{-\infty}^{+\infty} f(t)dt$ is defined and equal to 1. Let $\Omega = R$, let $\mathcal{F}$ be the smallest σ-algebra containing all intervals of real numbers, and define $P(A)$ for $A \in \mathcal{F}$ by putting

$$P((a, b)) = \int_a^b f(t)dt$$

for any interval (a, b). The integral on the right is also equal to $P((a, b])$, $P([a, b))$, and $P([a, b])$. Consider, for example, $P((a, b])$. Since

$$P((a, b)) \leq P((a, b]) \leq P\left(\left(a, b + \frac{1}{n}\right)\right) = \int_a^{b+(1/n)} f(t)dt$$

and the Riemann integral $\int_a^x f(t)dt$ is a continuous function of its upper limit,

$$P((a, b]) \leq \lim_{n \to \infty} \int_a^{b+(1/n)} f(t)dt = \int_a^b f(t)dt = P((a, b)).$$

Therefore, $P((a,b]) = P((a,b))$. In calculating probabilities $P(I)$ for an interval I, we can remove or adjoin endpoints to I without affecting the probabilities. Now define $X : \Omega \to R$ by putting $X(\omega) = \omega$ for all $\omega \in R$. Then $(X \le x) = \{\omega : X(\omega) \le x) = \{\omega : \omega \le x\} = (-\infty, x]$ and

$$F_X(x) = P(X \le x) = P((-\infty, x]) = P((-\infty, x)) = \int_{-\infty}^{x} f(t)dt.$$

Caveat: Generally speaking, endpoints can be removed or adjoined in this way only when probabilities are computed by integrating a Riemann integrable function.

EXAMPLE 6.3 Consider the function

$$f(x) = \begin{cases} 0 & \text{if } x < 0 \\ e^{-x} & \text{if } x \ge 0. \end{cases}$$

Since f is nonnegative and $\int_{-\infty}^{+\infty} f(t)dt = \int_0^{\infty} e^{-t}dt = \lim_{b\to+\infty} \int_0^b e^{-t}dt = \lim_{b\to+\infty}[-e^{-b} + 1] = 1$, there is a random variable X with distribution function

$$F_X(x) = \begin{cases} 0 & \text{if } x < 0 \\ 1 - e^{-x} & \text{if } x \ge 0. \end{cases}$$

The graphs of f and F_X are shown in Figure 6.1. The functions f and F_X are called the exponential density function and the distribution function, respectively. ■

As in the discrete case, it is necessary to perform various algebraic operations on random variables and deal with functions of random variables. Let $(\Omega, \mathcal{F}, P)$ be a probability space. Given random variables X and Y, we can define $X + Y$ by putting $(X + Y)(\omega) = X(\omega) + Y(\omega)$ for $\omega \in \Omega$ and define $X\,Y$ by putting $(X\,Y)(\omega) = X(\omega)Y(\omega)$ for $\omega \in \Omega$. More generally, if ϕ is a function of n real variables $x_1, \ldots, x_n$ and $X_1, \ldots, X_n$ are random variables, we can define $\phi(X_1, \ldots, X_n)(\omega) = \phi(X_1(\omega), \ldots, X_n(\omega))$; the above sum and product operations are special cases by taking $\phi(x, y) = x + y$ and $\phi(x, y) = xy$, respectively.

Lemma 6.2.1 *If X is a random variable, then $(X < x), (X \ge x), (X > x) \in \mathcal{F}$ for all $x \in R$. If $a, b \in R$ with $a < b$, then $(a < X \le b) \in \mathcal{F}$.*

PROOF: Let $Q = \{r_1, r_2, \ldots\}$ be the countable collection of rational numbers. Suppose $x \in R$ and $X(\omega) < x$. Then there is an $r_j \in Q$ such that $X(\omega) \le r_j < x$. Conversely, if $r_j < x$ and $X(\omega) \le r_j$, then $X(\omega) < x$. Thus,

$$(X < x) = \cup_{r_j \in Q, r_j < x}(X \le r_j) \in \mathcal{F}$$

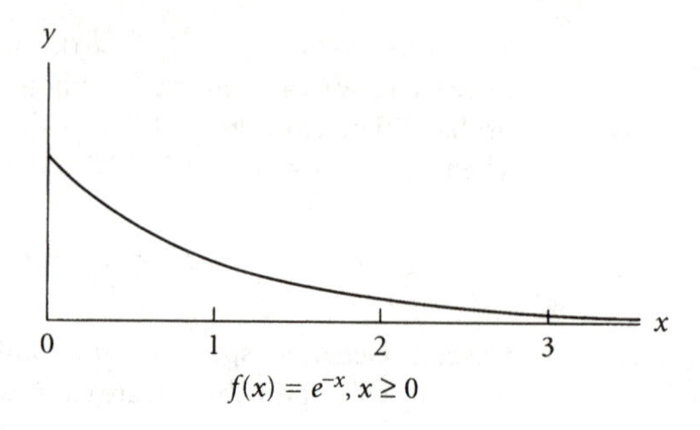

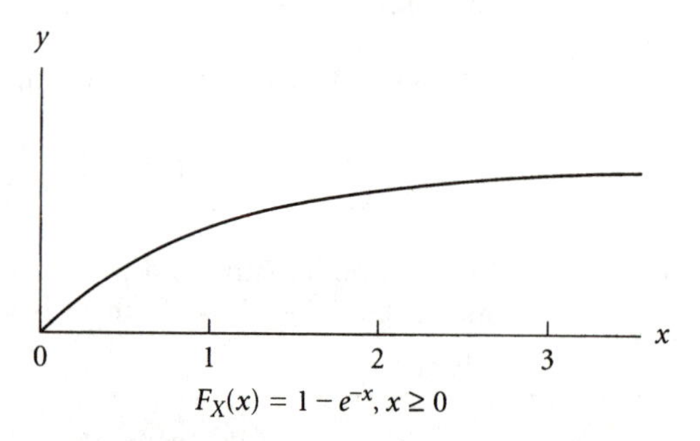

FIGURE 6.1 Exponential density and distribution functions.

by definition of a random variable. Since $(X \leq x) \in \mathcal{F}$,

$$(X > x) = (X \leq x)^c \in \mathcal{F}.$$

By the first part of the proof, $(X < x) \in \mathcal{F}$, and so $(X < x)^c = (X \geq x) \in \mathcal{F}$. If $a < b$, then $(a < X \leq b) = (X \leq b) \cap (X > a) \in \mathcal{F}$. ■

It is easily seen that $(a \leq X < b)$, $(a < X < b)$, and $(a \leq X \leq b)$ also belong to $\mathcal{F}$.

Theorem 6.2.2 *If X, Y are random variables and $a, b \in R$, then $aX + bY$, XY, and $|X|$ are all random variables.*

PROOF: We first show that aX is a random variable. If $a = 0$, then $aX \equiv 0$ and $(aX \leq x) = \emptyset$ if $x < 0$ and $(aX \leq x) = \Omega$ if $x \geq 0$ and aX is a random variable. If $a > 0$, then $(aX \leq x) = (X \leq x/a) \in \mathcal{F}$ for all $x \in R$, and if $a < 0$, then $(aX \leq x) = (X \geq x/a) \in \mathcal{F}$ for all $x \in R$. Thus, aX is a random variable. We now show that $X + Y$ is a random variable. Consider $(X + Y > z)$, $z \in R$. If ω is in this set, then $X(\omega) > z - Y(\omega)$ and there is a rational number r_j such that $X(\omega) > r_j > z - Y(\omega)$. The converse is also true. Therefore,

$$(X + Y > z) = \bigcup_{r_j \in Q} \Big((X > r_j) \cap (Y > z - r_j)\Big) \in \mathcal{F},$$

and therefore $(X + Y \leq z) \in \mathcal{F}$. To show that XY is a random variable, we first show that X^2 is a random variable. If $x < 0$, then $(X^2 \leq x) = \emptyset \in \mathcal{F}$; if $x \geq 0$, then $(X^2 \leq x) = (-\sqrt{x} \leq X \leq \sqrt{x}) \in \mathcal{F}$, as was to be proved. Since $XY = (1/4)((X + Y)^2 - (X - Y)^2)$, XY is a random variable by the previous steps. Since $(|X| \leq x) = \emptyset$ for $x < 0$ and $(|X| \leq x) = (-x \leq X \leq x)$ for $x \geq 0$, $|X|$ is a random variable. ■

It follows from this theorem that if n is a positive integer, X is a random variable, and $a_0, \ldots, a_n$ are constants, then $p(X) = a_0X^n + a_1X^{n-1} + \cdots + a_n$ is a random variable; i.e., a polynomial function of a random variable is again a random variable. This result can be extended to continuous functions. That is, if $\phi : R \to R$ is a continuous function and X is a random variable, then $\phi(X)$ is a random variable. Likewise, if ϕ is a continuous function of n variables $x_1, \ldots, x_n$ and $X_1, \ldots, X_n$ are random variables, then $\phi(X_1, \ldots, X_n)$ is a random variable. There is, however, trouble lurking beyond this point. In the case of a discrete random variable X, $\phi(X)$ is a random variable for *any* function $\phi : R \to R$. This fact need not be true for nondiscrete random variables. But since we will have no need to go beyond continuous functions of random variables, we will leave this matter where it belongs; namely, in a graduate course in real analysis.

One of the central problems we will take up has to do with finding the distribution function of $Y = \phi(X)$ knowing the distribution function of X.

EXAMPLE 6.4 Let X be a random variable having the distribution function

$$F_X(x) = \int_{-\infty}^{x} f(t)dt$$

where

$$f(t) = \begin{cases} 1 & \text{if } 0 < x < 1 \\ 0 & \text{otherwise.} \end{cases}$$

Then

$$F_X(x) = \begin{cases} 0 & \text{if } x < 0 \\ x & \text{if } 0 \leq x < 1 \\ 1 & \text{if } x \geq 1. \end{cases}$$

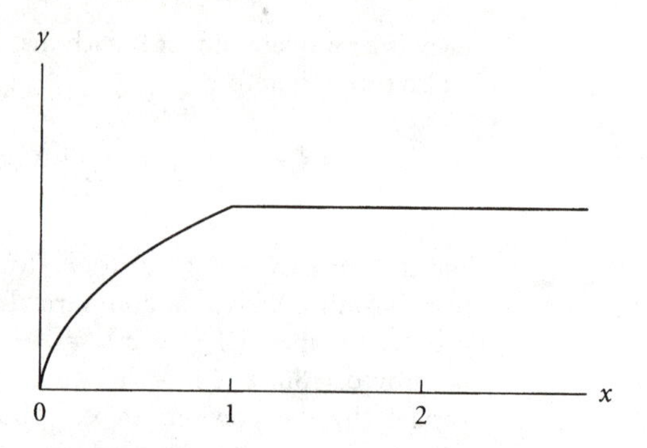

FIGURE 6.2 Distribution function of $Y = X^2$.

If $Y = X^2$, what is the distribution function of Y? If $y < 0$, then $F_Y(y) = P(Y \leq y) = P(X^2 \leq y) = P(\emptyset) = 0$. If $0 \leq y < 1$, then $F_Y(y) = P(Y \leq y) = P(X^2 \leq y) = P(-\sqrt{y} \leq X \leq \sqrt{y}) = \int_{-\sqrt{y}}^{\sqrt{y}} f(t)dt = \int_0^{\sqrt{y}} 1dt = \sqrt{y}$. If $y \geq 1$, then $F_Y(y) = P(Y \leq y) = P(X^2 \leq y) = \int_0^1 f(t)dt = 1$. Therefore,

$$F_Y(y) = \begin{cases} 0 & \text{if } y < 0 \\ \sqrt{y} & \text{if } 0 \leq y < 1 \\ 1 & \text{if } y \geq 1. \end{cases}$$

The graph of F_Y is shown in Figure 6.2. ■

Consider a conceptual experiment in which a point is chosen at random from a region S in the plane; e.g., a region encompassed by a simple closed curve. "At random" should mean that the probabilities that the chosen point will be in congruent subregions of S should be the same and that the probability that the chosen point will be in disjoint subregions should be the sum of the probabilities of being in each. These criteria suggest that probabilities should be determined by areas; e.g., if $A \subset S$ is the shaded region depicted in Figure 6.3, then the probability that the chosen point will be in A is given by

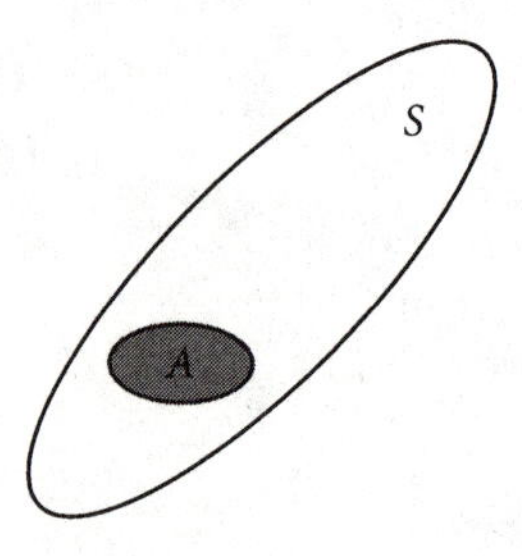

FIGURE 6.3 Geometric probabilities.

$$P(A) = \frac{|A|}{|S|}$$

where $|A|$ denotes the area of A. More generally, if S is a region in the n-dimensional space R^n and A is a subregion, then $P(A)$ is defined by the same equation, with $|A|$ representing the n-dimensional volume of A. Probabilities defined in this way are called *geometric probabilities.*

EXAMPLE 6.5 Suppose a point is chosen at random from a region S in the plane consisting of points (x, y) with $x^2 + y^2 \leq 1$; i.e., a point in a disk of radius 1 and having center at $(0, 0)$. Let A be the set of points in a disk of radius 1/2 and having the same center. In this case, $|A| = \pi(1/2)^2 = \pi/4$, $|S| = \pi$, and

$$P(A) = \frac{|A|}{|S|} = \frac{1}{4}. \blacksquare$$

EXERCISES 6.2

1. An experiment consists of choosing a point at random from a disk D in the plane with center at $(0, 0)$ of radius 1. If X is the distance of the point from the origin, find the distribution function of X.
2. An experiment consists of selecting a point at random from a ball in 3-space with center at the origin and radius 1. If X is the distance from the origin, find the distribution function of X.
3. By choosing a point X at random from the interval $[0, 1]$, the line segment $[0, 1]$ is broken into two line segments $[0, X]$ and $[X, 1]$. What is the probability that the length of the shorter segment will be less than or equal to one-fourth of the length of the longer segment?
4. A point X is chosen at random from the interval $[0, 1]$. What is the probability that the roots of the equation $4y^2 + 7Xy + 1 = 0$ will be real?
5. If

$$f(x) = \begin{cases} 0 & \text{if } x \leq -1 \\ x + 1 & \text{if } -1 < x < 0 \\ 1 - x & \text{if } 0 \leq x < 1 \\ 0 & \text{if } x \geq 1, \end{cases}$$

calculate $F(x) = \int_{-\infty}^{x} f(t)\,dt$ for each real number x.

6. Consider the function $g(x) = e^{-|x|}, x \in R$. Find a constant c such that $\int_{-\infty}^{+\infty} f(t)dt = 1$ where $f(t) = c\,g(t)$. If X is a random variable having distribution function $F_X(x) = \int_{-\infty}^{x} f(t)dt$, calculate $P(-1 \leq X \leq 1)$.
7. Let

$$f(t) = \begin{cases} 1/2 & \text{if } -1 < t < 1 \\ 0 & \text{otherwise,} \end{cases}$$

$F(x) = \int_{-\infty}^{x} f(t)dt, x \in R$, and X be a random variable having distribution function F. Calculate $F(x)$ for each $x \in R$. If $Y = X^2$, what is the distribution function of Y?

8. Let X be a random variable having the distribution function $F_X(x) = \int_{-\infty}^{x} f(t)dt$ where

$$f(t) = \begin{cases} 1 & \text{if } 0 \le x \le 1 \\ 0 & \text{otherwise.} \end{cases}$$

If $Y = X^3$, what is the distribution function of Y?

9. If

$$F(x) = \begin{cases} 0 & \text{if } x < 0 \\ x^2 & \text{if } 0 \le x < 1/2 \\ x - (1/4) & \text{if } 1/2 \le x \le 5/4 \\ 1 & \text{if } x > 5/4, \end{cases}$$

find a function $f(x), x \in R$, such that $F(x) = \int_{-\infty}^{x} f(t)\,dt$ for all $x \in R$.

10. If $X : \Omega \to R$, show that the following statements are equivalent.
   (a) $(X \le x) \in \mathcal{F}$ for all $x \in R$.
   (b) $(X < x) \in \mathcal{F}$ for all $x \in R$.
   (c) $(X \ge x) \in \mathcal{F}$ for all $x \in R$.
   (d) $(X > x) \in \mathcal{F}$ for all $x \in R$.

6.3 DISTRIBUTION FUNCTIONS

Let $(\Omega, \mathcal{F}, P)$ be a probability space and let X be a random variable. The distribution function F_X of X was defined in the previous section and is given by

$$F_X(x) = P(X \le x), \qquad x \in R.$$

If the random variable X is known from context, the subscript X will be suppressed. Before looking at properties of distribution functions, we review some definitions from the calculus. Consider a function $f : R \to R$ and let $a \in R$.

1. If there is a number L with the property that for each $\epsilon > 0$ there is a $\delta > 0$ such that

$$|f(x) - L| < \epsilon \text{ whenever } a < x < a + \delta,$$

we write $\lim_{x \to a+} f(x) = L$. L is usually denoted by $f(a+)$.

2. If there is a number l with the property that for each $\epsilon > 0$ there is a $\delta > 0$ such that

$$|f(x) - l| < \epsilon \text{ whenever } a - \delta < x < a,$$

we write $\lim_{x \to a-} f(x) = l$. l is usually denoted by $f(a-)$.

3. If there is a number L with the property that for each $\epsilon > 0$ there is an $M \in R$ such that

$$|f(x) - L| < \epsilon \text{ whenever } x > M,$$

we write $\lim_{x \to +\infty} f(x) = L$. L is usually denoted by $f(+\infty)$.

4. If there is a number l with the property that for each $\epsilon > 0$ there is an $m \in R$ such that

$$|f(x) - l| < \epsilon \text{ whenever } x < m,$$

we write $\lim_{x \to -\infty} f(x) = l$. l is usually denoted by $f(-\infty)$.

Theorem 6.3.1 *If F is a distribution function, then*

(i) $0 \leq F(x) \leq 1$ *for all* $x \in R$.

(ii) $F(x) \leq F(y)$ *whenever* $x \leq y$.

(iii) $F(-\infty) = 0$ *and* $F(+\infty) = 1$.

(iv) $F(x+) = \lim_{y \to x+} F(y)$ *exists for each* $x \in R$ *and* $F(x) = F(x+)$ $F(x-) = \lim_{y \to x-} F(y)$ *exists for each* $x \in R$.

(v) F *is right-continuous at each* $x \in R$*; i.e.,* $F(x) = F(x+)$ *for all* $x \in R$.

In addition, $F(x-) = P(X < x)$.

PROOF:

(i) Since $F(x)$ is a probability, *(i)* is trivially true.

(ii) If $x \leq y$, then $(X \leq x) \subset (X \leq y)$ and $F(x) = P(X \leq x) \leq P(X \leq y) = F(y)$, so that *(ii)* is true.

(iii) We will prove only the second part of *(iii)*, the proof of the first part being similar. We first prove that $\lim_{n \to \infty} F(n) = 1$. Note that $\{(X \leq n)\}_{n=1}^{\infty}$ is an increasing sequence of events. For any $\omega \in \Omega, X(\omega)$ is real number and there is an n such that $X(\omega) \leq n$; i.e., $\Omega \subset \cup_{n=1}^{\infty} (X \leq n)$. Since the opposite relation is always true, the events $(X \leq n)$ increase to Ω. Therefore, $\lim_{n \to \infty} F(n) = \lim_{n \to \infty} P(X \leq n) = P(\Omega) = 1$ by Theorem 2.5.3. Thus, for each $\epsilon > 0$ there is a positive integer N such that $|F(N) - 1| < \epsilon$, which

implies that $1 - \epsilon < F(N) \leq 1$. If $x \geq N$, then $1 - \epsilon < F(N) \leq F(x) \leq 1 < 1 + \epsilon$, and so $|F(x) - 1| < \epsilon$ whenever $x \geq N$. This proves that $F(+\infty) = \lim_{x \to +\infty} F(x) = 1$.

(*iv*) Fix $x \in R$. Since the sequence of events $\{(X \leq x+(1/n))\}_{n=1}^{\infty}$ is a decreasing sequence and $(X \leq x) = \cap_{n=1}^{\infty}(X \leq x + (1/n))$, $F(x) = \lim_{n \to \infty} F(x + (1/n))$ by Theorem 2.5.3. Thus, for each $\epsilon > 0$ there is a positive integer N such that $|F(x) - F(x + (1/N))| < \epsilon$, which implies that $F(x + (1/N)) < F(x) + \epsilon$. Suppose $x < y < x + (1/N)$. Then,

$$F(x) - \epsilon \leq F(x) \leq F(y) \leq F\left(x + \frac{1}{N}\right) < F(x) + \epsilon;$$

i.e., $|F(y) - F(x)| < \epsilon$ whenever $|y - x| < 1/N$. This shows that $F(x) = F(x+) = \lim_{y \to x+} F(y)$. In the second part of (*iv*), we can show only that the left limit $\lim_{y \to x-} F(y)$ exists; it need not be equal to $F(x)$. To show that the left limit at x exists, note that the sequence of events $\{(X \leq x - (1/n))\}_{n=1}^{\infty}$ is an increasing sequence with $\cup_{n=1}^{\infty}(X \leq x - (1/n)) = (X < x)$. By Theorem 2.5.3,

$$\lim_{n \to \infty} F\left(x - \frac{1}{n}\right) = \lim_{n \to \infty} P\left(X \leq x - \frac{1}{n}\right) = P(X < x).$$

Thus, given $\epsilon > 0$, there is a positive integer N such that

$$P(X < x) - \epsilon < F\left(x - \frac{1}{N}\right) \leq P(X < x).$$

Let $\delta = (1/N)$. Suppose $x - \delta = x - (1/N) < y < x$. Let M be a positive integer such that $y < x - (1/M) < x$. Then

$$P(X < x) - \epsilon < F\left(x - \frac{1}{N}\right) \leq F(y) \leq F\left(x - \frac{1}{M}\right) \leq P(X < x);$$

i.e., $|F(y) - P(X < x)| < \epsilon$. We have shown that for each $\epsilon > 0$ there is a $\delta > 0$ such that $|F(y) - P(X < x)| < \epsilon$ whenever $x - \delta < y < x$; i.e., we have shown not only that the left limit at x exists but also that $F(x-) = \lim_{y \to x-} F(y) = P(X < x)$.

(*v*) Statement (*v*) is just a restatement of the first part of (*iv*). ■

Corollary 6.3.2 *For each $x \in R$, $P(X = x) = F(x+) - F(x-) = F(x) - F(x-)$.*

PROOF: $P(X = x) = P((X \leq x) \cap (X < x)^c) = P(X \leq x) - P(X < x) = F(x) - F(x-)$. ■

Since $P(X < x) \leq P(X \leq x)$, we always have

$$F(x-) \leq F(x) = F(x+),$$

but there may not be equality on the left. Note that F is continuous at x if and only if $F(x-) = F(x)$ and that F can have jump discontinuities only as depicted in Figure 6.4, the magnitude of the jump at x being $F(x) - F(x-)$. How large is the set of points of discontinuity of F?

Theorem 6.3.3 *The set of points of discontinuity of a distribution function F is at most countable.*

PROOF: For each $n \geq 1$, let $D_n = \{x : F(x) - F(x-) \geq 1/n\}$. Each D_n is empty or finite, because otherwise the sum of the jumps of F at points in D_n would exceed 1, which cannot happen since the total increase in F is 1. Since $D = \cup_{n=1}^{\infty} D_n$ is the set of points at which F has a jump discontinuity, D is at most countable, by Theorem 2.3.1. ■

EXAMPLE 6.6 Consider a random variable X with distribution function

$$F(x) = \begin{cases} 0 & \text{if } x < 1 \\ (1/4)(x-1) & \text{if } 1 \leq x < 2 \\ 1/2 & \text{if } 2 \leq x < 4 \\ (1/2)(x-3) & \text{if } 4 \leq x < 5 \\ 1 & \text{if } x \geq 5. \end{cases}$$

Clearly, $F(2-) = 1/4$ and $F(2) = F(2+) = 1/2$. F is not continuous at 2, and $P(X = 2) = 1/4$. ■

Theorem 6.3.4 *Given a function $F : R \to R$ with properties (i)–(v) in Theorem 6.3.1, there is a probability space $(\Omega, \mathcal{F}, P)$ and a random variable X having F as its distribution function.*

Sketch of Proof: Take $\Omega = R$ and take $\mathcal{F}$ to be the smallest σ-algebra containing all intervals of real numbers. For each $\omega \in \Omega$, let $X(\omega) = \omega$. Since $(X \leq a) = \{\omega : X(\omega) \leq a\} = \{\omega : \omega \leq a\} = (-\infty, a] \in \mathcal{F}$, X is a ran-

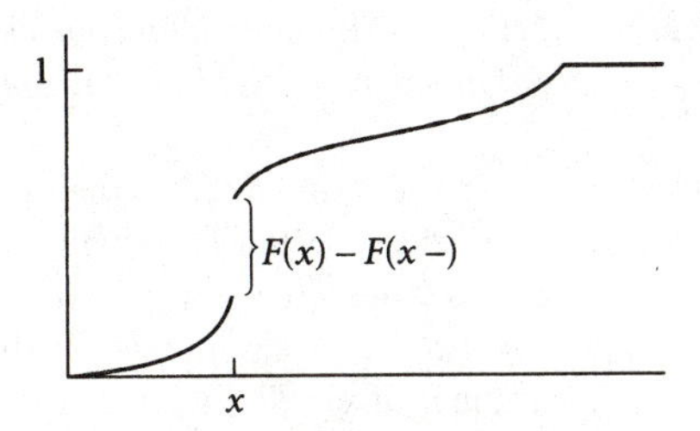

FIGURE 6.4 Jump discontinuity.

dom variable. For an interval $(a, b]$, define $P((a, b]) = F(b) - F(a)$. The function P can then be extended to all events in $\mathcal{F}$. Thus, $F_X(x) = P(X \leq x) = P((-\infty, x]) = \lim_{n \to -\infty} P((n, x]) = \lim_{n \to -\infty}(F(x) - F(n)) = F(x)$. ■

If $f : R \to R$ is nonnegative and Riemann integrable on every interval of real numbers with $\int_{-\infty}^{+\infty} f(t)dt = 1$, then Theorem 6.3.4 applies to the function

$$F(x) = \int_{-\infty}^{x} f(t)dt.$$

Thus, there is a probability space $(\Omega, \mathcal{F}, P)$ and a random variable X having F as its distribution function. The function f is called a *density function* for F and for X. Is the converse true? That is, given a random variable X with distribution function F, is there a nonnegative Riemann integrable function $f : R \to R$ such that the above equation holds? If there were such a function f, F would certainly have to be a continuous function, since the indefinite integral of f is a continuous function of its upper limit. In Example 6.6, the distribution function F is not continuous and consequently does not have a density function. A positive answer to the question just posed requires that F be at least continuous. But that is not enough to ensure that F has a density function. There is a criterion for determining if there is such a function.

Definition 6.3 *The function $F : R \to R$ is* absolutely continuous *if for each $\epsilon > 0$ there is a $\delta > 0$ such that*

$$\sum_{i=1}^{n} |F(\beta_i) - F(\alpha_i)| < \epsilon$$

whenever $n \geq 1$ and $(\alpha_1, \beta_1), \ldots, (\alpha_n, \beta_n)$ are nonoverlapping intervals with $\sum_{i=1}^{n} |\beta_i - \alpha_i| < \delta$. ■

If F is absolutely continuous, then F is continuous, as can be seen by taking $n = 1$ in this definition.

The following theorem settles the question asked above.

Theorem 6.3.5 *Let F be a distribution function on R. Then F is the indefinite integral of a function f if and only if F is absolutely continuous.*

The function f of this theorem can be identified, at most points of R, as the derivative $F'(x)$ of F. We will not elaborate on what is meant by "most points" at this stage. Such matters, as well as the proof of this theorem, are best left to advanced analysis courses. It should also be noted that the function f is not unique. If $g : R \to R$ agrees with f except at a finite number of points (or even at countably many points), then $F(x) = \int_{-\infty}^{x} g(t)dt$ also. In practice, the following fact can be used to establish that the distribution function F has a density function. If $f(x) = F'(x)$ wherever the derivative is defined and $\int_{-\infty}^{\infty} f(x)\,dx = 1$, then f is a density function for F.

Consider a situation in which it is known that the random variable X has a density function f and we want to find a density function, if there is one, for the random variable $Y = \phi(X)$ where $\phi : R \to R$ is continuous. The procedure for doing this is best illustrated by means of examples.

EXAMPLE 6.7 Suppose the random variable X has the density

$$f(x) = \begin{cases} e^{-x} & \text{if } x \geq 0 \\ 0 & \text{if } x < 0. \end{cases}$$

If $Y = X^2$, what is the density of Y? We first calculate the distribution function of X. For $x < 0$, $F_X(x) = \int_{-\infty}^{x} f(t)dt = 0$; for $x \geq 0$, $F_X(x) = \int_0^x e^{-t}dt = 1 - e^{-x}$. Thus,

$$F_X(x) = \begin{cases} 0 & \text{if } x < 0 \\ 1 - e^{-x} & \text{if } x \geq 0. \end{cases}$$

Let G be the distribution function of Y. For $y < 0$, $G(y) = P(Y \leq y) = 0$. Suppose $y \geq 0$. Then $G(y) = P(Y \leq y) = P(X^2 \leq y) = P(-\sqrt{y} \leq X \leq \sqrt{y}) = P(-\sqrt{y} < X \leq \sqrt{y}) = F_X(\sqrt{y}) - F_X(-\sqrt{y})$. Therefore,

$$G(y) = \begin{cases} 0 & \text{if } y < 0 \\ F_X(\sqrt{y}) - F_X(-\sqrt{y}) & \text{if } y \geq 0. \end{cases}$$

Since $-\sqrt{y} < 0$ for $y > 0$, $F_X(-\sqrt{y}) = 0$ and

$$G(y) = \begin{cases} 0 & \text{if } y < 0 \\ 1 - e^{-\sqrt{y}} & \text{if } y \geq 0. \end{cases}$$

The density $g(y) = G'(y)$ is therefore given by

$$g(y) = \begin{cases} 0 & \text{if } y < 0 \\ (1/2\sqrt{y})e^{-\sqrt{y}} & \text{if } y \geq 0. \end{cases} \quad \blacksquare$$

EXAMPLE 6.8 Suppose the random variable X has the density

$$f(x) = \begin{cases} 1 & \text{if } 0 \leq x \leq 1 \\ 0 & \text{otherwise} \end{cases}$$

and let $Y = \sqrt{X}$. Then

$$F_X(x) = \begin{cases} 0 & \text{if } x < 0 \\ x & \text{if } 0 \leq x < 1 \\ 1 & \text{if } x \geq 1. \end{cases}$$

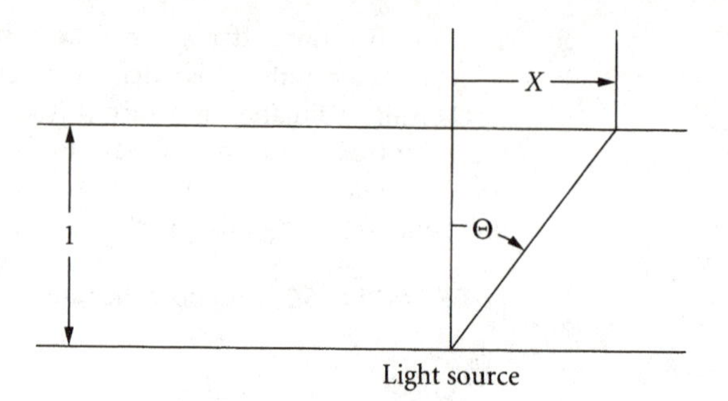

FIGURE 6.5 Cauchy density.

We can assume that X takes on values in $[0, 1]$. The same will be true of Y. Therefore, if G is the distribution function of Y, then $G(y) = 0$ if $y < 0$ and $G(y) = 1$ if $y > 1$. Suppose $0 \le y \le 1$. Then $G(y) = P(Y \le y) = P(0 \le \sqrt{X} \le y) = P(0 \le X \le y^2) = F_X(y^2) - F_X(0) = F_X(y^2) = y^2$. Hence,

$$G(y) = \begin{cases} 0 & \text{if } y < 0 \\ y^2 & \text{if } 0 \le y \le 1 \\ 1 & \text{if } y \ge 1. \end{cases}$$

The density $g(y) = G'(y)$ is therefore given by

$$g(y) = \begin{cases} 2y & \text{if } 0 \le y \le 1 \\ 0 & \text{otherwise.} \end{cases} \quad \blacksquare$$

In both of these examples, the distribution function of $Y = \phi(X)$ is obtained by converting $P(Y \le y)$ into a probability statement about X using the properties of the function ϕ.

EXAMPLE 6.9 (Cauchy Density) A source of light is mounted on one of two parallel walls, which are a unit distance apart as depicted in Figure 6.5. An angle Θ is chosen at random from the interval $(-\pi/2, \pi/2)$, measured from the perpendicular to the wall at the source, and a light beam is cast in that direction. The density function of Θ is then

$$f_\Theta(\theta) = \begin{cases} 1/\pi & \text{if } -\pi/2 < \theta < \pi/2 \\ 0 & \text{otherwise,} \end{cases}$$

and the distribution function $F_\Theta(\theta)$ is given by

$$F_\Theta(\theta) = \begin{cases} 0 & \text{if } \theta < -\pi/2 \\ (1/\pi)(\theta + (\pi/2)) & \text{if } -\pi/2 \le \theta < \pi/2 \\ 1 & \text{if } \theta \ge \pi/2. \end{cases}$$

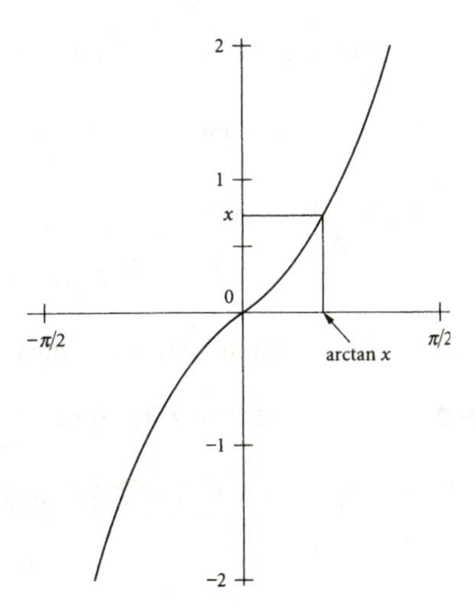

FIGURE 6.6 $X = \tan\Theta$.

Let X be the directed distance from the nearest point on the opposite wall to the point where the light beam hits the opposite wall. Then $X = \tan\Theta$. Since Θ can take on values between $-\pi/2$ and $\pi/2, X$ can take on values between $-\infty$ and $+\infty$. For $-\infty < x < +\infty$,

$$F_X(x) = P(X \le x) = P(\tan\Theta \le x).$$

To determine $P(\tan\Theta \le x)$, we must convert the statement $\tan\Theta \le x$ into a statement about Θ. Since $\arctan x$ is an increasing function on $(-\infty, +\infty)$, $\Theta = \arctan(\tan\Theta) \le \arctan x$ whenever $\tan\Theta \le x$ and conversely. Thus, $P(\tan\Theta \le x) = P(\Theta \le \arctan x)$, but we should keep in mind that Θ takes on values between $-\pi/2$ and $\pi/2$, so that

$$\begin{aligned} F_X(x) &= P(\tan\Theta \le x) = P\left(-\frac{\pi}{2} < \Theta \le \arctan x\right) \\ &= F_\Theta(\arctan x) - F_\Theta\left(-\frac{\pi}{2}\right) \\ &= \frac{1}{\pi}\left(\arctan x + \frac{\pi}{2}\right). \end{aligned}$$

Therefore,

$$f_X(x) = F'_X(x) = \frac{1}{\pi}\frac{1}{1+x^2}, \qquad -\infty < x < +\infty. \tag{6.3}$$

Equation 6.3 can be obtained by looking at the graph of the tangent function in Figure 6.6; for $\tan\Theta \le x, \Theta$ must be in the interval from $-\pi/2$ to $\arctan x$, and according to the definition of "random,"

$$F_X(x) = P(\tan\Theta \le x) = P\left(-\frac{\pi}{2} \le \Theta \le \arctan x\right) = \frac{\arctan x - (-\pi/2)}{\pi}$$

so that

$$f_X(x) = F'_X(x) = \frac{1}{\pi}\frac{1}{1+x^2}, \qquad -\infty < x < +\infty.$$

This function is known as the Cauchy density. ■

EXERCISES 6.3

1. Calculate $F'(x)$ for the distribution function of Example 6.6 and verify that F is not the indefinite integral of F'.
2. Let X be a random variable with distribution function

$$F(x) = \begin{cases} 0 & \text{if } x < 0 \\ x^2/4 & \text{if } 0 \le x < 1 \\ 1/2 & \text{if } 1 \le x < 2 \\ (1/2)(x-1) & \text{if } 2 \le x < 3 \\ 1 & \text{if } x \ge 3. \end{cases}$$

Calculate

(a) $P(0 \le X < 1)$.

(b) $P(0 \le X \le 1)$.

(c) $P(X = 1)$.

(d) $P(1/2 \le X \le 5/2)$.

3. Let X be a random variable having density function

$$f(x) = \begin{cases} 1/\pi & \text{if } -\pi/2 \le x \le \pi/2 \\ 0 & \text{otherwise} \end{cases}$$

and let $Y = \sin X$. Find a density function for Y.

4. Let X be a random variable with distribution function

$$F(x) = \begin{cases} 0 & \text{if } x < 0 \\ 1 - e^{-x} & \text{if } x \ge 0. \end{cases}$$

If $M > 0$, let $Y = \min(X, M)$. Determine the distribution function of Y. Does Y have a density function?

5. Let X be a random variable with distribution function

$$F(x) = \begin{cases} 0 & \text{if } x < 0 \\ 1 - e^{-x} & \text{if } x \ge 0 \end{cases}$$

and let $Y = \sqrt{X}$. Since $P(X \ge 0) = 1$, $\sqrt{X}$ is defined. What is the density function of Y?

6. Let X be a random variable having a density function

$$f_X(x) = \begin{cases} e^{-x} & \text{if } x \geq 0 \\ 0 & \text{if } x < 0 \end{cases}$$

and let $Y = \log X$. What is the density of Y?

7. Consider a searchlight that is mounted on a wall. An angle Θ is chosen at random between $-\pi/2$ and $\pi/2$, as measured from a perpendicular to the wall, and the light beam falls on an object 100 units away. If X denotes the distance of the object from the wall, what are the distribution and density functions of X?

6.4 JOINT DISTRIBUTION FUNCTIONS

It is not unusual in experimental situations to consider two or more numerical attributes of an outcome simultaneously. For the time being, we will consider only two attributes.

Let X, Y be two random variables on the probability space $(\Omega, \mathcal{F}, P)$. Then $P(X \leq x, Y \leq y)$ is meaningful and defines a function $F_{X,Y}$ of two real variables.

Definition 6.4 *If $x, y \in R$, the function of two real variables*

$$F_{X,Y}(x, y) = P(X \leq x, Y \leq y)$$

is called the joint distribution function *of the pair* (X, Y). ■

EXAMPLE 6.10 Suppose a point is chosen at random from the *unit square* U in the plane with opposite vertices at $(0, 0)$ and $(1, 1)$. Clearly, we should take $\Omega = U$ and "random" should mean that the probabilities that the outcome will be in two congruent regions in Ω will be the same. This suggests that the probability that the outcome will be in a region A within Ω should be equal to the area of the region divided by the area of the unit square, which is 1. Thus, $P(A) = \text{Area}(A)$. If $\omega = (x, y) \in \Omega$, let $X(\omega) = x$ and let $Y(\omega) = y$. Then X and Y should be random variables with

$$F_{X,Y}(x, y) = \begin{cases} 0 & \text{if } x < 0 \text{ or } y < 0 \\ x & \text{if } 0 \leq x \leq 1, y > 1 \\ y & \text{if } x > 1, 0 \leq y \leq 1 \\ xy & \text{if } 0 \leq x \leq 1, 0 \leq y \leq 1 \\ 1 & \text{if } x > 1, y > 1. \end{cases}$$

As an example of the computation of $F_{X,Y}(x, y)$, suppose $0 \leq x \leq 1, 0 \leq y \leq 1$. Then $F_{X,Y}(x, y)$ is the area of the shaded rectangle with opposite vertices at $(0, 0)$ and (x, y) as depicted in Figure 6.7. Since the area is xy, $F_{X,Y}(x, y) = xy$. ■

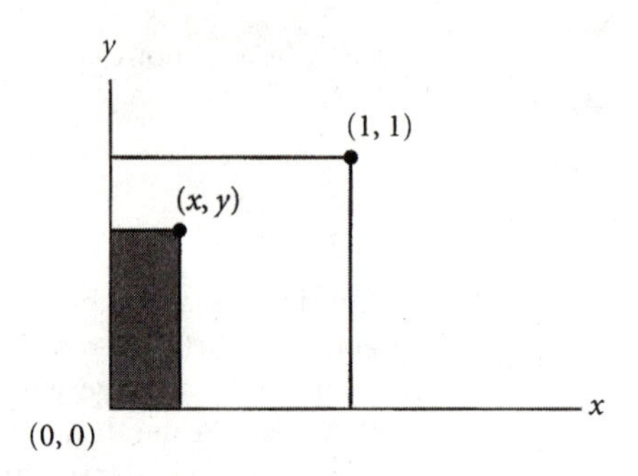

FIGURE 6.7 $F_{X,Y}(x, y), 0 \leq x \leq 1, 0 \leq y \leq 1.$

Joint distribution functions have properties similar to those listed in Theorem 6.3.1 for a distribution function of a single random variable.

Theorem 6.4.1 *If F is a joint distribution function, then*

(i) $0 \leq F(x, y) \leq 1$ *for all* $(x, y) \in R^2$.

(ii) *If* (x_1, y_1) *and* (x_2, y_2) *are any two points in* R^2 *with* $x_1 \leq x_2$ *and* $y_1 \leq y_2$, *then* $F(x_2, y_2) - F(x_1, y_2) - F(x_2, y_1) + F(x_1, y_1) \geq 0$.

(iii) $\lim_{x \to +\infty, y \to +\infty} F(x, y) = 1; \lim_{x \to -\infty, y \to -\infty} F(x, y) = 0;$
for each $y \in R$, $\lim_{x \to -\infty} F(x, y) = 0$; *and*
for each $x \in R$, $\lim_{y \to -\infty} F(x, y) = 0$.

(iv) *For each* $(a, b) \in R^2$, $\lim_{x \to a+, y \to b+} F(x, y) = F(a, b)$,
$\lim_{x \to a+} F(x, b) = F(a, b)$, *and* $\lim_{y \to b+} F(a, y) = F(a, b)$.

(v) *For each* $(a, b) \in R^2$, $\lim_{x \to a-} F(x, b)$ *and* $\lim_{y \to b-} F(a, y)$ *exist.*

The inequality of (ii) is easily reconstructed using Figure 6.8 by associating with each of the vertices a + or − sign, starting with a + at the upper right vertex and alternating signs, and then applying the signs to the value of F at the corresponding point. Except for (ii), proofs of these statements are similar to the proofs of the statements in Theorem 6.3.1.

Proof of (*ii*) Let (x_1, y_1) and (x_2, y_2) be two arbitrary points in the plane with $x_1 \leq x_2$ and $y_1 \leq y_2$. Then

$$F(x_2, y_2) - F(x_1, y_2) - F(x_2, y_1) + F(x_1, y_1) \geq 0. \tag{6.4}$$

This inequality follows from the fact that

$$\begin{aligned} 0 &\leq P(x_1 < X \leq x_2, y_1 < Y \leq y_2) \\ &= P((X \leq x_2) \cap (y_1 < Y \leq y_2) \cap (X \leq x_1)^c) \end{aligned}$$

$$
\begin{aligned}
&= P(X \le x_2, y_1 < Y \le y_2) - P(X \le x_1, y_1 < Y \le y_2) \\
&= P((X \le x_2) \cap (Y \le y_2) \cap (Y \le y_1)^c) \\
&- P((X \le x_1) \cap (Y \le y_2) \cap (Y \le y_1)^c) \\
&= P(X \le x_2, Y \le y_2) - P(X \le x_2, Y \le y_1) \\
&- P(X_1 \le x_1, Y \le y_2) + P(X \le x_1, Y \le y_1) \\
&= F_{X,Y}(x_2, y_2) - F_{X,Y}(x_2, y_1) - F_{X,Y}(x_1, y_2) + F_{X,Y}(x_1, y_1)
\end{aligned}
$$

since $(X \le x_2) \cap (Y \le y_2) \cap (Y \le y_1) = (X \le x_2) \cap (Y \le y_1)$, and so forth. ■

In the case of a single random variable X, Theorem 6.3.5 gives a necessary and sufficient condition that F_X have a density function; namely, that F_X be absolutely continuous. This concept can be extended to joint distribution functions, but it is best left to more advanced courses.

Definition 6.5 *A nonnegative Riemann integrable function $f : R^2 \to R$ is a* density function *for the joint distribution function F if*

$$F(x, y) = \iint_{A_{x,y}} f(u, v)da \qquad \textit{for all } x, y \in R$$

where $A_{x,y} = \{(u, v) : u \le x, v \le y\}$ and da denotes integration with respect to area. ■

In practice, the preceding double integral over $A_{x,y}$ is calculated using iterated integrals, since

$$\iint_{A_{x,y}} f(u, v)da = \int_{-\infty}^{x} \left(\int_{-\infty}^{y} f(u, v)dv\right)du = \int_{-\infty}^{y} \left(\int_{-\infty}^{x} f(u, v)du\right)dv.$$

We will assume in the remainder of this section that all joint distribution functions have density functions.

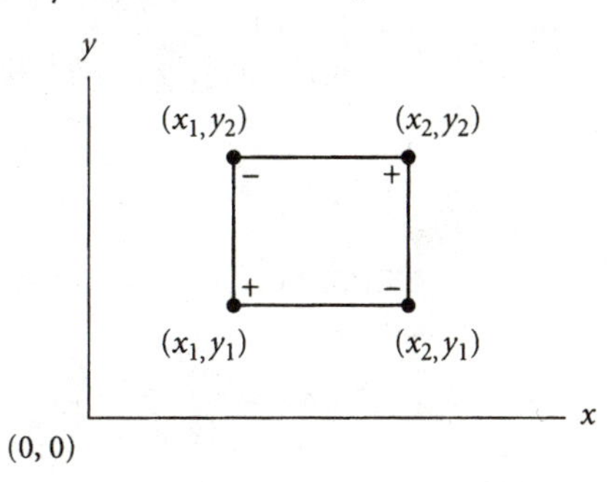

FIGURE 6.8 Alternating signs.

Most distribution functions come about by starting with a nonnegative Riemann integrable function $f : R^2 \rightarrow R$ with total integral 1 and constructing a probability space $(\Omega, \mathcal{F}, P)$ and a pair of random variables (X, Y) such that $F_{X,Y}$ has density function f by imitating the construction of the previous section as follows. Let $\Omega = R^2$, let $\mathcal{F}$ be the smallest σ-algebra containing all rectangles in R^2, and for $\omega = (x, y) \in \Omega$ let $X(\omega) = x$, $Y(\omega) = y$. Then X and Y are random variables. For any rectangle $I \subset R^2$, define

$$P(I) = \iint\limits_I f(u, v)\,da,$$

noting that including or excluding the edges of I has no effect on the value of the double integral. The probability function P can then be extended to all events in $\mathcal{F}$. Since $(X \leq x, Y \leq y) = \{\omega : X(\omega) \leq x, Y(\omega) \leq y) = \{(u, v) : u \leq x, v \leq y\}$,

$$F_{X,Y}(x, y) = P(X \leq x, Y \leq y) = \iint\limits_{A_{x,y}} f(u, v)\,da$$

where $A_{x,y} = \{(u, v) : u \leq x, v \leq y\}$. It follows that the pair (X, Y) has f as density function. It can be shown that for $A \in \mathcal{F}$,

$$P((X, Y) \in A) = \iint\limits_A f(u, v)\,da.$$

Without getting involved deeply in integration theory, for computational purposes we must limit the class of regions A for which this probability can be calculated. For example, if $a \leq b$, ϕ_1, ϕ_2 are continuous functions on $[a, b]$ with $\phi_1(x) \leq \phi_2(x), x \in [a, b]$, and $A = \{(u, v) : a \leq u \leq b, \phi_1(u) \leq v \leq \phi_2(u)\}$, then

$$P((X, Y) \in A) = \iint\limits_A f(u, v)\,da = \int_a^b \left(\int_{\phi_1(u)}^{\phi_2(u)} f(u, v)\,dv \right) du.$$

EXAMPLE 6.11 Let X and Y be random variables with joint density function

$$f_{X,Y}(x, y) = \begin{cases} 1 & \text{if } 0 \leq x \leq 1, 0 \leq y \leq 1 \\ 0 & \text{otherwise.} \end{cases}$$

Suppose we want to calculate $P(X^2 + Y^2 \leq 1)$. We must first define a region $A \subset R^2$ such that $(X^2 + Y^2 \leq 1) = ((X, Y) \in A)$. This is done by formally replacing X and Y by typical values u and v, respectively. In this case, we let

$$A = \{(u, v) : u^2 + v^2 \leq 1\},$$

as shown in Figure 6.9. It is easy to see that $(X(\omega), Y(\omega)) \in A$ if and only if $X^2(\omega) + Y^2(\omega) \leq 1$. Therefore,

$$P(X^2 + Y^2 \leq 1) = P((X, Y) \in A) = \iint_A f_{X,Y}(u, v)da.$$

Recalling that the integrand vanishes outside the unit square U and is equal to 1 inside, the integral is equal to Area$(A \cap U) = \pi/4$. Thus, $P(X^2 + Y^2 \leq 1) = \pi/4$. ■

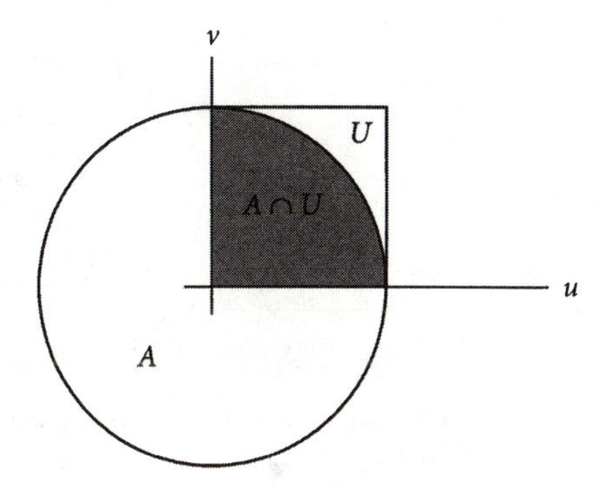

FIGURE 6.9 $P(X^2 + Y^2 \leq 1)$.

Knowing the joint distribution function or density function of the random variables X and Y, the corresponding individual distribution or density functions can be obtained.

Theorem 6.4.2 *If X and Y have the joint distribution function $F_{X,Y}$, then*

(i) $F_X(x) = \lim_{y \to +\infty} F_{X,Y}(x, y)$.

(ii) $F_Y(y) = \lim_{x \to +\infty} F_{X,Y}(x, y)$.

In addition, if X and Y have a joint density $f_{X,Y}$, then

(iii) $f_X(x) = \int_{-\infty}^{+\infty} f_{X,Y}(x, v)dv, \quad x \in R$, *and*

(iv) $f_Y(y) = \int_{-\infty}^{+\infty} f_{X,Y}(u, y)du, \quad y \in R$,

are densities for X and Y, respectively.

PROOF: The first step in proving (i) is to show that

$$F_X(x) = \lim_{n \to +\infty} F_{X,Y}(x, n).$$

Since the sequence of events $\{(X \leq x, Y \leq n)\}$ increases to the event $(X \leq x)$ as $n \to +\infty$ for each $x \in R$,

$$\lim_{n \to +\infty} F_{X,Y}(x, n) = \lim_{n \to +\infty} P(X \leq x, Y \leq n) = P(X \leq x) = F_X(x).$$

The rest of the proof of (i) is the same as the corresponding part of the proof of (iii) in Theorem 6.3.1. To prove (iii), let

$$g(x) = \int_{-\infty}^{+\infty} f_{X,Y}(x, v)dv.$$

Since

$$\begin{aligned} F_X(x) &= P(X \leq x, -\infty < Y < +\infty) \\ &= \int_{-\infty}^{x} \left(\int_{-\infty}^{+\infty} f_{X,Y}(u, v)dv \right) du \\ &= \int_{-\infty}^{x} g(u)du, \end{aligned}$$

it follows that g is a density for X. ■

When f_X and f_Y are obtained in this way, they are called *marginal density functions.*

EXAMPLE 6.12 Suppose a point is chosen at random from a disk D with center $(0, 0)$ and radius 1. Let X be the x-coordinate of the chosen point and Y the y-coordinate. The joint density $f_{X,Y}$ is then

$$f_{X,Y}(u, v) = \begin{cases} 1/\pi & \text{if } u^2 + v^2 \leq 1 \\ 0 & \text{otherwise.} \end{cases}$$

The density of X is then given by $f_X(x) = \int_{-\infty}^{+\infty} f_{X,Y}(x, v)dv, -\infty < x < +\infty$. To evaluate the integral, it is necessary to consider the cases $x < -1, -1 < x < 1$, and $x > 1$ separately. When $x < -1$, the function $f_{X,Y}(x, v)$ vanishes on the vertical line through x on the u-axis, and so $f_X(x) = 0$. The same is true when $x > 1$. When $-1 < x < 1, f_{X,Y}(x, \cdot)$ is equal to $1/\pi$ on the line segment joining $(x, -\sqrt{1 - x^2})$ to $(x, \sqrt{1 - x^2})$ as depicted in Figure 6.10 and equal to 0 at other points of the vertical line through x, so that

$$f_X(x) = \int_{-\sqrt{1-x^2}}^{\sqrt{1-x^2}} \frac{1}{\pi} dv = \frac{2}{\pi} \sqrt{1 - x^2}.$$

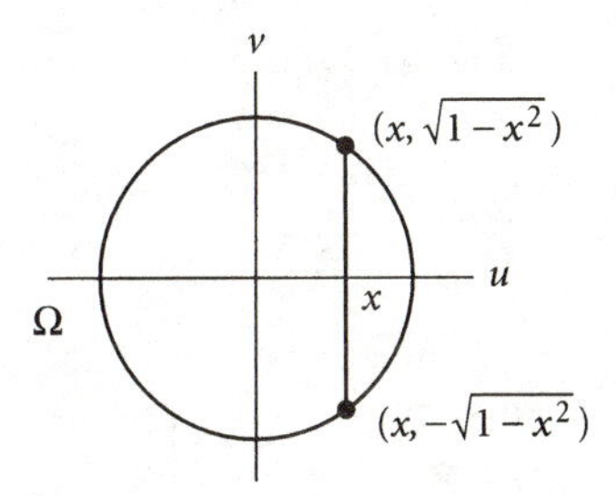

FIGURE 6.10 Marginal density.

Therefore,

$$f_X(x) = \begin{cases} (2/\pi)\sqrt{1-x^2} & \text{if } -1 < x < 1 \\ 0 & \text{otherwise.} \ \blacksquare \end{cases}$$

This example illustrates a technique for finding the density function of a random variable X. The consideration of a second random variable Y can result in the determination of the joint density $f_{X,Y}$, from which f_X can be obtained as above.

Suppose now that we are given the joint density function of two random variables X and Y and we would like to find the density f_Z of the sum $Z = X + Y$, assuming there is such a density.

Theorem 6.4.3 *Let X and Y be random variables having a joint density $f_{X,Y}$ and let $Z = X + Y$. The density of Z exists and is given by*

$$f_Z(z) = \int_{-\infty}^{+\infty} f_{X,Y}(u, z-u)\,du$$
$$= \int_{-\infty}^{+\infty} f_{X,Y}(z-v, v)\,dv, \qquad z \in R. \tag{6.5}$$

PROOF: Consider first the distribution function F_Z. Since $F_Z(z) = P(Z \leq z) = P(X + Y \leq z)$,

$$F_Z(z) = \iint_{\{u+v\leq z\}} f_{X,Y}(u, v)\,da = \int_{-\infty}^{+\infty}\left(\int_{-\infty}^{z-u} f_{X,Y}(u, v)\,dv\right)du.$$

With u fixed, let $w = v + u$ in the inner integral to obtain

$$F_Z(z) = \int_{-\infty}^{+\infty}\left(\int_{-\infty}^{z} f_{X,Y}(u, w-u)\,dw\right)du$$
$$= \int_{-\infty}^{z}\left(\int_{-\infty}^{+\infty} f_{X,Y}(u, w-u)\,du\right)dw.$$

If we let

$$g(z) = \int_{-\infty}^{+\infty} f_{X,Y}(x, z - x)dx = \int_{-\infty}^{+\infty} f_{X,Y}(u, z - u)du,$$

then $F_Z(z) = \int_{-\infty}^{z} g(w)dw$. Therefore, g is a density for the random variable Z. ■

The formula for the density of Z in Theorem 6.4.3 takes on a more usable form in the case of independent random variables.

Definition 6.6 *The random variables X and Y are* independent *if*

$$P(X \leq x, Y \leq y) = P(X \leq x)P(Y \leq y) \qquad \text{for all } x, y \in R. \;\blacksquare$$

This definition is clearly equivalent to the requirement that

$$F_{X,Y}(x, y) = F_X(x)F_Y(y) \qquad \text{for all } x, y \in R.$$

If X and Y are independent random variables and ϕ and ψ are continuous real-valued functions on R, then $\phi(X)$ and $\psi(Y)$ are also independent random variables. As in the discrete case, the proof of this fact will be omitted.

If X and Y have a joint density $f_{X,Y}$, then we know that X has a density f_X and Y has a density f_Y.

Theorem 6.4.4 *Let X and Y be random variables having a joint density. Then X and Y are independent if and only if $f_X(x)f_Y(y)$ is a density for X and Y.*

PROOF: Suppose $f_X(x)f_Y(y)$ is a joint density for X and Y. Then

$$\begin{aligned} P(X \leq x, Y \leq y) &= \int_{-\infty}^{x}\left(\int_{-\infty}^{y} f_X(u)f_Y(v)dv\right)du \\ &= \int_{-\infty}^{x} f_X(u)\left(\int_{-\infty}^{y} f_Y(v)dv\right)du \end{aligned}$$

$$= \int_{-\infty}^{x} f_X(u)P(Y \leq y)du$$
$$= P(Y \leq y)\int_{-\infty}^{x} f_X(u)\,du$$
$$= P(Y \leq y)P(X \leq x),$$

and therefore X and Y are independent random variables. On the other hand, if X and Y are independent, then

$$\begin{aligned} F_{X,Y}(x,y) &= P(X \leq x, Y \leq y) \\ &= P(X \leq x)P(Y \leq y) \\ &= \left(\int_{-\infty}^{x} f_X(u)du\right)\left(\int_{-\infty}^{y} f_Y(v)dv\right) \\ &= \int_{-\infty}^{x}\int_{-\infty}^{y} f_X(u)f_Y(v)da, \end{aligned}$$

so that $f_X(x)f_Y(y)$ is a joint density for X and Y. ■

Theorem 6.4.5 *If the random variables X and Y are independent with densities f_X and f_Y, respectively, and $Z = X + Y$, then Z has the density*

$$f_Z(z) = \int_{-\infty}^{+\infty} f_X(x)f_Y(z-x)dx = \int_{-\infty}^{+\infty} f_X(z-y)f_Y(y)dy, \qquad z \in R.$$

PROOF: Theorem 6.4.3 ■

The above equations take on simpler forms if, in addition, X and Y are nonnegative random variables. In this case,

$$f_Z(z) = \begin{cases} \int_0^z f_X(x)f_Y(z-x)dx & \text{if } z \geq 0 \\ 0 & \text{if } z < 0, \end{cases} \tag{6.6}$$

with a similar result holding for integration with respect to y instead of x. Since Z is nonnegative, $f_Z(z) = 0$ for $z < 0$. For $z > 0$, the integral over $(-\infty, +\infty)$ can be replaced by the integral over $(0, +\infty)$ since $f_X(x) = 0$ for $x < 0$; since $f_Y(z-x) = 0$ when $x > z$, the integral over $(0, +\infty)$ can be replaced by the integral over $(0, z)$.

Caveat: In real life, independence of random variables is the exception rather than the rule.

EXAMPLE 6.13 Let X and Y have the joint density

$$f_{X,Y}(x, y) = \begin{cases} e^{-x-y} & \text{if } x \geq 0, y \geq 0 \\ 0 & \text{otherwise.} \end{cases}$$

Since the joint density vanishes outside the first quadrant, the pair (X, Y) is in the first quadrant with probability 1 and outside with probability 0. Thus, $f_X(x) = 0$ if $x < 0$ and $f_Y(y) = 0$ if $y < 0$. Suppose $x > 0$. Then

$$f_X(x) = \int_{-\infty}^{+\infty} f_{X,Y}(x, v)dv = \int_0^{+\infty} e^{-x-v}dv = e^{-x}.$$

Thus,

$$f_X(x) = \begin{cases} e^{-x} & \text{if } x \geq 0 \\ 0 & \text{if } x < 0. \end{cases}$$

Similarly,

$$f_Y(y) = \begin{cases} e^{-y} & \text{if } y \geq 0 \\ 0 & \text{if } y < 0. \end{cases}$$

Since $f_{X,Y}(x, y) = f_X(x)f_Y(y)$, the random variables X and Y are independent. Now let $Z = X + Y$. Again $f_Z(z) = 0$ if $z < 0$. For $z \geq 0$,

$$f_Z(z) = \int_{-\infty}^{+\infty} f_X(x)f_Y(z-x)dx = \int_0^{+\infty} e^{-x}f_Y(z-x)dx.$$

Since $f_Y(z - x) = 0$ when $z - x < 0$,

$$f_Z(z) = \int_0^{z} e^{-x}e^{-(z-x)}dx = ze^{-z}.$$

Therefore,

$$f_Z(z) = \begin{cases} ze^{-z} & \text{if } z \geq 0 \\ 0 & \text{if } z < 0. \end{cases} \quad \blacksquare$$

EXERCISES 6.4

1. Let X and Y be independent random variables having density functions

$$f_X(x) = \begin{cases} 1 & \text{if } 0 \le x \le 1 \\ 0 & \text{otherwise} \end{cases} \qquad f_Y(y) = \begin{cases} 1 & \text{if } 0 \le y \le 1 \\ 0 & \text{otherwise.} \end{cases}$$

Calculate $P(Y \le X)$.

2. Let X and Y be independent random variables having density functions

$$f_X(x) = \begin{cases} 2e^{-2x} & \text{if } x \ge 0 \\ 0 & \text{if } x < 0 \end{cases} \qquad f_Y(y) = \begin{cases} 3e^{-3y} & \text{if } y \ge 0 \\ 0 & \text{if } y < 0. \end{cases}$$

Calculate $P(Y \le X)$.

3. Let

$$f_{X,Y}(x, y) = \begin{cases} c(1 - x^2 - y^2) & \text{if } 0 \le x^2 + y^2 \le 1, x \ge 0, y \ge 0 \\ 0 & \text{otherwise} \end{cases}$$

be a density function. Find the value of c and calculate $P(X \ge 1/2)$.

4. Let

$$f_{X,Y}(x, y) = \begin{cases} xe^{-x-xy} & \text{if } x \ge 0, y \ge 0 \\ 0 & \text{otherwise.} \end{cases}$$

Find $f_X(x)$ and $f_Y(y)$.

5. Let X and Y be independent random variables with densities

$$f_X(x) = \begin{cases} 1 & \text{if } 0 \le x \le 1 \\ 0 & \text{otherwise} \end{cases} \qquad f_Y(y) = \begin{cases} e^{-y} & \text{if } y \ge 0 \\ 0 & \text{if } y < 0 \end{cases}$$

What is the density of $Z = X + Y$?

6. Let X and Y be the random variables of Problem 2 and let $Z = \min(X, Y)$. Find the density of Z.

7. Let X and Y be the random variables of Problem 2 and let $Z = X + Y$. Find the density of Z.

8. Let X and Y be the random variables of Problem 1 and let $Z = X + Y$. Find the density of Z. (Hint: Calculate the distribution function by considering the intersection of the region $\{(u, v) : v \le z - u\}$ with the unit square U in the uv-plane.)

6.5 COMPUTATIONS WITH DENSITIES

We begin this section by cataloging several common density functions, special cases of which were seen in the previous section.

EXAMPLE 6.14 A density commonly used for many board games that employ a spinner is the *uniform density* on $[a, b]$ defined by

$$f(x) = \begin{cases} 1/(b-a) & \text{if } a \leq x \leq b \\ 0 & \text{otherwise.} \end{cases} \quad \blacksquare$$

EXAMPLE 6.15 The continuous version of the discrete geometric density is the *exponential density* with parameter $\lambda > 0$ defined by

$$f(x) = \begin{cases} \lambda e^{-\lambda x} & \text{if } x \geq 0 \\ 0 & \text{if } x > 0. \end{cases} \quad \blacksquare$$

Waiting times connected with continuously varying random processes sometimes have an exponential density.

One of the best known density functions is the *normal density* (or *Gauss density* or *Laplace density*). Consider the function $e^{-x^2/2}$ for $x \geq 0$ and let $c = \int_0^{+\infty} e^{-x^2/2} dx$. The constant c can be determined indirectly by calculating

$$\begin{aligned} c^2 &= \left(\int_0^{+\infty} e^{-x^2/2} dx\right)\left(\int_0^{+\infty} e^{-y^2/2} dy\right) \\ &= \iint_A e^{-(x^2+y^2)/2} da \end{aligned}$$

where $A = \{(x, y) : x \geq 0, y \geq 0\}$ is the first quadrant in the plane. Transforming to polar coordinates,

$$c^2 = \int_0^{\pi/2} \left(\int_0^{+\infty} e^{-r^2/2} r\, dr\right) d\theta = \int_0^{\pi/2} 1\, d\theta = \frac{\pi}{2}.$$

Thus, $c = \int_0^{+\infty} e^{-x^2/2} dx = \sqrt{\pi/2}$, and so

$$\int_{-\infty}^{+\infty} e^{-x^2/2} dx = \sqrt{2\pi}.$$

EXAMPLE 6.16 If we define

$$\phi(x) = \frac{1}{\sqrt{2\pi}} e^{-x^2/2}, \qquad x \in R, \tag{6.7}$$

then ϕ can serve as the density function of a random variable. The density ϕ is called the *standard normal density*. The graph of the standard normal density is depicted in Figure 6.11. $\blacksquare$

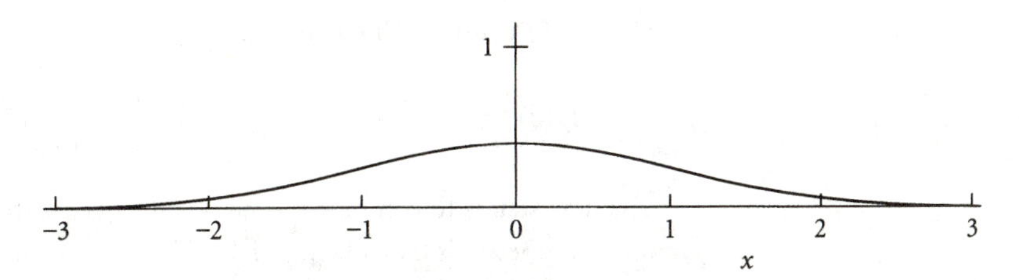

FIGURE 6.11 Standard normal density.

The corresponding distribution function defined by $\Phi(x) = \int_{-\infty}^{x} \phi(t)dt$, $x \in R$, is called the *standard normal distribution function.* Values of $\Phi(x)$ have been calculated using Maple V software and are given in the Standard Normal Distribution Function table (see p. 346). $\Phi(x)$ can be determined for negative values of x from this table by using the fact that ϕ is a symmetric function, so that $\Phi(x) = 1 - \Phi(-x)$. For example, $\Phi(-.75) = 1 - \Phi(.75) = .2266$.

If X is any random variable with density f_X and $Y = aX + b, a, b \in R$ $a \neq 0$, then f_Y can be expressed in terms of f_X as follows. Suppose first that $a > 0$. Since $F_Y(y) = P(Y \leq y) = P(X \leq (y-b)/a) = F_X((y-b)/a)$, $f_Y(y) = (1/a)f_X((y-b)/a)$. If $a < 0$, the $F_Y(y) = P(X \geq (y-b)/a) = P(X > (y-b)/a) = 1 - F_X((y-b)/a)$ and $f_Y(y) = (-1/a)f_X((y-b)/a)$. Since $-a = |a|$ when $a < 0$, the two cases can be combined by writing

$$f_Y(y) = \frac{1}{|a|} f_X\left(\frac{y-b}{a}\right) \tag{6.8}$$

EXAMPLE 6.17 Consider a random variable X having a standard normal density $\phi(x)$ and $Y = \sigma X + \mu$ where $\mu, \sigma \in R, \sigma > 0$. Then $f_Y(y) = (1/\sigma)\phi((y-\mu)/\sigma), y \in R$. Therefore,

$$f_Y(y) = \frac{1}{\sqrt{2\pi}\sigma} e^{-(y-\mu)^2/2\sigma^2}.$$

A random variable Y having this density is said to have a *normal density* $n(\mu, \sigma^2)$ with parameters μ and σ, called the *mean* and *standard deviation,* respectively. The latter terms will be justified later. For the time being, μ and σ are just parameters. ■

If X and Y are independent random variables having normal densities, what can we say about the density of the sum?

Theorem 6.5.1 *Let X and Y be independent random variables having normal densities $n(\mu_X, \sigma_X^2)$ and $n(\mu_Y, \sigma_Y^2)$, respectively. Then $Z = X + Y$ has a normal density $n(\mu_X + \mu_Y, \sigma_X^2 + \sigma_Y^2)$.*

Sketch of Proof: By Theorem 6.4.3,

$$f_Z(z) = \int_{-\infty}^{+\infty} \frac{1}{\sqrt{2\pi}\sigma_X} e^{-(x-\mu_X)^2/2\sigma_X^2} \frac{1}{\sqrt{2\pi}\sigma_Y} e^{-(z-x-\mu_Y)^2/2\sigma_Y^2}\, dx.$$

The next step is to combine the two exponents and then complete the square on x. The result is that a factor of $1/\sqrt{2\pi}$ along with an exponential function of a quadratic in z can be taken outside the integral, leaving the total integral of a normal density, which is 1. Although the algebra is tedious, readers should carry out these steps at least once in their lifetime. ■

If the random variable Y has a normal density with parameters μ and σ, then probabilities of the type $P(a < Y \leq b)$ can be expressed in terms of Φ as follows. Since $Y = \sigma X + \mu$, where X has a standard normal density,

$$\begin{aligned} P(a < Y \leq b) &= P(a < \sigma X + \mu \leq b) \\ &= P\left(\frac{a-\mu}{\sigma} < X \leq \frac{b-\mu}{\sigma}\right) \\ &= \Phi\left(\frac{b-\mu}{\sigma}\right) - \Phi\left(\frac{a-\mu}{\sigma}\right). \end{aligned}$$

EXAMPLE 6.18 Suppose the random variable X has a normal density with parameters $\mu = 100$ and $\sigma = 10$. According to the Standard Normal Distribution Function table (see p. 346) and the fact that $\Phi(x) = 1 - \Phi(-x)$, $x \in R$,

$$\begin{aligned} &P(75 < X \leq 125) \\ &= \Phi\left(\frac{125-100}{10}\right) - \Phi\left(\frac{75-100}{10}\right) = \Phi(2.5) - \Phi(-2.5) \\ &= \Phi(2.5) - (1 - \Phi(2.5)) = 2\Phi(2.5) - 1 = .9876. \ ■ \end{aligned}$$

If X is any random variable with density f_X and $Y = X^2$, then the density of Y can be obtained as follows. We first express F_Y in terms of F_X. Since Y is nonnegative, $F_Y(y) = 0$ if $y < 0$. Suppose $y \geq 0$. Then $F_Y(y) = P(Y \leq y) = P(X^2 \leq y) = P(-\sqrt{y} \leq X \leq \sqrt{y}) = P(-\sqrt{y} < X \leq \sqrt{y}) = F_X(\sqrt{y}) - F_X(-\sqrt{y})$. Since the derivative of the latter expression is $(1/(2\sqrt{y}))(F'_X(\sqrt{y}) + F'_X(-\sqrt{y}))$,

$$f_Y(y) = \begin{cases} 0 & \text{if } y < 0 \\ (1/(2\sqrt{y}))(f_X(\sqrt{y}) + f_X(-\sqrt{y})) & \text{if } y \geq 0. \end{cases} \tag{6.9}$$

EXAMPLE 6.19 Let X have a standard normal density and let $Y = X^2$. Applying the formula above,

$$f_Y(y) = \begin{cases} 0 & \text{if } y < 0 \\ (1/\sqrt{2\pi y})e^{-y/2} & \text{if } y \geq 0. \ ■ \end{cases} \tag{6.10}$$

The last density is a special case of a family of density functions having the form $x^{\alpha-1}e^{-\lambda x}$ for positive x except for a multiplicative constant where α and λ are positive parameters. To determine the multiplicative constant, we must evaluate the integral

$$\int_0^{+\infty} x^{\alpha-1}e^{-\lambda x}\,dx.$$

Letting $y = \lambda x$, this integral becomes $\lambda^{-\alpha}\int_0^{+\infty} y^{\alpha-1}e^{-y}dy$. Putting aside the evaluation of the latter integral for the time being, for each $\alpha > 0$ let

$$\Gamma(\alpha) = \int_0^{+\infty} y^{\alpha-1}e^{-y}dy.$$

Then

$$\int_0^{+\infty} x^{\alpha-1}e^{-\lambda x}dx = \frac{\Gamma(\alpha)}{\lambda^\alpha}.$$

The reciprocal of this constant is therefore the required multiplicative constant.

EXAMPLE 6.20 A random variable having the density $\Gamma(\alpha, \lambda)$ defined by

$$\Gamma(\alpha, \lambda)(x) = \begin{cases} (\lambda^\alpha/\Gamma(\alpha))x^{\alpha-1}e^{-\lambda x} & \text{if } x \geq 0 \\ 0 & \text{if } x < 0, \end{cases} \qquad (6.11)$$

is called a *gamma density* with parameters α and λ. ■

Returning to $\Gamma(\alpha)$ as a function of α, the recurrence relation

$$\Gamma(\alpha + 1) = \alpha\Gamma(\alpha), \qquad \alpha > 0, \qquad (6.12)$$

can be obtained by applying integration by parts to the integral

$$\int_0^{+\infty} x^{(\alpha+1)-1}e^{-x}dx.$$

Since $\Gamma(1) = \int_0^{+\infty} e^{-y}dy = 1$, it is easy to show by an induction argument that for every positive integer n,

$$\Gamma(n + 1) = n!.$$

Since the density given in Equation 6.10 is a $\Gamma(1/2, 1/2)$ density, it follows that the multiplicative constants in Equations 6.10 and 6.11 must be equal in this case. Therefore,

$$\Gamma\left(\frac{1}{2}\right) = \sqrt{\pi}.$$

From this result, $\Gamma(\alpha)$ can be calculated using Equation 6.12 for any $\alpha > 0$ that is an odd multiple of $1/2$. For example,

$$\Gamma\left(\frac{5}{2}\right) = \frac{3}{2}\Gamma\left(\frac{3}{2}\right) = \frac{3}{2}\frac{1}{2}\Gamma\left(\frac{1}{2}\right) = \frac{3}{4}\sqrt{\pi}.$$

EXAMPLE 6.21 Let X have a normal density $n(0, \sigma^2)$ and let $Y = X^2$. Then $f_Y(y) = 0$ if $y < 0$. Suppose $y > 0$. Then

$$\begin{aligned} f_Y(y) &= \frac{1}{2\sqrt{y}}(f_X(\sqrt{y}) + f_X(-\sqrt{y})) \\ &= \frac{1}{\sqrt{y}} \frac{1}{\sqrt{2\pi}\sigma} e^{-y/2\sigma^2} \end{aligned}$$

Since $\sqrt{\pi} = \Gamma(1/2)$,

$$f_Y(y) = \begin{cases} ((1/2\sigma^2)^{1/2}/\Gamma(1/2))y^{(1/2)-1}e^{-y/2\sigma^2} & \text{if } y > 0 \\ 0 & \text{if } y \le 0. \end{cases}$$

It follows that Y has a $\Gamma(1/2, 1/2\sigma^2)$ density. ■

The exponential density is a special case of the gamma density with $\alpha = 1$ and $\lambda = 1$. According to Example 6.13, if X and Y are independent and have exponential densities with the same parameter $\lambda = 1$, then $Z = X + Y$ has a $\Gamma(2, 1)$ density. This suggests the following theorem.

Theorem 6.5.2 *Let X and Y be independent random variables having gamma densities $\Gamma(\alpha_1, \lambda)$ and $\Gamma(\alpha_2, \lambda)$, respectively. If $Z = X + Y$, then Z has a $\Gamma(\alpha_1 + \alpha_2, \lambda)$ density.*

PROOF: By Equation 6.6 for $z > 0$,

$$\begin{aligned} f_Z(z) &= \int_0^z \frac{\lambda^{\alpha_1}}{\Gamma(\alpha_1)} x^{\alpha_1 - 1} e^{-\lambda x} \frac{\lambda^{\alpha_2}}{\Gamma(\alpha_2)} (z - x)^{\alpha_2 - 1} e^{-\lambda(z - x)} dx \\ &= \frac{\lambda^{\alpha_1 + \alpha_2}}{\Gamma(\alpha_1)\Gamma(\alpha_2)} e^{-\lambda z} \int_0^z x^{\alpha_1 - 1}(z - x)^{\alpha_2 - 1}\, dx. \end{aligned}$$

Making the substitution $y = x/z$ in the integral,

$$\int_0^z x^{\alpha_1 - 1}(z - x)^{\alpha_2 - 1} dx = z^{\alpha_1 + \alpha_2 - 1} \int_0^1 y^{\alpha_1 - 1}(1 - y)^{\alpha_2 - 1} dy.$$

The latter integral is a constant that will be denoted by $B(\alpha_1, \alpha_2)$. Therefore,

$$f_Z(z) = \begin{cases} \lambda^{\alpha_1 + \alpha_2}/(\Gamma(\alpha_1)\Gamma(\alpha_2))B(\alpha_1, \alpha_2) z^{\alpha_1 + \alpha_2 - 1} e^{-\lambda z} & \text{if } z > 0 \\ 0 & \text{if } z \le 0. \end{cases}$$

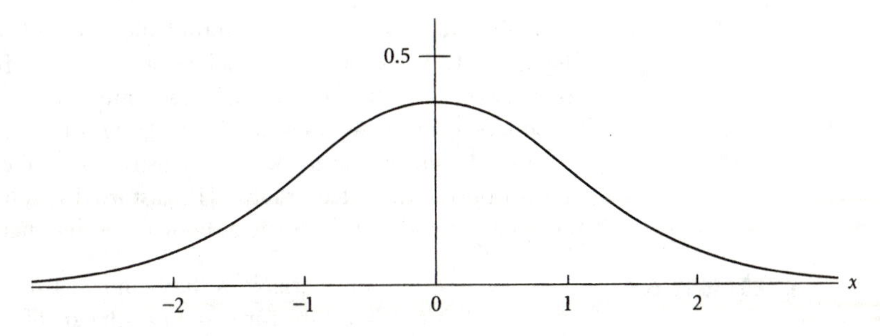

FIGURE 6.12 The bell curve.

Disregarding the fractional constant in the description of $f_Z(z)$, this function has the form of a gamma density and therefore must be a $\Gamma(\alpha_1 + \alpha_2, \lambda)$ density. This concludes the proof. ■

The fact that the above function is a gamma density permits us to determine the constant $B(\alpha_1, \alpha_2)$. Since the function is a gamma density $\Gamma(\alpha_1 + \alpha_2, \lambda)$, we must have

$$\frac{B(\alpha_1, \alpha_2)}{\Gamma(\alpha_1)\Gamma(\alpha_2)} = \frac{1}{\Gamma(\alpha_1 + \alpha_2)},$$

and therefore

$$B(\alpha_1, \alpha_2) = \int_0^1 y^{\alpha_1 - 1}(1 - y)^{\alpha_2 - 1} dy = \frac{\Gamma(\alpha_1)\Gamma(\alpha_2)}{\Gamma(\alpha_1 + \alpha_2)}.$$

The reader should observe the remarkable fact that we have evaluated a family of integrals using probability methods.

EXAMPLE 6.22

$$\int_0^1 y^{5/2}(1 - y)^{7/2} dy = \frac{\Gamma(7/2)\Gamma(9/2)}{\Gamma(8)} \approx .0077. \quad ■$$

Returning to the latter part of the above proof, consider the function

$$\beta(\alpha_1, \alpha_2)(x) = \begin{cases} (\Gamma(\alpha_1 + \alpha_2)/(\Gamma(\alpha_1)\Gamma(\alpha_2))x^{\alpha_1 - 1}(1 - x)^{\alpha_2 - 1} & \text{if } 0 < x < 1 \\ 0 & \text{otherwise.} \end{cases}$$

Since this function is nonnegative and the multiplicative constant has been chosen so that its total integral is 1, it is a density function called the *beta density* with parameters α_1 and α_2.

A final remark about the graph of the standard normal density shown in Figure 6.11. The standard normal density is shown there using the same units on the axes. The usual practice is to use a much larger unit on the y-axis, which produces the distorted view of the standard normal density seen in Figure 6.12. Figure 6.11 shows that the normal density is spread out more uniformly than shown on the distorted graph. The distorted graph is commonly called the "bell curve" and is the favorite of book cover designers.

EXERCISES 6.5

1. If the random variable X has a normal density $n(\mu, \sigma^2)$ and $Y = aX + b$, what is the form and parameters of the density of Y?
2. A random variable X has a normal density $n(\mu, \sigma^2)$ and it is known that $P(X \leq 100) = .9938$ and $P(X \leq 60) = .9332$. Find μ and σ.
3. Let X and Y be independent random variables with standard normal densities. Assuming that X^2 and Y^2 are independent, find the density of $Z = X^2 + Y^2$.
4. Let X and Y be independent random variables having standard normal densities. Calculate the probability that the pair (X, Y) will be between the lines through the origin making angles $\pi/6$ and $\pi/4$ radians with the x-axis and also in the first quadrant.
5. A point in the plane is chosen in such a way that its x-coordinate X is $n(0, \sigma^2)$, its y-coordinate Y is $n(0, \sigma^2)$, and the two are independent. Find the density of the distance $Z = \sqrt{X^2 + Y^2}$ of the point from the origin. The density of Z is called the *Rayleigh density.*
6. If the random variable X has a uniform density on the interval $[a, b], a < b$, find a function ϕ such that $Y = \phi(X)$ has a uniform density on $[0, 1]$.
7. Let X have a standard normal density $n(0, 1)$. Find a function ϕ such that $Y = \phi(X)$ has a uniform density on $(0, 1)$. (Hint: Φ is strictly increasing and has an inverse function Φ^{-1}.)

6.6 MULTIVARIATE AND CONDITIONAL DENSITIES

Given discrete random variables X and Y, the conditional density of Y given $X = x$ was defined in Section 4.5 to be the function

$$f_{Y|X}(y \mid x) = P(Y = y \mid X = x).$$

This is not possible for continuous random variables, because if x is a point of continuity of F_X, then $P(X = x) = 0$ and the above conditional probability is not defined. In particular, if F_X is continuous at every point, then the conditional probability is not defined for any x. Even if $P(X = x) > 0$, $P(Y = y \mid X = x)$ would still be equal to zero for most y since $P(Y = y) = 0$

whenever F_Y is continuous at y. In the case of continuous random variables, it is necessary to avoid events of the type $(Y = y)$ or $(X = x)$.

To keep the discussion at the introductory level, we will assume that X and Y are random variables with a joint density function. Consider the conditional probability $P(Y \leq y \mid x \leq X \leq x + \Delta x)$. Assuming that the given event has positive probability,

$$P(Y \leq y \mid x \leq X \leq x + \Delta x) = \frac{P(Y \leq y, x \leq X \leq x + \Delta x)}{P(x \leq X \leq x + \Delta x)}$$

$$= \frac{\int_{-\infty}^{y} \int_{x}^{x+\Delta x} f_{X,Y}(u, v)\,du\,dv}{\int_{x}^{x+\Delta x} f_X(u)\,du}.$$

We might try defining the conditional distribution of Y given $X = x$ by

$$F_{Y|X}(y \mid X = x) = \lim_{\Delta x \to 0} P(Y \leq y \mid x \leq X \leq x + \Delta x)$$

$$= \lim_{\Delta x \to 0} \int_{-\infty}^{y} \frac{\int_{x}^{x+\Delta x} f_{X,Y}(u, v)\,du}{\int_{x}^{x+\Delta x} f_X(u)\,du}\,dv.$$

Assuming that the limit can be taken under the first integral sign,

$$F_{Y|X}(y \mid X = x) = \int_{-\infty}^{y} \left(\lim_{\Delta x \to 0} \frac{\int_{x}^{x+\Delta x} f_{X,Y}(u, v)\,du}{\int_{x}^{x+\Delta x} f_X(u)\,du} \right) dv.$$

This suggests that the conditional density of Y given that $X = x$ should be defined as

$$f_{Y|X}(y \mid x) = \lim_{\Delta x \to 0} \frac{(1/\Delta x) \int_{x}^{x+\Delta x} f_{X,Y}(u, y)\,du}{(1/\Delta x) \int_{x}^{x+\Delta x} f_X(u)\,du}.$$

If the functions f_X and $f_{X,Y}$ are continuous at x and (x, y), respectively, then the denominator and numerator will have limits $f_X(x)$ and $f_{X,Y}(x, y)$, respectively, since they represent the average values of these functions near the points of continuity. In this case, we would have

$$f_{Y|X}(y \mid x) = \frac{f_{X,Y}(x, y)}{f_X(x)} \qquad \text{provided } f_X(x) > 0.$$

Definition 6.7 *If X and Y are random variables having a joint density, the* conditional density of Y given X *is defined for $x, y \in R$ by*

$$f_{Y|X}(y \mid x) = \begin{cases} f_{X,Y}(x, y)/f_X(x) & \text{if } f_X(x) > 0 \\ 0 & \text{if } f_X(x) = 0. \end{cases} \blacksquare$$

Note that the definition does not require any of the assumptions made in the above heuristic argument. Note also that the equation

$$f_{X,Y}(x,y) = f_{Y|X}(y \mid x) f_X(x), \qquad x, y \in R,$$

holds at all points x for which $f_X(x) > 0$.

As in the discrete case, experiments are sometimes defined in terms of densities and conditional densities.

EXAMPLE 6.23 A point X is chosen at random from the interval $[0, 1]$. Given that $X = x$, a point Y is then chosen at random from the interval $[0, x]$. Suppose we are required to find the density of Y. The density of X is clearly uniform on $[0, 1]$; i.e.,

$$f_X(x) = \begin{cases} 1 & \text{if } 0 \leq x \leq 1 \\ 0 & \text{otherwise.} \end{cases}$$

The information concerning Y is given in conditional form; namely,

$$f_{Y|X}(y \mid x) = \begin{cases} 1/x & \text{if } 0 \leq y \leq x \\ 0 & \text{otherwise.} \end{cases}$$

That is, given that $X = x$, Y is uniform on $[0, x]$. The joint density $f_{X,Y}(x, y)$ is given by

$$f_{X,Y}(x,y) = f_{Y|X}(y \mid x) f_X(x) = \begin{cases} 1/x & \text{if } 0 \leq y \leq x, 0 \leq x \leq 1 \\ 0 & \text{otherwise.} \end{cases}$$

Since Y takes on values in $[0, 1]$, $f_Y(y) = 0$ if $y < 0$ or $y > 1$. Suppose $0 \leq y \leq 1$. Then

$$f_Y(y) = \int_{-\infty}^{+\infty} f_{X,Y}(x,y)\,dx = \int_y^1 \frac{1}{x}\,dx = -\ln y.$$

Thus,

$$f_Y(y) = \begin{cases} -\ln y & \text{if } 0 \leq y \leq 1 \\ 0 & \text{otherwise.} \end{cases} \quad \blacksquare$$

EXAMPLE 6.24 To assess the reliability of a component of a system, fatigue or wear and tear must be taken into consideration. Suppose it is known that if a component has survived up to time t, then it will fail in a small time interval $(t, t+\Delta t)$ with probability approximately proportional to the length Δt of the interval where the constant of proportionality depends upon t; i.e., the conditional probability is approximately equal to $\beta(t)\Delta t$ for some nonnegative function $\beta(t)$ on $(0, \infty)$. Let T denote the time at which the component will fail. We will use an intuitive argument to determine the distribution function of T from the given data $\beta(t)$. According to the above assumption,

$$P(t < T \leq t + \Delta t \mid T > t) \approx \beta(t)\Delta t.$$

On the other hand, if we let F_T denote the distribution function of T, then

$$\begin{aligned} P(t < T \leq t + \Delta t \mid T > t) &= \frac{P(t < T \leq t + \Delta t)}{P(T > t)} \\ &= \frac{F_T(t + \Delta t) - F_T(t)}{1 - F_T(t)}. \end{aligned}$$

Assuming that F_T has a continuous density function f_T, by the mean value theorem of the integral calculus,

$$P(t < T \leq t + \Delta t \mid T > t) \approx \frac{f_T(t)\Delta t}{1 - F_T(t)}.$$

Thus,

$$\beta(t) = \lim_{\Delta t \to 0} \frac{P(t < T \leq t + \Delta t \mid T > t)}{\Delta t} = \frac{f_T(t)}{1 - F_T(t)}.$$

Therefore,

$$\beta(t) = -\frac{d}{dt}\big(\ln(1 - F_T(t))\big).$$

Integrating from 0 to t and using the fact that $F_T(0) = 0$,

$$\ln(1 - F_T(t)) = -\int_0^t \beta(s)\, ds,$$

so that

$$F_T(t) = 1 - e^{-\int_0^t \beta(s)\, ds}$$

and

$$f_T(t) = \beta(t)e^{-\int_0^t \beta(s)\, ds}.$$

Since any real-life component will eventually fail, we should require that $\lim_{t\to+\infty} F_T(t) = 1$ and therefore that

$$\int_0^{\infty} \beta(s)\,ds = +\infty. \blacksquare$$

EXAMPLE 6.25 Consider a system made up of two components that are connected in series with associated failure rates $\beta_1(t)$ and $\beta_2(t)$. Let T be the time of failure of the system and let T_1, T_2 be the times of failure of the two components. If the two components fail independently of each other, what is the failure rate $\beta(t)$ of the system? Since the components are connected in series, $T = \min(T_1, T_2)$. By the assumed independence,

$$\begin{aligned} P(T > t) &= P(\min(T_1, T_2) > t) \\ &= P(T_1 > t, T_2 > t) \\ &= P(T_1 > t)P(T_2 > t) \\ &= (1 - F_{T_1}(t))(1 - F_{T_{2(t)}}). \end{aligned}$$

Thus,

$$1 - F_T(t) = e^{-\int_0^t \beta_1(s)\,ds} e^{-\int_0 t\beta_2 s\,ds} = e^{-\int_0^t(\beta_1(s)+\beta_2(s))\,ds}$$

and

$$F_T(t) = 1 - e^{-\int_0^t(\beta_1(s)+\beta_2(s))\,ds},$$

and therefore $\beta(t) = \beta_1(t) + \beta_2(t)$. $\blacksquare$

We need not limit ourselves to just two random variables. If $X_1, \ldots, X_n$ are n random variables, we can define the *joint distribution function* $F_{X_1,\ldots,X_n}$ by the equation

$$F_{X_1,\ldots,X_n}(x_1, \ldots, x_n) = P(X_1 \le x_1, \ldots, X_n \le x_n).$$

The distribution function $F_{X_1,\ldots,X_n}$ and random variables $X_1, \ldots, X_n$ are said to have the Riemann integrable function $f_{X_1,\ldots,X_n} : R^n \to R$ as *joint density function* if

$$\begin{aligned} &P(a_1 < X_1 \le b_1, \ldots, a_n < X_n \le b_n) \\ &= \int_{a_1}^{b_1}\left(\cdots\left(\int_{a_n}^{b_n} f_{X_1,\ldots,X_n}(x_1, \ldots, x_n)dx_n\right)\cdots\right)dx_1. \end{aligned}$$

More generally, if $A \subset R^n$ and A is in the smallest σ-algebra of subsets of R^n containing all n-dimensional rectangles in R^n, then

$$P((X_1, \ldots, X_n) \in A) = \underset{A}{\int \cdots \int} f_{X_1,\ldots,X_n}(x_1, \ldots, x_n) dV_n$$

where dV_n denotes integration with respect to n-dimensional volume. To calculate a probability of the type $P(g(X_1, \ldots, X_n) \leq a)$ where g is some function of n-variables, the probability is put into the form

$$\begin{aligned} P(g(X_1, \ldots, X_n) \leq a) &= P((X_1, \ldots, X_n) \in A) \\ &= \underset{A}{\int \cdots \int} f_{X_1,\ldots,X_n}(x_1, \ldots, x_n) dV_n \end{aligned}$$

where $A = \{(x_1, \ldots, x_n) : g(x_1, \ldots, x_n) \leq a\}$.

EXAMPLE 6.26 Consider the unit cube $U \subset R^n$ defined by

$$U = \{(x_1, \ldots, x_n) : 0 \leq x_1 \leq 1, \ldots, 0 \leq x_n \leq 1\}.$$

An experiment consists of choosing a point at random from U. The latter statement means that the probability that the outcome will be in a region A, if A is in the σ-algebra described above, is taken to be the volume of $A \cap U$, $\mathrm{Vol}(A \cap U)$, divided by the volume of U, which is 1. If $X_1, \ldots, X_n$ denote the first, second, ..., nth-coordinates, respectively, of the chosen point, then

$$P((X_1, \ldots, X_n) \in A) = \mathrm{Vol}(A \cap U).$$

In particular, if we want to evaluate $P(X_1 \leq X_2 \leq \cdots \leq X_n)$, we put $A = \{(x_1, \ldots, x_n) : x_1 \leq x_2 \leq \cdots \leq x_n\}$, so that

$$P(X_1 \leq X_2 \leq \cdots \leq X_n) = \underset{A \cap U}{\int \cdots \int} 1 dV_n.$$

The multiple integral can be calculated using iterated integrals by fixing $x_1, \ldots, x_{n-1}$ and integrating with respect to x_n between x_{n-1} and 1, then integrating with respect to x_{n-1} between x_{n-2} and 1, and so forth. Thus,

$$\begin{aligned} P(X_1 \leq X_2 \leq \cdots \leq X_n) &= \int_0^1 \left(\cdots \int_{x_{n-2}}^1 \left(\int_{x_{n-1}}^1 dx_n \right) dx_{n-1} \ldots \right) dx_1 \\ &= \int_0^1 \left(\cdots \int_{x_{n-2}}^1 (1 - x_{n-1}) dx_{n-1} \cdots \right) dx_1 \end{aligned}$$

$$\vdots$$

$$= \int_0^1 \frac{(1 - x_1)^{n-1}}{(n-1)!} dx_1 = \frac{1}{n!}. \quad \blacksquare$$

Independence of continuous random variables is defined as before; i.e., the random variables $X_1, \ldots, X_n$ are independent if

$$P(X_1 \leq x_1, \ldots, X_n \leq x_n) = P(X_1 \leq x_1) \times \cdots \times P(X_n \leq x_n)$$

for all $x_1, \ldots x_n \in R$ or, equivalently,

$$F_{X_1,\ldots,X_n}(x_1, \ldots, x_n) = F_{X_1}(x_1) \times \cdots \times F_{X_n}(x_n)$$

for all $x_1, \ldots, x_n \in R$. Assuming that $X_1, \ldots, X_n$ have a joint density function $f_{X_1,\ldots,X_n}$, then each X_i has a density function $f_{X_i}, i = 1, \ldots, n$, and the random variables are independent if and only if

$$f_{X_1,\ldots,X_n}(x_1, \ldots, x_n) = f_{X_1}(x_1) \times \cdots \times f_{X_n}(x_n)$$

for all $x_1, \ldots, x_n \in R$. More precisely, if and only if $f_{X_1}(x_1) \times \cdots \times f_{X_n}(x_n)$ is a joint density for $X_1, \ldots, X_n$. As in the discrete case, if $X_1, \ldots, X_n$ are independent random variables and $\phi_1, \ldots, \phi_n$ are real-valued continuous functions on R, then the random variables $\phi_1(X_1), \ldots, \phi_n(X_n)$ are independent random variables, as are the two random variables $X_1 + \cdots + X_{n-1}$ and X_n.

EXAMPLE 6.27 Let X, Y, and Z be independent random variables with each having a uniform density on $[0, 1]$. The joint density is the product of the three densities and is given by

$$f_{X,Y,Z}(x, y, z) = \begin{cases} 1 & \text{if } 0 \leq x, y, z \leq 1 \\ 0 & \text{otherwise.} \end{cases}$$

Consider $P(Z \geq XY)$. This probability can be calculated by defining $A = \{(x, y, z) : z \geq xy\}$ so that

$$P(Z \geq XY) = \iiint_A f_{X,Y,Z}(x, y, z)\, dV_3$$

$$= \iiint_{z \geq xy} f_{X,Y,Z}(x, y, z)\, dV_3.$$

The region of integration consists of all points $(x, y, z) \in R^3$ that lie above the surface $z = xy$. Since the integrand vanishes outside the unit cube, we can integrate over the region consisting of points (x, y, z) that are above the surface

$z = xy$, below the surface $z = 1$, and above the unit square in the xy-plane. By fixing x and y with $0 \le x, y \le 1$, we can integrate with respect to z from xy to 1, so that

$$\begin{aligned} P(Z \ge XY) &= \int_0^1 \int_0^1 \left(\int_{xy}^1 1\, dz \right) dx dy \\ &= \int_0^1 \int_0^1 (1 - xy)\, dx dy \\ &= \frac{3}{4}. \quad \blacksquare \end{aligned}$$

The proofs of the following theorems will be reserved as exercises.

Theorem 6.6.1 *Let $X_1, \ldots, X_n$ be independent random variables. If X_i has a gamma density $\Gamma(\alpha_i, \lambda)$, $i = 1, \ldots, n$, then $X_1 + \cdots + X_n$ has the gamma density $\Gamma(\alpha_1 + \cdots + \alpha_n, \lambda)$.*

Theorem 6.6.2 *Let $X_1, \ldots, X_n$ be independent random variables. If X_i has a normal density $n(\mu_i, \sigma_i^2)$, then $X_1 + \cdots + X_n$ has a $n(\mu, \sigma^2)$ density where $\mu = \mu_1 + \cdots + \mu_n$ and $\sigma^2 = \sigma_1^2 + \cdots + \sigma_n^2$.*

Let $X_1, \ldots, X_n$ be random variables having a joint density function $f_{X_1, \ldots, X_n}$. If $1 \le m < n$, define

$$f_{X_{m+1}, \ldots, X_n | X_1, \ldots, X_m}(x_{m+1}, \ldots, x_n \mid x_1, \ldots, x_m) = \frac{f_{X_1, \ldots, X_n}(x_1, \ldots, x_n)}{f_{X_1, \ldots, X_m}(x_1, \ldots, x_m)},$$

provided the denominator is positive. As before,

$$f_{X_1, \ldots, X_n}(x_1, \ldots, x_n) = f_{X_{m+1}, \ldots, X_n | X_1, \ldots, X_m}(x_{m+1}, \ldots, x_n \mid x_1, \ldots, x_m) f_{X_1, \ldots, X_m}(x_1, \ldots, x_m),$$

provided the second factor on the right is positive.

EXAMPLE 6.28 A point X is chosen at random from the interval $[0, 1]$. Given that $X = x$, a point Y is chosen at random from the interval $[0, x]$. Finally, given that $Y = y$, a point Z is chosen at random from the interval $[0, y]$. What is the density of Z? The given information specifies the following densities and conditional densities:

$$f_X(x) = \begin{cases} 1 & \text{if } 0 \le x \le 1 \\ 0 & \text{otherwise.} \end{cases}$$

$$f_{Y|X}(y \mid x) = \begin{cases} 1/x & \text{if } 0 \le y \le x \le 1 \\ 0 & \text{otherwise.} \end{cases}$$

$$f_{Z|X,Y}(z \mid x, y) = f_{Z|Y}(z \mid y) = \begin{cases} 1/y & \text{if } 0 \le z \le y \le x \le 1 \\ 0 & \text{otherwise.} \end{cases}$$

Since

$$f_{X,Y,Z}(x, y, z) = f_{Z|X,Y}(z \mid x, y) f_{X,Y}(x, y) = f_{Z|Y}(z \mid y) f_{Y|X}(y \mid x) f_X(x)$$

provided $0 \le z \le y \le x \le 1$,

$$f_{X,Y,Z}(x, y, z) = \begin{cases} 1/xy & \text{if } 0 \le z \le y \le x \le 1 \\ 0 & \text{otherwise.} \end{cases}$$

For $0 \le z \le 1$,

$$\begin{aligned} f_Z(z) &= \int_{-\infty}^{+\infty} \left(\int_{-\infty}^{+\infty} f_{X,Y,Z}(x, y, z)\,dx \right) dy \\ &= \int_z^1 \left(\int_y^1 \frac{1}{xy}\,dx \right) dy \\ &= \int_z^1 \left(-\frac{1}{y} \ln y \right) dy \\ &= \frac{1}{2} (\ln z)^2, \end{aligned}$$

and $f_Z(z) = 0$ otherwise. ■

EXERCISES 6.6

1. Let $X_1, \ldots, X_n$ be independent random variables each having an exponential density with parameter $\lambda > 0$. Find the density of $Z = X_1 + \cdots + X_n$.
2. Let $X_1, \ldots, X_n$ be independent random variables each having a standard normal density. Find the density of $Z = \sqrt{X_1^2 + X_2^2 + \cdots + X_n^2}$.
3. Write out complete proofs of Theorems 6.6.1 and 6.6.2.
4. Consider a disk D with center at $(0, 0)$ and radius 1. A point with x-coordinate X is chosen at random from the line segment joining $(-1, 0)$ to $(1, 0)$, and then a point is chosen at random from the line segment joining $(x, -\sqrt{1 - x^2})$ to $(x, \sqrt{1 - x^2})$. Let Y be the y-coordinate of the latter point. What is the density of Y?
5. Let X_1, X_2, X_3 be independent random variables each having an exponential density with the same parameter $\lambda = 1$. Calculate the probability $P(X_1 \le 2X_2 \le 3X_3)$.

6. Random variables $X_1, X_2, \ldots, X_n$ are defined as follows. A point X_1 is selected at random from the interval $[0, 1]$; given that $X_1 = x_1$, a point X_2 is selected at random from the interval $[0, x_1]$; and at the last step, given that $X_{n-1} = x_{n-1}$, a point X_n is selected at random from the interval $[0, x_{n-1}]$. What is the density of X_n?

7. Let $X_1, \ldots, X_n$ be independent random variables each having a uniform density on $[0, 1]$. Let $U = \min(X_1, \ldots, X_n)$ and $V = \max(X_1, \ldots, X_n)$. Find the joint distribution function of U and V and verify that

$$f_{U,V}(u, v) = \begin{cases} n(n-1)(v-u)^{n-2} & \text{if } 0 \le u \le v \le 1 \\ 0 & \text{otherwise} \end{cases}$$

is the joint density of U and V.

8. Let X, Y, and Z be random variables with joint density

$$f_{X,Y,Z}(x, y, z) = \begin{cases} z^2 e^{-z(1+x+y)} & \text{if } x \ge 0, y \ge 0, z \ge 0 \\ 0 & \text{otherwise.} \end{cases}$$

Find $f_X, f_Y, f_Z, f_{X,Y}$, and $f_{X,Y|Z}$.

9. A system consists of two components operating in parallel with associated failure times T_1 and T_2 and failure rates $\beta_1(t)$ and $\beta_2(t)$, respectively. Let T be the time of failure of the system. Assuming that the two components fail independently of each other, what is the density of T?

CHAPTER 7

EXPECTATION REVISITED

7.1 INTRODUCTION

Although this chapter discusses most of the concepts introduced in Chapter 4, the introduction of the Riemann-Stieltjes integral makes it possible to formulate a definition of expected value that combines the discrete and continuous cases. An additional condition, however, must be imposed to have an effective means of calculating expected values using the calculus.

One of the hallmarks of probability theory is a classic theorem known as the DeMoivre-Laplace limit theorem, which states that a sum of binomial probabilities can be approximated by the integral of a function known as the "bell-shaped" normal density. Although the proof of this result is tedious, a complete proof is given with enough detail that a postcalculus student can follow the arguments. Reading the proof at this stage is not essential for learning probability.

The chapter concludes with applications to certain types of sequences of random variables called stationary processes, which occur in filtering theory and prediction theory. Processes of this type have their origin in the works of G. U. Yule and E. Slutsky during the period 1920–1940.

7.2 RIEMANN-STIELTJES INTEGRAL

If X is a discrete random variable with range $\{x_1, x_2, \ldots\}$ and density function f_X, the expected value of X was defined to be $E[X] = \sum_j x_j f_X(x_j)$, provided the series converges absolutely. This definition cannot be used for continuous random variables for several reasons. What is needed is a formulation of expected value that applies to continuous random variables and that reduces to the definition above in the case of discrete random variables. Fortunately, there is a type of integral that can do both, called the Riemann-Stieltjes integral.

Let us quickly review the Riemann integral. Let $[a, b] \subset R, a < b$, be a finite interval and let $\phi : [a, b] \to R$. A collection of points $\{x_0, x_1, \ldots, x_n\}$ with $a = x_0 \leq x_1 \leq x_2 \leq \cdots \leq x_n = b$ is called a partition of $[a, b]$ and is denoted by π. The norm of the partition π is denoted by $|\pi|$ and defined by $|\pi| = \max\{x_1 - x_0, x_2 - x_1, \ldots, x_n - x_{n-1}\}$. For $i = 1, \ldots, n$, let ξ_i be any point of the interval $[x_{i-1}, x_i]$ and let $\Delta x_i = x_i - x_{i-1}$. The Riemann integral of ϕ over $[a, b]$ is then defined as

$$\int_a^b \phi(t)\, dt = \lim_{|\pi| \to 0} \sum_{i=1}^{n} \phi(\xi_i)\, \Delta x_i$$

provided the limit on the right exists. There is nothing sacred about using the weight Δx_i for the subinterval $[x_{i-1}, x_i]$. We could just as well use some other weighting scheme.

Let F be any distribution function on R. Using the same notation as above, we could replace Δx_i by $\Delta F_i = F(x_i) - F(x_{i-1})$ and define an integral of ϕ over $[a, b]$ with respect to F by

$$\int_a^b \phi(t)\, dF(t) = \lim_{|\pi| \to 0} \sum_{i=1}^{n} \phi(\xi_i)\, \Delta F_i$$

provided the limit on the right exists. To be more precise about the existence of the limit, if π_1 and π_2 are any two partitions of $[a, b]$, we say that π_2 is finer than π_1 if $\pi_1 \subset \pi_2$.

Definition 7.1 *The function $\phi : [a, b] \to R$ is* Riemann-Stieltjes integrable *over $[a, b]$ with respect to F if there is a real number L such that for every $\varepsilon > 0$ there is a partition π_ε of $[a, b]$ for which*

$$\left| \sum_{i=1}^{n} \phi(\xi_i)\, \Delta F_i - L \right| < \varepsilon$$

for all partitions π finer than π_ε. The number L is denoted by

$$\int_a^b \phi(t)\, dF(t) \quad \text{or} \quad \int_a^b \phi\, dF. \ \blacksquare$$

The Riemann-Stieltjes integral shares many of the properties of the Riemann integral:

1. If ϕ_1, ϕ_2 are Riemann-Stieltjes integrable over $[a, b]$ with respect to F and $c_1, c_2 \in R$, then $c_1\phi_1 + c_2\phi_2$ is Riemann-Stieltjes integrable over $[a, b]$ with respect to F and

$$\int_a^b (c_1\phi_1 + c_2\phi_2)\,dF = c_1\int_a^b \phi_1\,dF + c_2\int_a^b \phi_2\,dF.$$

2. Let $a < c < b$ and $\phi : [a, b] \to R$. If two of the following three integrals exist, then so does the third and

$$\int_a^c \phi\,dF + \int_c^b \phi\,dF = \int_a^b \phi\,dF.$$

3. If $a < b$ and ϕ is Riemann-Stieltjes integrable over $[a, b]$ with respect to F, then

$$\int_a^b \phi\,dF = -\int_b^a \phi\,dF.$$

4. If ϕ is continuous on $[a, b]$, then ϕ is Riemann-Stieltjes integrable over $[a, b]$ with respect to F.

There is also an integration by parts formula that will be omitted because it will not be needed here. Proofs of the above statements can be found in the book by Apostol listed at the end of this chapter.

EXAMPLE 7.1 Let ϕ be continuous on $[a, b]$ and let

$$F(x) = \begin{cases} 0 & \text{if } x < c \\ 1 & \text{if } x \geq c \end{cases}$$

where $a < c < b$. Then

$$\int_a^b \phi(t)\,dF(t) = \phi(c).$$

This can be seen as follows. Since ϕ is continuous at c, given any $\varepsilon > 0$ there is a $\delta > 0$ such that

$$|\phi(x) - \phi(c)| < \varepsilon \text{ whenever } |x - c| < \delta, x \in [a, b].$$

Fix a partition π_ε of $[a, b]$ with $|\pi_\varepsilon| < \delta$. Consider any partition $\pi = \{x_0, x_1, \dots, x_n\}$ finer than π_ε. Then $c \in [x_{i_0-1}, x_{i_0}]$ for some $i_0 = 1, 2, \dots, n$. Since $F(x_i) - F(x_{i-1}) = 0$ for $i \neq i_0$ and $F(x_{i_0}) - F(x_{i_0-1}) = 1$,

$$\sum_{i=1}^{n} \phi(\xi_i)\,\Delta F_i = \phi(\xi_{i_0})$$

where $\xi_{i_0} \in [x_{i_0-1}, x_{i_0}]$. Since $|\xi_{i_0} - c| \leq |\pi| \leq |\pi_\varepsilon| < \delta, |\phi(\xi_{i_0}) - \phi(c)| < \varepsilon$, and therefore

$$\left|\sum_{i=1}^{n} \phi(\xi_i)\,\Delta F_i - \phi(c)\right| = |\phi(\xi_{i_0}) - \phi(c)| < \varepsilon$$

for any partition π finer than π_ε. Thus, $L = \phi(c)$ satisfies the above definition and $\int_a^b \phi(t)\,dF(t) = \phi(c)$. ■

Note that continuity of ϕ was required only at the point c in this example.

The same type of argument can be used to show that if $\phi : [a, b] \to R$ is continuous and there are a finite number of points $c_1, c_2, \ldots, c_n$ with $a \leq c_1 < c_2 < \cdots < c_n \leq b$ such that F only increases by jumps at $c_1, c_2, \ldots, c_n$, then

$$\int_a^b \phi(t)\,dF(t) = \sum_{i=1}^{n} \phi(c_i)(F(c_i) - F(c_i-)).$$

In particular, if X is a discrete random variable with range $\{c_1, \ldots, c_n\}$ and $F = F_X$, then $P(X = c_i) = F_X(c_i) - F_X(c_i-) = f_X(c_i)$ and

$$E[X] = \sum_{i=1}^{n} \phi(c_i) f_X(c_i) = \int_a^b \phi(t)\,dF(t).$$

Improper integrals of the type $\int_a^{+\infty} \phi\,dF$ and $\int_{-\infty}^{a} \phi\,dF$ can be defined in the usual way as the limits

$$\lim_{b\to+\infty} \int_a^b \phi\,dF \qquad \text{and} \qquad \lim_{c\to-\infty} \int_c^a \phi\,dF,$$

respectively, provided the limits exist in R. If both limits do exist, then $\int_{-\infty}^{+\infty} \phi\,dF$ is defined by the equation

$$\int_{-\infty}^{+\infty} \phi\,dF = \int_{-\infty}^{a} \phi\,dF + \int_a^{+\infty} \phi\,dF, \qquad a \in R.$$

Definition 7.2 *The improper integral* $\int_{-\infty}^{+\infty} \phi\,dF$ *is* absolutely convergent *if the improper integral* $\int_{-\infty}^{+\infty} |\phi|\,dF$ *is defined.* ■

We can now formulate the definition of expected value of any random variable.

Definition 7.3 *The expected value* $E[X]$ *of the random variable* X *is defined by*

$$E[X] = \int_{-\infty}^{+\infty} x\, dF_X(x)$$

provided the improper integral is absolutely convergent; i.e., $\int_{-\infty}^{+\infty} |x|\, dF_X(x) < +\infty$. ■

If X is a discrete random variable, this definition of $E[X]$ agrees with the definition given in Section 4.2.

We have seen that if the distribution function F_X is absolutely continuous, then there is a density function f_X such that

$$F_X(x) = \int_{-\infty}^{x} f_X(t)\, dt, \qquad x \in R,$$

and

$$F_X(b) - F_X(a) = \int_a^b f_X(t)\, dt.$$

Suppose in addition, that f_X is continuous on the interval $[a, b]$. Then by the mean value theorem of the integral calculus, there is a number c with $a \leq c \leq b$ such that

$$F_X(b) - F_X(a) = \int_a^b f_X(t)\, dt = f_X(c)(b - a).$$

Suppose now that the function $\phi : [a, b] \to R$ is continuous on $[a, b]$ and let $\pi = \{x_0, \ldots, x_n\}$ be any partition of $[a, b]$. Then

$$\sum_{i=1}^{n} \phi(\xi_i)\, \Delta F_i = \sum_{i=1}^{n} \phi(\xi_i) \int_{x_{i-1}}^{x_i} f_X(t)\, dt = \sum_{i=1}^{n} \phi(\xi_i) f_X(\xi_i^*)\, \Delta x_i$$

where $\xi_i^* \in [x_{i-1}, x_i], i = 1, \ldots, n$. Then

$$\int_a^b \phi(t)\, dF(t) = \lim_{|\pi| \to 0} \sum_{i=1}^{n} \phi(\xi_i)\, \Delta F_i = \lim_{|\pi| \to 0} \sum_{i=1}^{n} \phi(\xi_i) f_X(\xi_i^*)\, \Delta x_i.$$

If the ξ_i and ξ_i^* were the same in the second sum, the latter limit would exist since ϕ is Riemann-Stieltjes integrable with respect to F_X and would be equal to $\int_a^b \phi(t) f_X(t)\, dt$; the same result is valid even if ξ_i and ξ_i^* are not the same, by a result known as Duhamel's principle. Therefore, assuming that F_X has a density f_X that is continuous on $[a, b]$ and ϕ is also continuous on $[a, b]$,

$$\int_a^b \phi(x)\, dF_X(x) = \int_a^b \phi(x) f_X(x)\, dx.$$

Theorem 7.2.1 *Let X be a random variable with $E[X]$ defined and having a continuous density. Then*

$$E[X] = \int_{-\infty}^{+\infty} x f_X(x)\, dx.$$

PROOF: Since $\int_0^b |x| f_X(x)\, dx = \int_0^b |x|\, dF_X(x)$ for all $b \geq 0$, $\int_0^{+\infty} |x| f_X(x)\, dx = \int_0^{+\infty} |x|\, dF_X(x) < +\infty$. Similarly, $\int_{-\infty}^0 |x| f_X(x)\, dx = \int_{-\infty}^0 |x|\, dF_X(x) < +\infty$. Since the integrals $\int_{-\infty}^0 f_X(x)\, dx$ and $\int_0^{\infty} x f_X(x)\, dx$ are absolutely convergent,

$$\begin{aligned} E[X] = \int_{-\infty}^{+\infty} x\, dF_X(x) &= \int_{-\infty}^0 x\, dF_X(x) + \int_0^{+\infty} x\, dF_X(x) \\ &= \int_{-\infty}^0 x f_X(x)\, dx + \int_0^{+\infty} x f_X(x)\, dx \\ &= \int_{-\infty}^{+\infty} x f_X(x)\, dx. \quad \blacksquare \end{aligned}$$

If $Z = \phi(X)$ and we want to calculate $E[Z]$, we must first find the density function of Z, assuming there is one. The following theorem allows us to bypass this step by using f_X rather than f_Z as a weight function. The proof will be omitted. Proofs require approximating the random variable X by a discrete random variable and applying Theorem 4.2.1.

Theorem 7.2.2 *Let X be a random variable having a continuous density and let $\phi : R \to R$ be a continuous function. If the integral $\int_{-\infty}^{+\infty} \phi(x) f_X(x)\, dx$ converges absolutely, then $E[\phi(X)]$ is defined and*

$$E[\phi(X)] = \int_{-\infty}^{+\infty} \phi(x) f_X(x)\, dx.$$

EXAMPLE 7.2 Let X have a uniform density on $[a, b]$, $a < b$. Then

$$E[X] = \int_{-\infty}^{+\infty} x f_X(x)\, dx = \int_a^b x \frac{1}{b-a}\, dx = \frac{a+b}{2}.$$

Note that $E[X]$ is the midpoint of the interval $[a, b]$. If we take $\phi(x) = x^2$, then

$$E[X^2] = \int_{-\infty}^{+\infty} x^2 f_X(x)\,dx = \int_a^b x^2 \frac{1}{b-a}\,dx = \frac{1}{3}(b^2 + ab + a^2). \blacksquare$$

To calculate $E[\phi(X)]$, the integral is set up by replacing X in the argument of ϕ by a typical value x, multiplying by the density of X, and integrating with respect to x.

EXAMPLE 7.3 Let X be a random variable having a $\Gamma(1/2, 1/2)$ density. Then

$$\begin{aligned} E[X^2] &= \int_{-\infty}^{+\infty} x^2 \Gamma\left(\frac{1}{2}, \frac{1}{2}\right)(x)\,dx \\ &= \int_0^{+\infty} x^2 \frac{1}{\sqrt{2\pi}} x^{(1/2)-1} e^{-x/2}\,dx \\ &= \frac{1}{\sqrt{2\pi}} \int_0^{+\infty} x^{(5/2)-1} e^{-x/2}\,dx. \end{aligned}$$

Note that the integrand looks like a $\Gamma(5/2, 1/2)$ density. It would be if it were multiplied by the constant

$$\frac{(1/2)^{5/2}}{\Gamma(5/2)} = \frac{1}{3\sqrt{2\pi}}.$$

Thus,

$$E[X^2] = \frac{1}{\sqrt{2\pi}} 3\sqrt{2\pi} \int_0^{+\infty} \frac{1}{3\sqrt{2\pi}} x^{(5/2)-1} e^{-x/2}\,dx.$$

Since the $\Gamma(5/2, 1/2)(x)$ density is zero for $x < 0$, the last integral is the total integral of $\Gamma(5/2, 1/2)$ over $(-\infty, +\infty)$ and has value 1. Therefore, $E[X^2] = 3$. $\blacksquare$

There are random variables for which $E[X]$ is not defined according to the criterion in Definition 7.3.

EXAMPLE 7.4 Let X be a random variable having the Cauchy density

$$f_X(x) = \frac{1}{\pi} \frac{1}{1+x^2}, \qquad x \in R.$$

Since

$$\int_{-\infty}^{+\infty} |x| \frac{1}{\pi} \frac{1}{1+x^2}\,dx = \frac{2}{\pi} \int_0^{+\infty} \frac{x}{1+x^2}\,dx = +\infty,$$

the integral

$$\int_{-\infty}^{+\infty} x \frac{1}{\pi} \frac{1}{1+x^2}\, dx$$

is not absolutely convergent, and therefore $E[X]$ is not defined according to Definition 7.3. ■

If the random variable X takes on only nonnegative values with probability 1, then $f_X(x) = 0$ for $x < 0$ and the integral defining $E[X]$ has the form $\int_0^{+\infty} x f_X(x)\, dx$. Absolute convergence and convergence are the same in this case. If the integral is not convergent, then as in calculus we write $E[X] = \int_0^{+\infty} x f_X(x)\, dx = +\infty$. In addition, if T is a waiting time that takes on the value $+\infty$ with positive probability, we then put $E[T] = +\infty$.

EXERCISES 7.2

1. Let X be a random variable having a uniform density on $[0, \pi]$. Calculate $E[\sin X]$.
2. Let X be a random variable having the standard normal density $\phi(x) = (1/\sqrt{2\pi})e^{-x^2/2}$, $-\infty < x < \infty$. Calculate $E[|X|]$.
3. Let X be a random variable having a uniform density on $[0, 1]$. Calculate $E[\min(X, 1/2)]$ and $E[\max(X, 1/2)]$.
4. Let $X_1, \ldots, X_n$ be independent random variables each having a uniform density on $[0, 1]$, let $U = \min(X_1, \ldots, X_n)$, and let $V = \max(X_1, \ldots, X_n)$. Calculate $E[U]$, $E[U^2]$, $E[V]$, and $E[V^2]$.
5. Let X be a random variable having an exponential density with parameter $\lambda > 0$. Calculate $E[X]$ and $E[X^2]$, if defined.
6. Let X be a random variable having a gamma density $\Gamma(\alpha, \lambda)$ where $\alpha, \lambda > 0$. If defined, calculate $E[X^r]$ where r is a positive integer.
7. Let X be a random variable having a beta density $\beta(\alpha_1, \alpha_2)$ where $\alpha_1, \alpha_2 > 0$. If defined, calculate $E[X^r]$ where r is a positive integer.
8. Let F be a distribution function that increases only by jumps at the points $c_1 < c_2 < \cdots < c_m$ and let $\phi : R \to R$ be continuous at $c_1, \ldots, c_m$. Prove that

$$\int_{-\infty}^{+\infty} \phi(t)dF(t) = \sum_{i=1}^{m} \phi(c_i)(F(c_i) - F(c_i-)).$$

9. Let X be a random variable having a density function f_X for which $f_X(x) = 0$ if $x < 0$. Assuming that $E[X]$ is defined, show that

$$E[X] = \int_0^{+\infty} P(X > x)\, dx = \int_0^{+\infty} (1 - F_X(x))\, dx.$$

(See Exercise 4.2.6.)

10. If X and Y are random variables having densities with $Y \geq X \geq 0$ with probability 1, show that $E[Y] \geq E[X] \geq 0$.

7.3 EXPECTATION AND CONDITIONAL EXPECTATION

The reason for defining the expected value of a random variable in terms of a Riemann-Stieltjes integral in the previous section was to convince the reader that there is a way of treating the discrete and continuous cases simultaneously and also of treating random variables that are neither discrete nor continuous. Operationally, the discrete case uses summation and the continuous case uses integration.

Let $X_1, \ldots, X_n$ be n random variables, let $\phi : R^n \to R$ be a real-valued continuous function of n variables, and let $Z = \phi(X_1, \ldots, X_n)$. Then the expected value $E[Z] = \int_{-\infty}^{+\infty} z\, dF_Z(z)$ is defined, provided the integral is absolutely convergent. The definition, however, requires the determination of F_Z knowing the probability characteristics of $X_1, \ldots, X_n$. The expected value of Z can be calculated using the following theorem.

Theorem 7.3.1 *Let $X_1, \ldots, X_n$ be n random variables having a joint density $f_{X_1,\ldots,X_n}$ and let ϕ be a real-valued continuous function of n variables. If $Z = \phi(X_1, \ldots, X_n)$, then*

$$E[Z] = \int \cdots \int_{R^n} \phi(x_1, \ldots, x_n) f_{X_1,\ldots,X_n}(x_1, \ldots, x_n)\, dV_n,$$

where dV_n denotes integration with respect to volume, provided the integral converges absolutely.

In a slightly more advanced course dealing with probability measures, this theorem is relatively easy to prove.

EXAMPLE 7.5 Let (X, Y) be the coordinates of point chosen at random from the unit square $U = \{(x,y) : 0 \leq x \leq 1, 0 \leq y \leq 1\}$ in the plane. Then

$$f_{X,Y}(x,y) = \begin{cases} 1 & \text{if } 0 \leq x \leq 1, 0 \leq y \leq 1 \\ 0 & \text{otherwise} \end{cases}$$

and

$$E[XY] = \iint_{R^2} xy f_{X,Y}(x,y)\, dA = \int_0^1 \left(\int_0^1 xy\, dy \right) dx = \frac{1}{4}. \quad \blacksquare$$

In this example, note that the integral for calculating $E[XY]$ is obtained by replacing X and Y by x and y, respectively, to form the integrand xy, multiplying by the joint density, and then integrating with respect to area.

Granted Theorem 7.3.1, we can establish properties of the expected value.

Theorem 7.3.2 *Let X and Y be random variables having a joint density with $E[X]$ and $E[Y]$ defined and let $a, b \in R$.*

(*i*) *If $P(X \geq 0) = 1$, then $E[X] \geq 0$.*

(*ii*) *The expected value of $aX + bY$ is defined, and $E[aX + bY] = aE[X] + bE[Y]$.*

(*iii*) *If $P(X \geq Y) = 1$, then $E[X] \geq E[Y]$.*

(*iv*) *$|E[X]| \leq E[|X|]$.*

PROOF:

(*i*) If $P(X \geq 0) = 1$, then $f_X(x) = 0$ for $x < 0$ and

$$E[X] = \int_{-\infty}^{+\infty} x f_X(x)\, dx = \int_0^{+\infty} x f_X(x)\, dx \geq 0.$$

(*ii*) Putting aside the question of whether or not $E[aX + bY]$ is defined,

$$\begin{aligned} E[aX + bY] &= \iint_{R^2} (ax + by) f_{X,Y}(x, y)\, dA \\ &= a \int_{-\infty}^{+\infty} \left(\int_{-\infty}^{+\infty} x f_{X,Y}(x, y)\, dy \right) dx \\ &\quad + b \int_{-\infty}^{+\infty} \left(\int_{-\infty}^{+\infty} y f_{X,Y}(x, y)\, dx \right) dy \\ &= a \int_{-\infty}^{+\infty} x f_X(x)\, dx + b \int_{-\infty}^{+\infty} y f_Y(y)\, dy \\ &= aE[X] + bE[Y]. \end{aligned}$$

The steps involved in showing that $E[aX + bY]$ is defined are similar using the inequality $|ax + by| \leq |a||x| + |b||y|$.

(*iii*) Since $P(X \geq Y) = 1, P(X - Y \geq 0) = 1$ and $E[X] - E[Y] = E[X - Y] \geq 0$ by (*ii*) and (*i*).

(*iv*) Since $-|X| \leq X \leq |X|, -E[|X|] \leq E[X] \leq E[|X|]$ by (*iii*), and therefore $|E[X]| \leq E[|X|]$. ■

EXAMPLE 7.6 Let (X, Y) be the coordinates of a point chosen at random from the part of a disk D with center $(0, 0)$ and radius 1 in the first quadrant. The joint density is

$$f_{X,Y}(x, y) = \begin{cases} (4/\pi) & \text{if } x^2 + y^2 \leq 1, x \geq 0, y \geq 0 \\ 0 & \text{otherwise,} \end{cases}$$

and the marginal densities are

$$f_X(x) = \begin{cases} (4/\pi)\sqrt{1 - x^2} & \text{if } 0 \leq x \leq 1 \\ 0 & \text{otherwise} \end{cases}$$

$$f_Y(y) = \begin{cases} (4/\pi)\sqrt{1 - y^2} & \text{if } 0 \leq y \leq 1 \\ 0 & \text{otherwise.} \end{cases}$$

Since $f_{X,Y}(x, y) \neq f_X(x)f_Y(y)$ for all x, y in a small disk centered at $(0, 0)$, X and Y are not independent random variables. Since

$$E[X] = E[Y] = \int_0^1 y\frac{4}{\pi}\sqrt{1 - y^2}\, dy = \frac{4}{3\pi},$$

$E[X + Y] = (4/3)\pi + (4/3)\pi = (8/3)\pi$. ■

Theorem 7.3.3 *Let X and Y be independent random variables for which $E[X]$ and $E[Y]$are defined. Then $E[XY]$ is defined and*

$$E[XY] = E[X]E[Y].$$

PROOF: Since

$$\begin{aligned}\iint_{R^2} |xy| f_X(x) f_Y(y)\, dA &= \iint_{R^2} |x||y| f_X(x) f_Y(y)\, dA \\ &= \int_{-\infty}^{+\infty} |x| f_X(x) \left(\int_{-\infty}^{+\infty} |y| f_Y(y)\, dy \right) dx \\ &= \left(\int_{-\infty}^{+\infty} |y| f_Y(y)\, dy \right) \left(\int_{-\infty}^{+\infty} |x| f_X(x)\, dx \right) \\ &< +\infty,\end{aligned}$$

$E[XY]$ is defined. The same calculation with $|xy|$ replaced by xy shows that $E[XY] = E[X]E[Y]$. ■

EXAMPLE 7.7 Let X and Y be independent random variables having exponential densities with parameters λ_1 and λ_2, respectively. By Exercise 7.2.5, $E[X] = 1/\lambda_1$ and $E[Y] = 1/\lambda_2$. By Theorem 7.3.3, $E[XY] = 1/(\lambda_1\lambda_2)$. ■

The preceding theorem can be extended to functions of X and Y as follows.

Theorem 7.3.4 *Let X and Y be independent random variables and let ϕ and ψ be continuous real-valued functions with $E[\phi(X)]$ and $E[\psi(Y)]$ defined. Then $E[\phi(X)\psi(Y)]$ is defined, and*

$$E[\phi(X)\psi(Y)] = E[\phi(X)]E[\psi(Y)].$$

All of the theorems and corollaries of Section 4.4 hold for arbitrary random variables. The proof of Schwarz's inequality (Equation 4.8) is precisely the same as in Section 4.4 as soon as it is established that $E[X^2] = 0$ implies that $P(X = 0) = 1$. Consider the case that X has a density function f_X. For every positive integer n,

$$0 = E[X^2] = \int_{-\infty}^{+\infty} x^2 f_X(x)\,dx \geq \int_{|x| \geq \frac{1}{n}} x^2 f_X(x)\,dx \geq \frac{1}{n^2}P\left(|X| \geq \frac{1}{n}\right)$$

Thus, $P(|X| \geq 1/n) = 0$ for every positive integer n, and since the events $(|X| \geq 1/n)$ increase to the event $(|X| > 0)$ as $n \to \infty$, $P(X = 0) = P(|X| = 0) = 1$. This is all that is needed to replicate the proof of Schwarz's inequality. It is not essential that X have a density function.

The following example can also be interpreted as a random walk in the plane if the bonds are thought of as instantaneous displacements of a randomly moving particle. For a more comprehensive treatment of chain molecules, see the book by P. J. Flory listed at the end of the chapter.

EXAMPLE 7.8 (Chain Molecules) Consider a chain molecule formed in the following way. Let ℓ be a fixed positive number representing the distance between two molecules or the length of the bond between the two. Starting with an initial molecule at the origin of the xy-plane, a bond is formed between it and molecule #1 of length ℓ making an angle Θ_1 with the positive x-axis, where Θ_1 is chosen at random from the interval $[0, 2\pi]$; starting from the position of molecule #1, a bond is formed between it and molecule #2 of length ℓ making an angle Θ_2 with the positive x-axis, where Θ_2 is chosen at random from the interval $[0, 2\pi]$, independently of Θ_1, and so forth, as shown in Figure 7.1. If there are n bonds in the chain with an initial molecule at the origin, what is the expected value of the square of the distance of the nth molecule from the origin? For $i = 1, \ldots, n$, let (X_i, Y_i) be the change in the coordinates in going from the $(i-1)$st molecule to the ith molecule. Then

$$X_i = \ell\cos\Theta_i \qquad Y_i = \ell\sin\Theta_i.$$

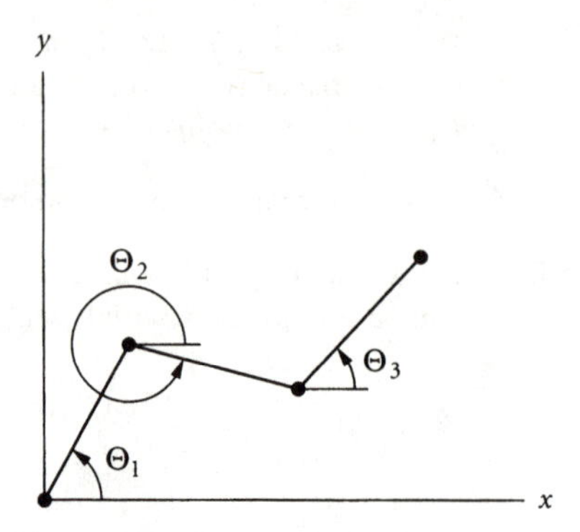

FIGURE 7.1 Chain molecule.

The position of molecule $\#n$ is then

$$\left(\sum_{i=1}^{n} X_i, \sum_{i=1}^{n} Y_i\right).$$

If D denotes the square of the distance of molecule $\#n$ from the initial molecule, then

$$\begin{aligned} D^2 &= \left(\sum_{i=1}^{n} X_i\right)^2 + \left(\sum_{i=1}^{n} Y_i\right)^2 \\ &= \sum_{i=1}^{n} X_i^2 + \sum_{i \neq j} X_i X_j + \sum_{i=1}^{n} Y_i^2 + \sum_{i \neq j} Y_i Y_j. \end{aligned}$$

Note that

$$\begin{aligned} E[X_i^2] &= E[\ell^2 \cos^2 \Theta_i] \\ &= \frac{\ell^2}{2\pi} \int_0^{2\pi} \cos^2 \theta_i \, d\theta_i \\ &= \frac{\ell^2}{2\pi} \int_0^{2\pi} \frac{1 + \cos 2\theta_i}{2} \, d\theta_i \\ &= \frac{\ell^2}{2} \end{aligned}$$

and that

$$\begin{aligned} E[X_i X_j] &= \ell^2 E[\cos \Theta_i \cos \Theta_j] \\ &= \ell^2 E[\cos \Theta_i] E[\cos \Theta_j] \\ &= \ell^2 \left(\frac{1}{2\pi} \int_0^{2\pi} \cos \theta_i \, d\theta_i\right)\left(\frac{1}{2\pi} \int_0^{2\pi} \cos \theta_j \, d\theta_j\right) \\ &= 0 \end{aligned}$$

by the independence of Θ_i and Θ_j for $i \neq j$ and Theorem 7.3.4. Similarly, $E[Y_i^2] = \ell^2/2$ and $E[Y_i Y_j] = 0$. Therefore,

$$E[D^2] = n\frac{\ell^2}{2} + n\frac{\ell^2}{2} = n\ell^2.$$

It should be emphasized that this is not the square of the expected value but the expected value of the square. By Schwarz's inequality, $(E[D])^2 \leq E[D^2] = n\ell^2$, from which it follows that $E[D] \leq \ell\sqrt{n}$. ■

In considering a single random variable X, most of the information concerning X is embodied in its density function. For some purposes, we would like to summarize that information in a few parameters.

Consider a random variable X that has a finite second moment and density function f_X; i.e., $E[X^2] = \int_{-\infty}^{+\infty} x^2 f_X(x)\,dx < +\infty$. In Section 4.3, we saw that $|x| \leq x^2 + 1$ for all $x \in R$, so that

$$\int_{-\infty}^{+\infty} |x| f_X(x)\,dx \leq \int_{-\infty}^{+\infty} (x^2+1) f_X(x)\,dx = \int_{-\infty}^{+\infty} x^2 f_X(x)\,dx + 1 < +\infty,$$

and it follows that $E[X]$ is defined. Thus,

$$\mu_X = E[X]$$

is finite and is called the *mean* of X. Since $(x - \mu_X)^2 = x^2 - 2\mu_X x + \mu_X^2$,

$$\begin{aligned}
&\int_{-\infty}^{+\infty} (x-\mu_X)^2 f_X(x)\,dx \\
&\quad = \int_{-\infty}^{+\infty} x^2 f_X(x)\,dx - 2\mu_X \int_{-\infty}^{+\infty} x f_X(x)\,dx + \int_{-\infty}^{+\infty} \mu_X^2 f_X(x)\,dx \\
&\quad = E[X^2] - \mu_X^2 < +\infty.
\end{aligned}$$

Thus, $E[(X - \mu_X)^2]$ is finite, and we define a second parameter

$$\sigma_X^2 = E[(X-\mu_X)^2] = E[X^2] - \mu_X^2 = E[X^2] - (E[X])^2,$$

called the *variance* of X. The variance of X is also denoted by var X. It is easily checked that

$$\begin{aligned}
\text{var}\,(cX) &= c^2\,\text{var}\,X \\
\text{var}\,(X + c) &= \text{var}\,X
\end{aligned}$$

EXAMPLE 7.9 Let X have a uniform density on $[a, b], a < b$. It was shown in the previous section that $E[X] = (a+b)/2$ and $E[X^2] = (1/3)(b^2+ab+a^2)$. Thus, $\mu_X = (a+b)/2$ and $\sigma_X^2 = \text{var}\,X = (b-a)^2/12$. ■

EXAMPLE 7.10 Let X have a $\Gamma(\alpha, \lambda)$ density. By Exercise 7.2.6, for each positive integer r,

$$E[X^r] = \frac{(\alpha + r - 1)(\alpha + r - 2) \times \cdots \times \alpha}{\lambda^r}.$$

In particular, $E[X] = \alpha/\lambda$ and $E[X^2] = ((\alpha + 1)\alpha)/\lambda^2$, so that $\mu_X = \alpha/\lambda$ and $\operatorname{var} X = \alpha/\lambda^2$. ■

EXAMPLE 7.11 Let X have an exponential density with parameter $\lambda > 0$. Since this density is the same as the $\Gamma(1, \lambda)$ density, $\mu_X = 1/\lambda$ and $\operatorname{var} X = 1/\lambda^2$ by Example 7.10. ■

EXAMPLE 7.12 Let X be a random variable having a standard normal density. Since

$$\int_{-\infty}^{+\infty} |x| \frac{1}{\sqrt{2\pi}} e^{-x^2/2}\, dx = 2 \int_0^{+\infty} x \frac{1}{\sqrt{2\pi}} e^{-x^2/2}\, dx < +\infty,$$

$E[X]$ is defined and

$$E[X] = \int_{-\infty}^{+\infty} x \frac{1}{\sqrt{2\pi}} e^{-x^2/2}\, dx = 0,$$

since the integrand is an odd function. To find $\operatorname{var} X = E[X^2] - (E[X])^2 = E[X^2]$, we need only determine $E[X^2]$. Letting $Y = X^2$, by Example 6.19 Y has a $\Gamma(1/2, 1/2)$ density. By Example 7.10, $E[Y] = 1$. Thus, $\mu_X = E[X] = 0$ and $\sigma_X = \operatorname{var} X = 1$. ■

Now let Y be a random variable Y having an $n(\mu, \sigma^2)$ density. By definition, $Y = \sigma X + \mu$ where X has a standard normal density. It follows that $\mu_Y = E[Y] = \sigma E[X] + \mu = \mu$ and $\sigma_Y^2 = \operatorname{var} Y = \operatorname{var}(\sigma X + \mu) = \sigma^2 \operatorname{var} X = \sigma^2$. From Example 6.17,

$$f_Y(y) = \frac{1}{\sqrt{2\pi}\sigma} e^{-(y-\mu)^2/2\sigma^2},$$

and μ_Y and σ_Y can be readily identified by examining the parameters μ and σ in the function f_Y.

Although the random variable X is not required to have a density in the next lemma, the proof will assume a density.

Lemma 7.3.5 (Markov's Inequality) *If X is any random variable for which $E[X]$ is defined and $t > 0$, then*

$$P(|X| \geq t) \leq \frac{E[|X|]}{t}.$$

PROOF: $E[|X|] = \int_{-\infty}^{+\infty} |x| f_X(x)\,dx \geq \int_{|x|\geq t} |x| f_X(x)\,dx \geq tP(|X| \geq t)$. ■

As in the discrete case, Chebyshev's inequality is an easy consequence of Markov's inequality.

Theorem 7.3.6 (Chebyshev's Inequality) *Let X be a random variable with mean μ and finite variance σ^2. Then*

$$P(|X - \mu| \geq \delta) \leq \frac{\sigma^2}{\delta^2}$$

for all $\delta > 0$.

Suppose now that the random variables $Y, X_1, \ldots, X_n$ have a joint density. We can then consider the conditional density of Y given $X_1, \ldots, X_n$ and define the *conditional expectation* of Y given $X_1, \ldots, X_n$ by the equation

$$E[Y \mid X_1 = x_1, \ldots, X_n = x_n] = \int_{-\infty}^{+\infty} y f_{Y|X_1,\ldots,X_n}(y|x_1, \ldots, x_n)\,dy,$$

provided the integral is absolutely convergent.

Theorem 7.3.7 *If $Y, X_1, \ldots, X_n$ have a joint density and $E[Y]$ is defined, then $E[Y \mid X_1 = x_1, \ldots, X_n = x_n]$ is defined, and*

$$E[Y] = \int \cdots \int_{R^n} E[Y|X_1 = x_1, \ldots, X_n = x_n] f_{X_1,\ldots,X_n}(x_1, \ldots, x_n)\,dx_1 \ldots dx_n.$$

Sketch of Proof: Since $E[Y]$ is defined,

$$\begin{aligned} +\infty &> \int_{-\infty}^{+\infty} |y| f_Y(y)\,dy \\ &= \int_{-\infty}^{+\infty} |y| \int \cdots \int_{R^n} f_{Y,X_1,\ldots,X_n}(y, x_1, \ldots, x_n)\,dx_1 \ldots dx_n\,dy \\ &= \int \cdots \int_{R^n} \left(\int_{-\infty}^{+\infty} |y| f_{Y|X_1,\ldots,X_n}(y|x_1, \ldots, x_n)\,dy \right) \\ &\times f_{X_1,\ldots,X_n}(x_1, \ldots, x_n)\,dx_1 \ldots dx_n. \end{aligned}$$ ■

Thus, the integral within parentheses is finite for "most" points $(x_1, \ldots, x_n)$ in R^n. The same calculation with $|y|$ replaced by y will establish the final result.

EXAMPLE 7.13 Let (X, Y) be the coordinates of a point chosen at random from a triangle with vertices at $(0, 0)$, $(1, 0)$, and $(1, 1)$. Intuitively, given that $X = x$ with $0 \leq x \leq 1$, the point (x, Y) is then chosen at random from the line segment joining $(x, 0)$ to (x, x); i.e., given that $X = x$, Y has a uniform density on $(0, x)$, and therefore $E[Y|X = x] = x/2$ for $0 \leq x \leq 1$. This intuitive argument can be justified by the following formal calculations. The joint density of X and Y is

$$f_{X,Y}(x, y) = \begin{cases} 2 & \text{if } 0 \leq y \leq x, 0 \leq x \leq 1 \\ 0 & \text{otherwise.} \end{cases}$$

Clearly, $f_X(x) = 0$ for $x < 0$ and $x > 1$. For $0 \leq x \leq 1$,

$$f_X(x) = \int_{-\infty}^{+\infty} f_{X,Y}(x, y)\, dy = \int_0^x 2\, dy = 2x.$$

Therefore,

$$f_X(x) = \begin{cases} 2x & \text{if } 0 \leq x \leq 1 \\ 0 & \text{otherwise} \end{cases}$$

and

$$f_{Y|X}(y|x) = \begin{cases} 1/x & \text{if } 0 \leq y \leq x, 0 \leq x \leq 1 \\ 0 & \text{otherwise.} \end{cases}$$

For $0 \leq x \leq 1$,

$$E[Y|X = x] = \int_{-\infty}^{+\infty} y f_{Y|X}(y|x) dy = \int_0^x y \frac{1}{x} dy = \frac{1}{x}\frac{x^2}{2} = \frac{x}{2}. \quad \blacksquare$$

Care must be taken in making heuristic arguments of the tyne appearing in this example. Such arguments can allow one to discover facts, but they should be formally verified as above before staking one's reputation on the result.

The last theorem can be stated in a more general context. Suppose $Y_1, \ldots, Y_m, X_1, \ldots, X_n$ are random variables having a joint density and ψ is a continuous real-valued function of m variables. If $E[\psi(Y_1, \ldots, Y_m)]$ is defined, then

$$E[\psi(Y_1, \ldots, Y_m)| X_1 = x_1, \ldots, X_n = x_n] =$$
$$\int_{R^m} \cdots \int \psi(y_1, \ldots, y_m) f_{Y_1,\ldots,Y_m|X_1,\ldots,X_n}(y_1, \ldots, y_m|x_1, \ldots, x_n)\, dy_1 \ldots dy_m.$$

Starting with this result, it is possible to develop such concepts as conditional variance of Y given $X_1 = x_1, \ldots, X_n = x_n$ denoted by $\text{var}(Y|X_1 = x_1, \ldots, X_n = x_n)$, and so forth.

EXERCISES 7.3

1. Let X and Y be independent random variables, both having an exponential density with parameter $\lambda = 1$. If $Z = \max(X, Y)$, calculate $E[Z]$ without using the density of Z.
2. A point (X, Y) is selected at random from the unit square $\{(x, y) : 0 \leq x \leq 1, 0 \leq y \leq 1\}$. Given that $X = x$ and $Y = y$, a point (U, V) is selected at random from the rectangle $\{(u, v) : 0 \leq u \leq x, 0 \leq v \leq y\}$. Calculate $E[U|X = x, Y = y]$.
3. Let Λ be a random variable having an exponential density with parameter 1. Given that $\Lambda = \lambda$, the random variables $X_1, \ldots, X_n$ are independent random variables each having an exponential density with parameter λ. Find $E[\Lambda|X_1 = x_1, \ldots, X_n = x_n]$.
4. Let X and Y be random variables having the joint density
$$f_{X,Y}(x, y) = \begin{cases} 8xy & \text{if } 0 \leq y \leq x \leq 1 \\ 0 & \text{otherwise.} \end{cases}$$
Find $E[Y|X = x]$ and $E[X|Y = y]$.
5. Let Λ be a random variable having a gamma density $\Gamma(\alpha, \beta)$ and given that $\Lambda = \lambda, X$ has an exponential density with parameter λ. Find the density of X and the conditional density of Λ given $X = x$.
6. Let U and V be the random variables of Exercises 6.6.7 and 7.2.4 and let $R = V - U$ be the range of the $X_1, \ldots, X_n$. Calculate $E[R]$ and $\text{var } R$.
7. Consider the random variables U and V of the previous problem. Determine $f_{U|V}(u|v)$ and calculate $E[U|V = v]$.
8. Consider the random variables X, Y, and Z having the joint density of Exercise 6.6.8. Determine $f_{Z|X,Y}(z|x, y)$ and calculate $E[Z|X = x, Y = y]$.
9. Let X and Y be the coordinates of a point chosen at random from the triangle in the plane with vertices at $(-1, 0)$, $(0, 1)$, and $(1, 0)$. Determine $E[Y|X = x]$ without calculations.
10. A number X is chosen at random from the interval $[0, 1]$. Given that $X = x$, a number Y is chosen at random from the interval $[0, x]$. Calculate $E[X|Y = y]$.

7.4 NORMAL DENSITY

Consider a sequence $\{X_j\}_{j=1}^n$ of n Bernoulli trials with probability of success $p = 1/2$ and let $S_n = \sum_{j=1}^n X_j$. We know that S_n has the binomial density function

$$b\left(j; n, \frac{1}{2}\right) = \binom{n}{j} 2^{-n}, \qquad j = 0, \ldots, n,$$

with $E[S_n] = n/2$ and $\sigma_{S_n} = \sqrt{n}/2$. If we were to examine a bar graph of this density, it would be centered about the point $x = n/2$ on the x-axis, and the variance $n/4$ would be large for large n, indicating that the density is spread out far from the mean. As n becomes large, the individual probabilities would also become small. To eliminate the spreading effect, we consider the normalized sum

$$S_n^* = \frac{S_n - (n/2)}{\sqrt{n}/2},$$

which is centered about $E[S_n^*] = 0$ and has variance var $S_n^* = \text{var}\,((S_n - (n/2))/(\sqrt{n}/2)) = (4/n)\,\text{var}\,S_n = 1$. The bar graph of S_{36}^* is depicted in Figure 7.2. The area of the rectangle centered above 0 represents the probability

$$P(S_{36}^* = 0) = P(S_n = 18) = \binom{36}{18} 2^{-36} \approx .132.$$

If we compare Figure 7.2 with the graph of the standard normal density depicted in Figure 6.11, it might appear that one of the two could be used to approximate the other. At one time it was impractical to calculate $P(\alpha < S_n < \beta)$ because of the large number of arithmetic operations required even when n is moderately large, and so the normal density was used to approximate

$$P(\alpha < S_n < \beta) = P\left(\frac{\alpha - n/2}{\sqrt{n}/2} < S_n^* < \frac{\beta - n/2}{\sqrt{n}/2}\right)$$

With the advent of fast computers, such calculations can now be done in fractions of a second for moderate values of n. Even though aṗproximation of

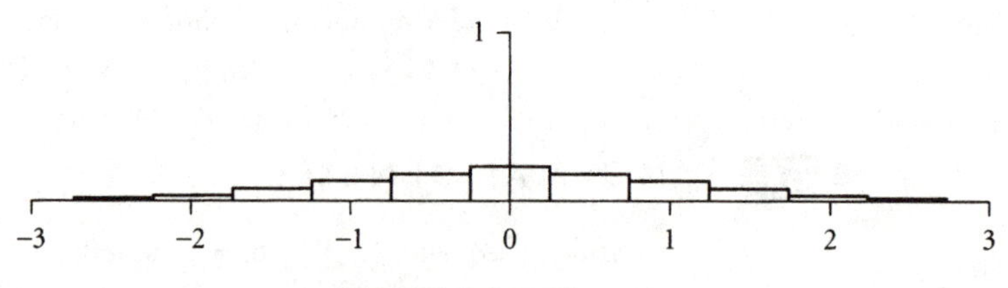

FIGURE 7.2 $P(S_{36}^* = i)$.

binomial probabilities is not as important as it once was, there are valid reasons for looking into normal approximations to the binomial density.

The following theorem was first proved by DeMoivre (1667–1754). The proof of the theorem involves only elementary facts from the calculus. Reading the proof is not essential for learning probability at this stage.

Theorem 7.4.1 (Central Limit Theorem) *Let* $\{X_j\}_{j=1}^n$ *be a sequence of n Bernoulli trials with probability of success* $p = 1/2$ *and let* $S_n = \sum_{i=1}^n X_i$. *Then*

$$\lim_{n\to\infty} P\left(\left|S_n - \frac{n}{2}\right| < x\frac{\sqrt{n}}{2}\right) = \frac{1}{\sqrt{2\pi}}\int_{-x}^{x} e^{-t^2/2}\,dt.$$

The probability on the left is the same as $P(|S_n^*| < x)$. The proof of the theorem requires the following approximations.

Lemma 7.4.2 *There are functions* γ *and* δ *such that*

(i) $\ln(1+x) = x(1+\gamma(x))$ if $|x| < 1$.

(ii) $|\gamma(x)| \le \delta(|x|)$ if $|x| < 1$.

(iii) δ *is nondecreasing on* $[0, 1)$ *and* $\lim_{x\to 0+}\delta(x) = 0$.

PROOF: The function $\ln(1+x)$ has the Maclaurin series expansion

$$\begin{aligned}\ln(1+x) &= x - \frac{x^2}{2} + \frac{x^3}{3} - \frac{x^4}{4} + \cdots \\ &= x\left(1 - \frac{x}{2} + \frac{x^2}{3} - \frac{x^3}{4} + \cdots\right) \\ &= x(1+\gamma(x))\end{aligned}$$

for $|x| < 1$ where $\gamma(x) = -(x/2) + (x^2/3) - (x^3/4) + \cdots$. Let

$$\delta(x) = \frac{x}{2} + \frac{x^2}{3} + \frac{x^3}{4} + \cdots.$$

Since the series defining δ converges absolutely in $(-1, +1)$ and the sum of a power series is a continuous function on its interval of convergence, $\lim_{x\to 0+}\delta(x) = \delta(0) = 0$. It is clear that δ is increasing on $[0, 1)$. ■

We will also need the following discrete version of the mean value theorem of the integral calculus.

Lemma 7.4.3 *If $a_1, \ldots, a_n$ are nonnegative real numbers and $b_1, \ldots, b_n$ are any real numbers, then there is an m with $|m| \le \max_{1 \le j \le n} |b_j|$ such that*

$$\sum_{j=1}^{n} a_j b_j = m \sum_{j=1}^{n} a_j.$$

PROOF: We can assume that some $a_j > 0$ because otherwise we can take $m = 0$. Since

$$\left|\sum_{j=1}^{n} a_j b_j\right| \le \sum_{j=1}^{n} |a_j||b_j| \le (\max_{1 \le j \le n} |b_j|) \sum_{j=1}^{n} a_j,$$

we can take

$$m = \frac{\sum_{j=1}^{n} a_j b_j}{\sum_{j=1}^{n} a_j}. \blacksquare$$

We will also need to approximate the exponential function. It follows from the Maclaurin series expansion of e^y that $e^y = 1 + \theta(y)$ where $\theta(y) = y + y^2/2! + y^3/3! + \cdots$ satisfies (1) $\lim_{y \to 0} \theta(y) = 0$, (2) $|\theta(y)| \le \theta(|y|)$, and (3) $\theta(y)$ is nondecreasing on $[0, +\infty)$.

Proof of Theorem 7.4.1 We prove the theorem for an even number of trials first. The number x will be fixed throughout the proof. Consider only those n for which $x\sqrt{n/2} < n$. Let

$$P_n(x) = P(|S_{2n} - n| < x\sqrt{n/2}).$$

Since S_{2n} has the binomial density $b(\cdot; 2n, 1/2)$,

$$P_n(x) = \sum_{|k-n| < x\sqrt{n/2}} \binom{2n}{k} 2^{-2n}.$$

Substituting j for $k - n$,

$$P_n(x) = \sum_{|j| < x\sqrt{n/2}} \binom{2n}{n+j} 2^{-2n}.$$

Note that $\binom{2n}{n+j} = \binom{2n}{n-j}$, so that the term in the sum corresponding to j is equal to the term corresponding to $-j$. Consider the terms of the sum for which $j \geq 0$. Since

$$\binom{2n}{n+j} = \frac{(2n)!}{(n+j)!(n-j)!}$$
$$(n+j)! = (n+j) \times \cdots \times (n+1)n!$$
$$(n-j)! = \frac{n!}{n(n-1) \times \cdots \times (n-j+1)},$$

we can write

$$\binom{2n}{n+j}2^{-2n} = \frac{(2n)!}{n!n!}2^{-2n}\frac{n(n-1) \times \cdots \times (n-j+1)}{(n+j)(n+j-1) \times \cdots \times (n+1)}.$$

By the above remark, this equation holds for $j < 0$. Therefore,

$$P_n(x) = \sum_{|j|<x\sqrt{n/2}} \frac{(2n)!}{n!n!}2^{-2n}\frac{n(n-1) \times \cdots \times (n-j+1)}{(n+j)(n+j-1) \times \cdots \times (n+1)}.$$

Letting

$$P_n = \frac{(2n)!}{n!n!}2^{-2n}$$

and applying Stirling's formula, Equation 5.10, to the factorials, $P_n \sim 1/\sqrt{\pi n}$. This means that $P_n/(1/\sqrt{\pi n}) \to 1$ as $n \to \infty$; i.e.,

$$\frac{P_n}{1/\sqrt{\pi n}} = 1 + \delta_n$$

where $\lim_{n\to\infty} \delta_n = 0$. Thus, we can write

$$P_n = \frac{1}{\sqrt{\pi n}}(1 + \delta_n).$$

Therefore,

$$P_n(x) = \sum_{|j|<x\sqrt{n/2}} \frac{1}{\sqrt{\pi n}}(1+\delta_n)\frac{n(n-1) \times \cdots \times (n-j+1)}{(n+j)(n+j-1) \times \cdots \times (n+1)}. \quad (7.1)$$

For $|j| < x\sqrt{n/2}$, let

$$D_{n,j} = \frac{n(n-1) \times \cdots \times (n-j+1)}{(n+j)(n+j-1) \times \cdots \times (n+1)}$$
$$= \frac{1}{(1+(j/n))(1+(j/(n-1))) \times \cdots \times (1+(j/(n-j+1)))}.$$

Taking natural logarithms and approximating the ln function using the γ function of Lemma 7.4.2,

$$\begin{aligned}\ln D_{n,j} &= -\sum_{i=0}^{j-1} \ln\left(1 + \frac{j}{n-i}\right)\\ &= -\sum_{i=0}^{j-1} \frac{j}{n-i}\left(1 + \gamma\left(\frac{j}{n-i}\right)\right).\end{aligned}$$

Writing $j/(n-i) = (j/n)(1 + (i/(n-i)))$,

$$\begin{aligned}\ln D_{n,j} &= -\sum_{i=0}^{j-1} \frac{j}{n}\left(1 + \frac{i}{n-i}\right)\left(1 + \gamma\left(\frac{j}{n-i}\right)\right)\\ &= -\sum_{i=0}^{j-1} \frac{j}{n} - \sum_{i=0}^{j-1} \frac{j}{n}\left(\gamma\left(\frac{j}{n-i}\right) + \frac{i}{n-i} + \frac{i}{n-i}\gamma\left(\frac{j}{n-i}\right)\right).\end{aligned}$$

Applying Lemma 7.4.3 to the second sum on the right,

$$\ln D_{n,j} = -\frac{j^2}{n} - \gamma_{n,j}\sum_{i=0}^{j-1} \frac{j}{n} = -\frac{j^2}{n} - \gamma_{n,j}\frac{j^2}{n}$$

where

$$|\gamma_{n,j}| \leq \max\left\{\left|\gamma\left(\frac{j}{n-i}\right) + \frac{i}{n-i} + \frac{i}{n-i}\gamma\left(\frac{j}{n-i}\right)\right|; i = 0, \ldots, j-1\right\}.$$

Recalling the δ function of Lemma 7.4.2,

$$\left|\gamma\left(\frac{j}{n-i}\right)\right| \leq \delta\left(\left|\frac{j}{n-i}\right|\right) \leq \delta\left(\frac{|j|}{n-|j|}\right).$$

Noting that

$$\frac{|j|}{n-|j|} \leq \frac{x\sqrt{n/2}}{n - x\sqrt{n/2}}$$

and letting $\Delta_{n,x}$ denote the quantity on the right, $\lim_{n\to\infty} \Delta_{n,x} = 0$ and

$$\left|\gamma\left(\frac{j}{n-i}\right)\right| \leq \delta(\Delta_{n,x}).$$

Also note that $|i/(n-i)| \leq |j|/(n-|j|) \leq \Delta_{n,x}$. Thus,

$$|\gamma_{n,j}| \leq \delta(\Delta_{n,x}) + \Delta_{n,x} + \Delta_{n,x}\delta(\Delta_{n,x}).$$

Letting $\tilde{\Delta}_{n,x}$ denote the quantity on the right,

$$|\gamma_{n,j}| \leq \tilde{\Delta}_{n,x} \qquad \text{for all } |j| < x\sqrt{n/2}$$

where $\lim_{n\to\infty} \tilde{\Delta}_{n,x} = 0$. Thus,

$$D_{n,j} = e^{-j^2/n} e^{-(\gamma_{n,j})(j^2/n)} \qquad \text{for all } |j| < x\sqrt{n/2}.$$

Using the approximation $e^y = 1 + \theta(y)$ discussed after Lemma 7.4.3,

$$D_{n,j} = e^{-j^2/n}\left(1 + \theta\left(-\gamma_{n,j}\frac{j^2}{n}\right)\right) \qquad \text{for all } |j| < x\sqrt{n/2}.$$

Thus, Equation 7.1 can be written

$$\begin{aligned}
P_n(x) &= \sum_{|j|<x\sqrt{n/2}} \frac{1}{\sqrt{n\pi}} e^{-j^2/n}(1+\delta_n)\left(1+\theta\left(-\gamma_{n,j}\frac{j^2}{n}\right)\right) \\
&= \sum_{|j|<x\sqrt{n/2}} \frac{1}{\sqrt{n\pi}} e^{-j^2/n} \\
&+ \sum_{|j|<x\sqrt{n/2}} \frac{1}{\sqrt{n\pi}} e^{-j^2/n}\left(\delta_n + \theta\left(-\gamma_{n,j}\frac{j^2}{n}\right) + \delta_n\theta\left(-\gamma_{n,j}\frac{j^2}{n}\right)\right).
\end{aligned}$$

We will now show that the second sum on the right has the limit zero as $n \to \infty$. Applying Lemma 7.4.3 to the second sum, it can be written

$$\Lambda_{n,x} \sum_{|j|<x\sqrt{n/2}} \frac{1}{\sqrt{n\pi}} e^{-j^2/n} \tag{7.2}$$

where

$$|\Lambda_{n,x}| \leq \max\left\{\left|\delta_n + \theta\left(-\gamma_{n,j}\frac{j^2}{n}\right) + \delta_n\theta\left(-\gamma_{n,j}\frac{j^2}{n}\right)\right|; |j| < x\sqrt{n/2}\right\}.$$

Since $|\theta(y)| \leq \theta(|y|)$ and $\theta(y)$ is nondecreasing on $[0,\infty)$,

$$\left|\theta\left(-\gamma_{n,j}\frac{j^2}{n}\right)\right| \leq \theta\left(|\gamma_{n,j}|\frac{j^2}{n}\right) \leq \theta\left(\tilde{\Delta}_{n,x}\frac{x^2}{2}\right)$$

and

$$|\Lambda_{n,x}| \leq \delta_n + \theta\left(\tilde{\Delta}_{n,x}\frac{x^2}{2}\right) + \delta_n\theta\left(\tilde{\Delta}_{n,x}\frac{x^2}{2}\right) \to 0 \text{ as } n \to \infty,$$

and therefore $\lim_{n\to\infty} \Lambda_{n,x} = 0$. Note also that

$$\sum_{|j|<x\sqrt{n/2}} \frac{1}{\sqrt{n\pi}} e^{-j^2/n} \le \frac{\sqrt{2}x}{\sqrt{\pi}},$$

and therefore the quantity in 7.2 has the limit zero as $n \to \infty$. It follows that

$$\lim_{n\to\infty} P_n(x) = \lim_{n\to\infty} \sum_{|j|<x\sqrt{n/2}} \frac{1}{\sqrt{n\pi}} e^{-j^2/n}$$

provided the limit on the right exists. If we let $x_j = j\sqrt{2/n}$, then the points $\{x_j : |j| < x\sqrt{n/2}\}$ constitute a partition of the interval $[-x, x]$ into subintervals $[x_{j-1}, x_j]$ of length $\Delta x_j = \sqrt{2/n}$, and

$$\sum_{|j|<x\sqrt{n/2}} \frac{1}{\sqrt{n\pi}} e^{-j^2/n} = \sum_{|j|<x\sqrt{n/2}} \frac{1}{\sqrt{2\pi}} e^{-x_j^2/2} \Delta x_j.$$

Since the sum on the right is just a Riemann sum defining the integral $\int_{-x}^{x} (1/\sqrt{2\pi}) e^{-t^2/2}\, dt$,

$$\lim_{n\to\infty} P_n(x) = \lim_{n\to\infty} \sum_{|j|<x\sqrt{n/2}} \frac{1}{\sqrt{n\pi}} e^{-j^2/n} = \int_{-x}^{x} \frac{1}{\sqrt{2\pi}} e^{-t^2/2}\, dt.$$

This completes the proof for even n. To take care of odd n it is necessary to do an epsilon argument. We will use $\phi(x)$ in place of $(1/\sqrt{2\pi})e^{-x^2/2}$ for the rest of the proof. Since $\int_{-x}^{x} \phi(t)\, dt$ is a continuous function of x, given $\varepsilon > 0$ there is an $h > 0$ such that

$$\begin{aligned} \int_{-x}^{x} \phi(t)\, dt - \varepsilon < \int_{-x+h}^{x-h} \phi(t)\, dt &< \int_{-x}^{x} \phi(t)\, dt \\ &< \int_{-x-h}^{x+h} \phi(t)\, dt < \int_{-x}^{x} \phi(t) dt + \varepsilon. \end{aligned}$$

Since

$$\lim_{n\to\infty} P(|S_{2n} - n| < (x - h)\sqrt{n/2}) = \int_{-x+h}^{x-h} \phi(t)\, dt$$

and

$$\lim_{n\to\infty} P(|S_{2n} - n| < (x + h)\sqrt{n/2}) = \int_{-x-h}^{x+h} \phi(t)\, dt,$$

there is a positive integer N such that for all $n \ge N$

(*i*) $(1/2) - (h/2)\sqrt{2n} < 0.$

(*ii*) $P(|S_{2n} - n| < (x-h)\sqrt{n/2}) > \int_{-x}^{x} \phi(t)\,dt - \varepsilon.$

(*iii*) $P(|S_{2n} - n| < (x+h)\sqrt{n/2}) < \int_{-x}^{x} \phi(t)\,dt + \varepsilon.$

Since $X_i(\omega) = 0$ or $1, |X_i(\omega) - (1/2)| = 1/2$ for all i, and therefore

$$\begin{aligned}\left|S_{2n+1}(\omega) - \frac{2n+1}{2}\right| &= \left|S_{2n}(\omega) + X_{2n+1}(\omega) - n - \frac{1}{2}\right| \\ &\leq |S_{2n}(\omega) - n| + \left|X_{2n+1}(\omega) - \frac{1}{2}\right| \\ &= |S_{2n}(\omega) - n| + \frac{1}{2}.\end{aligned}$$

Suppose $|S_{2n}(\omega) - n| < (x-h)\sqrt{n/2}$. By (*i*), for $n \geq N$,

$$\begin{aligned}\left|S_{2n+1}(\omega) - \frac{2n+1}{2}\right| &< (x-h)\sqrt{n/2} + \frac{1}{2} \\ &= x\sqrt{n/2} + \left(\frac{1}{2} - h\sqrt{n/2}\right) \\ &< x\sqrt{\frac{2n+1}{4}};\end{aligned}$$

i.e.,

$$\left\{\omega : |S_{2n}(\omega) - n| < (x-h)\sqrt{\frac{2n}{4}}\right\} \subset \left\{\omega : \left|S_{2n+1}(\omega) - \frac{2n+1}{2}\right| < x\sqrt{\frac{2n+1}{4}}\right\}.$$

Similarly,

$$\left\{\omega : \left|S_{2n+1}(\omega) - \frac{2n+1}{2}\right| < x\sqrt{\frac{2n+1}{4}}\right\} \subset \left\{\omega : |S_{2(n+1)}(\omega) - (n+1)| < (x+h)\sqrt{\frac{2(n+1)}{4}}\right\}.$$

It follows from these relations and (*ii*) and (*iii*) above that for $n \geq N$,

$$\int_{-x}^{x} \phi(t)\,dt - \varepsilon < P\left(|S_{2n} - n| < (x-h)\sqrt{\frac{n}{2}}\right)$$

$$\leq P\left(\left|S_{2n+1} - \frac{2n+1}{2}\right| < x\sqrt{\frac{2n+1}{4}}\right)$$

$$\leq P\left(|S_{2(n+1)} - (n+1)| < (x+h)\sqrt{\frac{n+1}{2}}\right)$$

$$< \int_{-x}^{x} \phi(t)\,dt + \varepsilon.$$

This shows that

$$\lim_{n\to\infty} P\left(|S_{2n+1} - \frac{2n+1}{2}| < x\sqrt{\frac{2n+1}{4}}\right) = \int_{-x}^{x} \phi(t)\,dt. \quad \blacksquare$$

The original central limit theorem has a tendency to underestimate the binomial probabilities, as can be seen from Table 7.1 in the $n = 36$, $p = 1/2$ case.

The central limit theorem was improved upon by Laplace (1749–1827). Let $\{X_j\}_{j=1}^{n}$ be a sequence of n Bernoulli random variables with probability of success p and let $S_n = \sum_{j=1}^{n} X_j$. The following result gives a better approximation of binomial probabilities. The proof belongs in a more advanced text.

$$P(\alpha \leq S_n \leq \beta) \approx \Phi\left(x_\beta + \frac{h}{2}\right) - \Phi\left(x_\alpha - \frac{h}{2}\right) \tag{7.3}$$

where $h = 1/\sqrt{npq}$ and $x_t = (t - np)h$.

EXAMPLE 7.14 Suppose $n = 36$ and $p = 1/2$. According to Table 7.1, $P(13 \leq S_n \leq 23) = .9347$. If we use Equation 7.3 to approximate this probability, then $h = 1/3$, $x_{13} = -5/3$, $x_{23} = 5/3$, and

$$P(13 \leq S_n \leq 23) \approx \Phi\left(\frac{5}{3} + \frac{1}{6}\right) - \Phi\left(-\frac{5}{3} - \frac{1}{6}\right) \approx .9332,$$

Number of Successes	Probability	Normal Approximation
$17 \leq S_n \leq 19$	.3833	.2611
$16 \leq S_n \leq 20$	.5950	.4950
$15 \leq S_n \leq 21$	.7570	.6827
$14 \leq S_n \leq 22$	.8675	.8176
$13 \leq S_n \leq 23$	.9347	.9044
$12 \leq S_n \leq 24$	.9711	.9545
$11 \leq S_n \leq 25$	.9887	.9804
$10 \leq S_n \leq 26$	.9960	.9923
$9 \leq S_n \leq 27$	.9988	.9973

TABLE 7.1 Normal Approximation of Binomial Probabilities

a much better approximation than that given in Table 7.1. ■

The following theorem is a weakened version of the last approximation but has the advantage of being easier to apply in some situations.

Theorem 7.4.4 (DeMoivre-Laplace Limit Theorem) *Let $\{X_j\}_{j=1}^n$ be a sequence of n Bernoulli random variables with probability of success p and let $S_n = \sum_{j=1}^n X_j$. Then for fixed $a < b$,*

$$P(a \le S_n^* \le b) \approx \Phi(b) - \Phi(a). \tag{7.4}$$

EXAMPLE 7.15 A survey is undertaken to determine how many voters in a population of eligible voters favor candidate A. Assume that the unknown proportion of voters who favor A is p and that voters act independently of one another. Suppose we want to determine how many should be polled so that the observed proportion of favorable voters is within .05 of p with probability at least .95. We can look upon the polling as a sequence of Bernoulli trials $\{X_j\}_{j=1}^n$ with unknown probability of success p. The observed proportion of favorable voters will then be S_n/n, and we want to choose n so that

$$P\left(\left|\frac{S_n}{n} - p\right| \le .05\right) \ge .95.$$

We could proceed as in Section 4.3 by using Inequality 4.6 to require that

$$P\left(\left|\frac{S_n}{n} - p\right| \ge .05\right) \le \frac{1}{4n(.05)^2} \le .05;$$

i.e., that $n \ge 2000$. Since Inequality 4.6 assumes virtually nothing about the density of S_n/n, a better result might be obtained by invoking Theorem 7.4.4. Note that

$$\begin{aligned} P\left(\left|\frac{S_n}{n} - p\right| \le .05\right) &= P\left(\left|\frac{S_n - np}{\sqrt{npq}}\right| \le (.05)\sqrt{n/pq}\right) \\ &= P(|S_n^*| \le (.05)\sqrt{n/pq}). \end{aligned}$$

By the Standard Normal Distribution Function table (see page 346), the approximate solution of the equation $\Phi(x) - \Phi(-x) = 2\Phi(x) - 1 = .95$ is $x = 1.96$. Since $P(|S_n^*| \le x) \approx \Phi(x) - \Phi(-x) = .95$, we should choose n so that

$$(.05)\sqrt{n/pq} \ge 1.96,$$

in which case we would have $P(|S_n^*| \le (.05)\sqrt{n/pq}) \ge .95$. This requires that

$$n \ge \left(\frac{1.96}{.05}\right)^2 pq.$$

Since $pq = p(1-p) \leq 1/4$, if we choose n so that

$$n \geq \left(\frac{1.96}{.05}\right)^2 \frac{1}{4} \geq \left(\frac{1.96}{.05}\right)^2 pq,$$

we would then have $P(|(S_n/n) - p| \leq .05) \geq .95$. We therefore take n to be the smallest integer for which

$$n \geq \left(\frac{1.96}{.05}\right)^2 \frac{1}{4} = 384.16;$$

i.e., we take $n = 385$. By polling 385 eligible voters and using S_n/n to estimate the unknown p, we know that our estimate will be within .05 of p 19 times out of 20. ■

The central limit theorem is valid in much more general situations than those dealt with here. For example, if $\{X_j\}$ is a sequence of independent random variables having the same distribution function with $\mu = E[X_1], \sigma^2 = \text{var } X_1 < +\infty$, and $S_n = \sum_{j=1}^{n} X_j$, then

$$\lim_{n\to\infty} P\left(a \leq \frac{S_n - n\mu}{\sigma\sqrt{n}} \leq b\right) = \Phi(b) - \Phi(a). \tag{7.5}$$

EXERCISES 7.4

Equations 7.4 and 7.5 were used to obtain answers to the following problems.

1. Approximate $\sum_{j=26}^{36} \binom{64}{j} 2^{-64}$.
2. Approximate $\sum_{j=30}^{39} \binom{128}{j} (1/4)^j (3/4)^{128-j}$.
3. A jumbo jet with a seating capacity of 360 passengers is allowed a maximum weight of 59,000 pounds for passengers. If the average weight of a passenger is 160 pounds with a standard deviation of $\sigma = 48$ pounds, what is the approximate probability that the weight limit will be exceeded, assuming the 360 passengers that board are a random sample from the population?
4. A national polling agency would like to determine the percentage of eligible voters who favor their client within 3 percentage points with 90 percent confidence. How many eligible voters should be polled?
5. Consider a particle taking a random walk on the integers starting at 0 with $p = 1/2$. What is the approximate probability that the particle will be within 30 units of 0 after 1000 steps?
6. Consider a particle taking a random walk on the integers starting at 0 with $p = .45$. What is the approximate probability that the particle will be to the right of -50 after 1000 steps?

7. The n real numbers $a_1, \ldots, a_n$ are rounded off to the nearest integers $a_1 + X_1, \ldots, a_n + X_n$, respectively, where the round-off errors $X_1, \ldots, X_n$ are assumed to be independent and have a uniform density on $[-1/2, 1/2]$. Use the central limit theorem to find a number $\lambda > 0$, depending upon n, such that $P(|\sum_{j=1}^{n} X_j| < \lambda) \approx .99$. (See the final paragraph of this section.)
8. A programmer decides to carry m significant figures to the right of the decimal point and round off the result of any addition, multiplication, or division operation to that many figures. Assume that 10^6 elementary operations are performed, that successive round-off errors are independent and have a uniform density on $[-(1/2)10^{-m}, (1/2)10^{-m}]$, and that the final error is the sum of all the round-off errors. Find an upper bound, which does not depend upon m, for the probability that the final error will be less than $5 \times 10^{-m+2}$ in absolute value.

7.5 COVARIANCE AND COVARIANCE FUNCTIONS

The covariance cov (X, Y) between two random variables with finite second moments was defined in Section 4.4. All the definitions, lemmas, and theorems of that section are valid for any random variables—discrete, continuous, or a mixture of the two. We can and will use the properties of cov (X, Y) described in Section 4.4.

Assuming that the random variables X, Y have finite second moments with $\sigma_X > 0$ and $\sigma_Y > 0$, the correlation between X and Y is defined just as in the discrete case by the equation

$$\rho(X, Y) = \frac{\text{cov}(X, Y)}{\sigma_X \sigma_Y} = \frac{E[(X - \mu_X)(Y - \mu_Y)]}{\sqrt{\text{var } X}\sqrt{\text{var } Y}}.$$

As pointed out above, Inequality 4.8, Schwarz's inequality, holds for any random variables with finite second moment.

Theorem 7.5.1 (Schwarz's Inequality) *If X and Y are any random variables with finite second moments, then*

$$(E[XY])^2 \leq E[X^2]E[Y^2] \tag{7.6}$$

with equality holding if and only if $P(X = 0) = 1$ or $P(Y = aX) = 1$ for some constant a.

As in the discrete case, $|\rho(X, Y)| \leq 1$ with equality if and only if there are constants $a, b \in R$ such that $P(Y = aX + b) = 1$.

If $\vec{x} = (x_1, \ldots, x_n)$ is a point in R^n, the length of the vector $\vec{x}$ is the quantity $(\sum_{i=1}^n x_i^2)^{1/2}$. By analogy, if $\{x_1, \ldots, x_n\}$ is the range of a random variable X, we could define the length of X, written $\|X\|$, by the equation

$$\|X\| = \left(\sum_{i=1}^{n} x_i^2 f_X(x_i)\right)^{1/2} = \sqrt{E[X^2]}.$$

Since the quantity on the right makes sense for any random variable with finite second moments, we can extend this notion as follows.

Definition 7.4 *If X is any random variable with finite second moment, define $\| X \| = \sqrt{E[X^2]}$; $\| X \|$ is called the* norm *of X.* ■

What is to be made of the equation $\|X\| = \sqrt{E[X^2]} = 0$? Putting $Y = 1$ in Inequality 7.6,

$$(E[X])^2 \leq E[X^2]$$

and $E[X] = 0$. According to the discussion following Theorem 7.3.4, $\text{var } X = 0$, and therefore $P(X = 0) = 1$. Note that this does not mean that $X(\omega) = 0$ for every $\omega \in \Omega$. If we have two random variables X and Y with $\|X - Y\| = \sqrt{E[(X - Y)^2]} = 0$, we can conclude only that $P(X = Y) = 1$. Two random variables X and Y with this property are said to be equal in the probability sense and will be regarded as the same in this section.

As in vector calculus, once we have a concept of length, we can go on to distances.

Definition 7.5 *If X and Y are random variables with finite second moments, the* mean square distance *between X and Y is the quantity*

$$\|X - Y\| = \sqrt{E[(X - Y)^2]}. \blacksquare$$

The following inequality is the analog of the geometrical fact that the length of a side of a triangle is less than or equal to the sum of the lengths of the other two sides.

Lemma 7.5.2 (Triangle Inequality) *If X and Y are any random variables with finite second moments, then*

$$\|X + Y\| \leq \|X\| + \|Y\|. \tag{7.7}$$

PROOF: Since $\|X + Y\|^2 = E[(X + Y)^2] = E[X^2] + 2E[XY] + E[Y^2]$ and

$E[XY] \leq \sqrt{E[X^2]}\sqrt{E[Y^2]}$ by Schwarz's inequality,

$$\begin{aligned}\|X+Y\|^2 &\leq E[X^2] + 2\sqrt{E[X^2]}\sqrt{E[Y^2]} + E[Y^2] \\ &= \left(\sqrt{E[X^2]} + \sqrt{E[Y^2]}\right)^2 \\ &= (\|X\| + \|Y\|)^2,\end{aligned}$$

and therefore $\|X+Y\| \leq \|X\| + \|Y\|$. ■

Replacing X by $X - Z$ and Y by $Z - Y$ in 7.7, we obtain the inequality

$$\|X - Y\| \leq \|X - Z\| + \|Z - Y\|, \tag{7.8}$$

which is also referred to as the *triangle inequality.* Another inequality can be obtained from Inequality 7.7 by replacing X by $X - Y$ to obtain $\|X\| \leq \|X - Y\| + \|Y\|$ or $\|X - Y\| \geq \|X\| - \|Y\|$. Interchanging X and Y in the latter inequality and using the fact that $\|X - Y\| = \|Y - X\|$, $\|X - Y\| \geq \|Y\| - \|X\|$. We thus obtain another version of the triangle inequality:

$$\|X - Y\| \geq \big|\|X\| - \|Y\|\big|. \tag{7.9}$$

With the above definitions in mind, we can now discuss convergence of sequences of random variables.

Definition 7.6 *Let $X, X_1, X_2, \ldots$ be random variables having finite second moments. The sequence $\{X_n\}_{n=1}^{\infty}$* converges in mean square *to X, written* ms-$\lim_{n\to\infty} X_n = X$, *if*

$$\lim_{n\to\infty} \|X_n - X\| = 0$$

or, equivalently,

$$\lim_{n\to\infty} E[(X_n - X)^2] = 0. \quad ■$$

We can use Inequality 7.8 to show that the mean square limit of a sequence $\{X_n\}$ is unique in the probability sense if it exists. Suppose

$$\lim_{n\to\infty} \|X_n - X\| = 0 \quad \text{and} \quad \lim_{n\to\infty} \|X_n - X'\| = 0.$$

Since

$$0 \leq \|X - X'\| \leq \|X - X_n\| + \|X_n - X'\| \to 0$$

as $n \to \infty$, $\|X - X'\| = 0$, and therefore $X = X'$ with probability 1.

Now that we have a means for taking limits, we can deal with infinite series. If $\{X_j\}_{j=1}^{\infty}$ is an infinite sequence of random variables having finite second moments, we can form the infinite series $\sum_{j=1}^{\infty} X_j$. Letting $S_n = \sum_{j=1}^{n} X_j, n \geq 1$, if there is a random variable S with finite second moment such that $S = \text{ms-lim}_{n\to\infty} S_n$, we write $S = \text{ms-lim}_{n\to\infty} \sum_{j=1}^{n} X_j = \text{ms-}\sum_{j=1}^{\infty} X_j$ and say that the series $\sum_{j=1}^{\infty} X_j$ converges in the mean square sense.

Convergence in the mean also implies convergence of means.

Lemma 7.5.3 *Let $X, X_1, X_2, \ldots$ and $Y, Y_1, Y_2 \ldots$ be random variables with finite second moments. If* $\text{ms-lim}_{n\to\infty} X_n = X$ *and* $\text{ms-lim}_{n\to\infty} Y_n = Y$, *then*

(i) $\lim_{n\to\infty} E[X_n] = E[X]$.

(ii) $\lim_{n\to\infty} E[X_n Y] = E[XY]$.

(iii) $\lim_{n\to\infty} E[X_n^2] = E[X^2]$.

(iv) $\lim_{n\to\infty} E[X_n Y_n] = E[XY]$.

PROOF: Replacing X and Y in Inequality 7.6 by 1 and $|X_n - X|$, respectively, by Theorem 7.3.2:

$$|E[X_n] - E[X]|^2 \leq E[|X_n - X|]^2 \leq E[(X_n - X)^2] \to 0$$

as $n \to \infty$. Thus, $\lim_{n\to\infty} E[X_n] = E[X]$ and (i) is proved. By Schwarz's inequality, $0 \leq |E[X_n Y] - E[XY]|^2 = |E[(X_n - X)Y]|^2 \leq E[(X_n - X)^2]E[Y^2] \to 0$ as $n \to \infty$, and $\lim_{n\to\infty} E[X_n Y] = E[XY]$, so that (ii) is true. Part (iii) is the same as the statement that $\lim_{n\to\infty} \|X_n\|^2 = \|X\|^2$. If we can show that $\lim_{n\to\infty} \|X_n\| = \|X\|$, then (iii) would be proved by continuity of the function $f(x) = x^2$. By Inequality 7.9, $|\|X_n\| - \|X\|| \leq \|X_n - X\| \to 0$ as $n \to \infty$ and (iii) is true. To prove (iv), by Schwarz's inequality:

$$\begin{aligned}
&|E[X_n Y_n] - E[XY]| \\
&= |E[(X_n - X)Y_n] + E[X(Y_n - Y)]| \\
&\leq E[|(X_n - X)Y_n|] + E[|X(Y_n - Y)|] \\
&\leq \sqrt{E[(X_n - X)^2]}\sqrt{E[Y_n^2]} + \sqrt{E[X^2]}\sqrt{E[(Y_n - Y)^2]}.
\end{aligned}$$

Since $\lim_{n\to\infty} E[(X_n - X)^2] = 0, \lim_{n\to\infty} E[(Y_n - Y)^2] = 0$ by hypothesis, and $\lim_{n\to\infty} E[Y_n^2] = E[Y^2]$ by (iii), $\lim_{n\to\infty} E[X_n Y_n] = E[XY]$. ■

In the remainder of this section, we will consider a family of random variables $\{X_t : t \in T\}$ having finite second moments where T is the set of all real numbers R or the set of integers Z, assuming that such a family exists.

Definition 7.7 *The family or process $\{X_t : t \in T\}$ is* weakly stationary *if*

1. $E[X_s] = E[X_t]$ for all $s, t \in T$.

2. $\operatorname{cov}(X_s, X_t) = \operatorname{cov}(X_{s+h}, X_{t+h})$ for all $h, s, t \in T$. ■

The function

$$r(h) = \operatorname{cov}(X_t, X_{t+h}) = \operatorname{cov}(X_0, X_h), \qquad h \in T,$$

is called the *covariance function* of the process. To avoid trivialities, we assume that $r(0) = \operatorname{cov}(X_0, X_0) = \operatorname{var} X_0 = \sigma^2 > 0$. Note that for $h > 0$, $r(-h) = \operatorname{cov}(X_t, X_{t-h}) = \operatorname{cov}(X_{t-h}, X_t) = r(h)$, and therefore

$$r(h) = r(-h) = r(|h|), \qquad h \in T.$$

The normalized covariance function

$$\rho(h) = \frac{r(h)}{r(0)}, \qquad h \in T,$$

is called the *correlation function* of the $\{X_t : t \in T\}$ process.

EXAMPLE 7.16 Let $\{X_j\}_{j=-\infty}^{\infty}$ be a sequence of independent random variables having the same distribution function, finite second moments, and $\sigma^2 = \operatorname{var} X_0$. Then for any $\nu \in Z$,

$$r(\nu) = E[(X_j - E[X_j])(X_{j+\nu} - E[X_{j+\nu}])] = \begin{cases} \sigma^2 & \text{if } \nu = 0 \\ 0 & \text{if } \nu \neq 0. \end{cases}$$ ■

EXAMPLE 7.17 Let U and V be uncorrelated random variables (i.e., $\rho(U, V) = 0$) with zero means and unit variances. For $\lambda \in R$, let

$$X_t = U \cos \lambda t + V \sin \lambda t, \qquad t \in R.$$

Since $E[X_t] = E[U] \cos \lambda t + E[V] \sin \lambda t = 0$, the covariance function is given by

$$\begin{aligned} r(h) &= \operatorname{cov}(X_t, X_{t+h}) = E[X_t X_{t+h}] \\ &= E[(U \cos \lambda t + V \sin \lambda t)(U \cos \lambda(t+h) + V \sin \lambda(t+h))] \\ &= \cos \lambda t \cos \lambda(t+h) E[U^2] + \cos \lambda t \sin \lambda(t+h) E[UV] \\ &+ \sin \lambda t \cos \lambda(t+h) E[UV] + \sin \lambda t \sin \lambda(t+h) E[V^2]. \end{aligned}$$

Since $E[U^2] = 1, E[V^2] = 1$, and $E[UV] = 0$,

$$r(h) = \cos \lambda t \cos \lambda(t+h) + \sin \lambda t \sin \lambda(t+h) = \cos \lambda h.$$

It follows that the process $\{X_t : t \in T\}$ is weakly stationary. Since we can write

$$c_1 \cos x + c_2 \sin x = \sqrt{c_1^2 + c_2^2} \cos(x + \theta)$$

where $\theta = \arctan(c_1/c_2)$, X_t can be written

$$X_t = \sqrt{U^2 + V^2} \cos(\lambda t + \Theta)$$

where $\Theta = \arctan(U/V)$, a random variable. Thus, X_t is a periodic function with random amplitude $\sqrt{U^2 + V^2}$, random phase shift Θ, and fixed frequency $\lambda/2\pi$. ■

We now consider an example that can serve as a model for the sound produced by n different tuning forks that are struck at random times.

EXAMPLE 7.18 Let $U_0, \ldots, U_n, V_0, \ldots V_n$ be uncorrelated random variables with zero means. Assume that the U_i and V_i have common variances $\sigma_i^2, i = 0, \ldots, n$, and let $\sigma^2 = \sigma_0^2 + \cdots + \sigma_n^2$. Also let $\lambda_0 \ldots, \lambda_n$ be distinct real numbers. Set

$$X_t = \sum_{j=0}^{n} (U_j \cos \lambda_j t + V_j \sin \lambda_j t), \qquad t \in R.$$

Note that $E[U_i U_j] = E[V_i V_j] = E[U_i V_j] = 0$ whenever $i \neq j$ and $E[U_i^2] = E[V_i^2] = \sigma_i^2$. Clearly,

$$E[X_t] = \sum_{j=0}^{n} (E[U_j] \cos \lambda_j t + E[V_j] \sin \lambda_j t) = 0$$

and

$$\begin{aligned} E[X_t X_{t+h}] &= E\left[\left\{\sum_{j=0}^{n} (U_j \cos \lambda_j t + V_j \sin \lambda_j t)\right\}\right. \\ &\quad \times \left.\left\{\sum_{k=0}^{n} (U_k \cos \lambda_k (t+h) + V_k \sin \lambda_k (t+h))\right\}\right] \\ &= \sum_{j=0}^{n} \left\{ E[U_j^2] \cos \lambda_j t \cos \lambda_j (t+h) \right. \\ &\quad \left. + E[V_j^2] \sin \lambda_j t \sin \lambda_j (t+h) \right\} \\ &= \sum_{j=0}^{n} \sigma_j^2 \cos \lambda_j h. \end{aligned}$$

The process $\{X_t : t \in R\}$ is therefore weakly stationary with covariance function

$$r(h) = \sum_{j=0}^{n} \sigma_j^2 \cos \lambda_j h.$$

Since $r(0) = \sum_{j=0}^{n} \sigma_j^2 = \sigma^2$, the correlation function is given by

$$\rho(h) = \sum_{j=0}^{n} \frac{\sigma_j^2}{\sigma^2} \cos \lambda_j h, \qquad h \in R. \tag{7.10}$$

As in Example 7.17, X_t can be thought of as a mixture of $n+1$ sound waves with random amplitudes $\sqrt{U_j^2 + V_j^2}$, random phase shifts $\Theta_j = \arctan(U_j/V_j)$, and fixed frequencies $\lambda_j/2\pi, j = 0, \ldots, n$. ■

Equation 7.10 suggests that the correlation function $\rho(h)$ of a weakly stationary process $\{X_t : t \in T\}$ has a representation

$$\rho(h) = \int_{-\infty}^{\infty} \cos \lambda h \, dF(\lambda), \qquad h \in R, \tag{7.11}$$

where F is a distribution function on R. Without going into details, such a function exists and is called the *spectral distribution function.* In Example 7.18, the function F increases only at the points λ_j by jumps of $\sigma_j^2/\sigma^2, j = 0, \ldots, n$.

For a weakly stationary process of the type $\{X_j\}_{j=-\infty}^{\infty}$, the correlation function $\rho(\nu)$ is defined only for $\nu \in Z$, and in this case the integrand $\cos \lambda\nu$ in Equation 7.11 is a periodic function of λ of period 2π since $\cos(\lambda + 2n\pi)\nu = \cos \lambda\nu$. In this case, Equation 7.11 can be written

$$\rho(\nu) = \sum_{j=-\infty}^{\infty} \int_{(2j-1)\pi}^{(2j+1)\pi} \cos \lambda\nu \, dF(\lambda).$$

It can be shown, but will not be done here, that the latter equation can be written

$$\rho(\nu) = \int_{(-\pi,\pi]} \cos \lambda\nu \, d\hat{F}(\lambda), \qquad \nu \in Z,$$

where $\hat{F}$ is a distribution function with $\hat{F}(-\pi) = 0$ and $\hat{F}(\pi) = 1$. In the particular case that $\sum_{\nu=-\infty}^{\infty} |\rho(\nu)| < \infty$, the spectral distribution function $\hat{F}$ has a density function f, called the *spectral density function,* so that

$$\rho(\nu) = \int_{-\pi}^{\pi} f(\lambda) \cos \lambda\nu \, d\lambda;$$

moreover,

$$f(\lambda) = \frac{1}{2\pi} + \frac{1}{\pi}\sum_{\nu=1}^{\infty} \rho(\nu)\cos\lambda\nu, \qquad -\pi < \lambda < \pi. \quad (7.12)$$

It is sometimes useful to smooth data by using a moving average, as in the next example.

EXAMPLE 7.19 Consider a sequence $\{X_j\}_{j=-\infty}^{\infty}$ of uncorrelated random variables having the same means and variances σ^2. Fix $m \geq 1$ and for each integer n, let

$$Y_n = a_1X_n + a_2X_{n-1} + \cdots + a_mX_{n-m+1} = \sum_{j=1}^{m} a_jX_{n-j+1}$$

where $a_1, \ldots, a_m$ are constants. Put $a_j = 0$ if $j \notin \{1, 2, \ldots, m\}$. The sequence $\{Y_j\}_{j=-\infty}^{\infty}$ is called a *moving average process.* Note that

$$E[Y_n] = \mu(a_1 + \cdots + a_m) \quad \text{and} \quad \operatorname{var} Y_n = \sigma^2(a_1^2 + \cdots + a_m^2).$$

To show that the Y_n process is weakly stationary, fix $\nu \geq 0$ and consider

$$\begin{aligned} &E\left[\left(Y_n - \mu\sum_{i=1}^{m} a_i\right)\left(Y_{n+\nu} - \mu\sum_{i=1}^{m} a_i\right)\right] \\ &= E\left[\left(\sum_{i=1}^{m} a_i(X_{n-i+1} - \mu)\right)\left(\sum_{j=1}^{m} a_j(X_{n+\nu-j+1} - \mu)\right)\right] \\ &= \sum_{i=1}^{m}\sum_{j=1}^{m} a_ia_jE[(X_{n-i+1} - \mu)(X_{n+\nu-j+1} - \mu)]. \end{aligned}$$

Since the terms of this sum are nonzero only when the subscripts of the X_j's are equal, the sum is equal to

$$\sigma^2\sum_{j=1}^{m} a_{j-\nu}a_j.$$

Since this quantity does not depend upon n, the $\{Y_j\}_{j=-\infty}^{\infty}$ sequence is weakly stationary with covariance function

$$r(\nu) = \sigma^2\sum_{j=1}^{m} a_{j-\nu}a_j.$$

Since $j - \nu \geq 1$ is required for the first factor to be nonzero, we must have $\nu \leq j - 1 \leq m - 1$, and therefore

$$\begin{aligned} r(\nu) &= \sigma^2 \sum_{j=\nu+1}^{m} a_{j-\nu} a_j \\ &= \sigma^2 (a_{m-\nu} a_m + a_{m-1-\nu} a_{m-1} + \cdots + a_1 a_{\nu+1}). \end{aligned}$$

Therefore,

$$r(\nu) = \begin{cases} \sigma^2 (a_{m-\nu} a_m + a_{m-1-\nu} a_{m-1} + \cdots + a_1 a_{\nu+1}) & \text{if } \nu \leq m - 1 \\ 0 & \text{if } \nu \geq m. \end{cases}$$

In particular, if $a_j = 1/\sqrt{m}, j = 1, \ldots, m$, then for $\nu \geq 0$,

$$r(\nu) = \begin{cases} \sigma^2 (1 - (\nu/m)) & \text{if } 0 \leq \nu \leq m - 1 \\ 0 & \text{if } \nu \geq m. \end{cases}$$

Since $r(-\nu) = r(\nu) = r(|\nu|)$,

$$r(\nu) = \begin{cases} \sigma^2 (1 - (|\nu|/m)) & \text{if } |\nu| \leq m - 1 \\ 0 & \text{if } |\nu| \geq m. \end{cases} \quad \blacksquare$$

Let $\{X_j\}_{j=-\infty}^{\infty}$ be a weakly stationary process. Suppose there is a real number λ with $|\lambda| < 1$ such that

$$X_n = \lambda X_{n-1} + N_n, \qquad n \in Z,$$

where the N_j's, representing a "noise" component, are uncorrelated with zero means and variance σ^2. By iteration,

$$\begin{aligned} X_n &= \lambda(\lambda X_{n-2} + N_{n-1}) + N_n \\ &= \lambda^2 X_{n-2} + \lambda N_{n-1} + N_n \\ &\;\;\vdots \\ &= \lambda^j X_{n-j} + \sum_{i=0}^{j-1} \lambda^i N_{n-i}. \end{aligned} \tag{7.13}$$

Thus,

$$E\left[\left(X_n - \sum_{i=0}^{j-1} \lambda^i N_{n-i}\right)^2\right] = E[(\lambda^j X_{n-j})^2] = \lambda^{2j} E[X_{n-j}^2].$$

Since $E[X_{n-j}^2] = \text{var } X_{n-j} + (E[X_{n-j}])^2$ and the $\{X_j\}_{j=-\infty}^{\infty}$ process is weakly stationary, the quantities on the right are independent of $n - j$, and so

$$E\left[\left(X_n - \sum_{i=0}^{j-1} \lambda^i N_{n-i}\right)^2\right] = c\lambda^{2j}$$

for some constant $c \geq 0$. Since $|\lambda| < 1$, $\lim_{j\to\infty} \lambda^{2j} = 0$, and so

$$X_n = \text{ms-}\lim_{j\to\infty} \sum_{i=0}^{j-1} \lambda^i N_{n-i} = \text{ms-}\sum_{i=0}^{\infty} \lambda^i N_{n-i}.$$

The significance of this equation lies in the fact that the $\{X_j\}$ process has a "representation" in terms of a sequence of random variables that are much easier to analyze. This is apparent in the following calculation of the covariance function of the $\{X_j\}_{j=-\infty}^{\infty}$ process. By Lemma 7.5.3,

$$\begin{aligned}
E[X_n] &= E\left[\text{ms-}\sum_{i=0}^{\infty} \lambda^i N_{n-i}\right] \\
&= \lim_{k\to\infty} E\left[\sum_{i=0}^{k} \lambda^i N_{n-i}\right] = \lim_{k\to\infty} \sum_{i=0}^{k} \lambda^i E[N_{n-i}] = 0
\end{aligned}$$

and

$$\begin{aligned}
E[X_n^2] &= E\left[\left(\text{ms-}\sum_{i=0}^{\infty} \lambda^i N_{n-i}\right)^2\right] \\
&= \lim_{k\to\infty} E\left[\left(\sum_{i=0}^{k} \lambda^i N_{n-i}\right)\left(\sum_{j=0}^{k} \lambda^j N_{n-j}\right)\right] \\
&= \lim_{k\to\infty} E\left[\sum_{i=0}^{k} \lambda^{2i} N_{n-i}^2\right] + \lim_{k\to\infty} E\left[\sum_{i\neq j} \lambda^{i+j} N_{n-i} N_{n-j}\right] \\
&= \lim_{k\to\infty} \sum_{i=0}^{k} \lambda^{2i} E[N_{n-i}^2] + \lim_{k\to\infty} \sum_{i\neq j} \lambda^{i+j} E[N_{n-i} N_{n-j}] \\
&= \sigma^2 \sum_{i=0}^{\infty} \lambda^{2i} = \frac{\sigma^2}{1-\lambda^2},
\end{aligned}$$

since $E[N_{n-i}^2] = \sigma^2$ and $E[N_{n-i}N_{n-j}] = 0$ when $i \neq j$. To calculate $\text{cov}(X_n, X_{n+k})$, note that $\text{cov}(X_n, X_{n+k}) = E[X_n X_{n+k}]$, since $E[X_n] = 0$ for all $n \in Z$. By Equation 7.13, for $k \geq 1$,

$$X_{n+k} = \lambda^k X_n + \sum_{i=0}^{k-1} \lambda^i N_{n+k-i},$$

so that

$$\begin{aligned} E[X_n X_{n+k}] &= \lambda^k E[X_n^2] + E\left[X_n \left(\sum_{i=0}^{k-1} \lambda^i N_{n+k-i}\right)\right] \\ &= \lambda^k E[X_n^2] + \lim_{l \to \infty} E\left[\left(\sum_{j=0}^{l} \lambda^j N_{n-j}\right)\left(\sum_{i=0}^{k-1} \lambda^i N_{n+k-i}\right)\right] \\ &= \lambda^k E[X_n^2] + \lim_{l \to \infty} \sum_{j=0}^{l} \sum_{i=0}^{k-1} \lambda^{j+i} E[N_{n-j} N_{n+k-i}]. \end{aligned}$$

Since $n+k-i \geq n+1$ for $i = 0, \ldots, k-1$ and $n-j \leq n$ for $j = 0, \ldots, l$, all of the terms in the double sum are zero. Thus,

$$E[X_n X_{n+k}] = \lambda^k \frac{\sigma^2}{1-\lambda^2}.$$

The covariance function of the $\{X_j\}_{j=-\infty}^{\infty}$ process is

$$r(k) = \frac{\sigma^2}{1-\lambda^2} \lambda^{|k|}, \qquad k \in Z,$$

and the correlation function is given by

$$\rho(k) = \lambda^{|k|}, \qquad k \in Z.$$

EXERCISES 7.5

1. Let $a_1, \ldots, a_m, b_1, \ldots, b_m$ be positive constants, let $Z_1, \ldots, Z_m$ be independent random variables having a uniform density on $[0, 2\pi]$, and let

$$X_n = \sum_{j=1}^{m} a_j \cos(n b_j + Z_j).$$

Show that the sequence $\{X_j\}_{j=-\infty}^{\infty}$ is weakly stationary and determine its correlation function.

2. Let $\{X_j\}_{j=-\infty}^{\infty}$ be a weakly stationary process, let $a_1, \ldots, a_m$ be constants, and let

$$Y_n = \sum_{j=1}^{m} a_j X_{n-j+1} = a_1 X_n + a_2 X_{n-1} + \cdots + a_m X_{n-m+1}.$$

Show that the process $\{Y_j\}_{j=-\infty}^{\infty}$ is weakly stationary and determine its correlation function in terms of the correlation function of the X_j process.

3. Consider a sequence of random variables $\{X_j\}_{j=-\infty}^{\infty}$ defined as the moving average

$$X_j = N_j + \alpha N_{j-1}, \qquad -\infty < j < \infty,$$

where $\{N_j\}_{j=-\infty}^{\infty}$ is a weakly stationary sequence of uncorrelated random variables having unit variances. Find the spectral density function of the X_j process.

4. Consider a sequence of random variables $\{X_j\}_{j=-\infty}^{\infty}$ defined as the moving average

$$X_j = N_j + \alpha N_{j-1} + \beta N_{j-2}, \qquad -\infty < j < \infty,$$

where $\{N_j\}_{j=-\infty}^{\infty}$ is a weakly stationary sequence having unit variances. Find the spectral density of the X_j process.

5. Let Λ and Θ be independent random variables where Λ takes on the values $\lambda_1, \ldots, \lambda_m$ with probabilities $p_1, \ldots, p_m$ and Θ has a uniform density on $[0, 2\pi]$, and let

$$X_t = \cos(\Lambda t + \Theta), \qquad t \in R.$$

Show that the process $\{X_t, t \in R\}$ is weakly stationary and determine its correlation function.

SUPPLEMENTAL READING LIST

1. T. M. Apostol (1957). *Mathematical Analysis.* Reading, Mass.: Addison-Wesley.
2. P. J. Flory (1969). *Statistical Mechanics of Chain Molecules.* New York: Wiley/Interscience.

CHAPTER 8

CONTINUOUS PARAMETER MARKOV PROCESSES

8.1 INTRODUCTION

In this chapter, we will pursue a path that is less dependent upon the structure of the probability space and more dependent upon macro properties of non-random functions.

In the next section, starting with a few heuristic principles governing the probabilistic behavior of an evolving system in small time intervals, a system of differential equations is derived and solved, resulting in time dependent probability functions that describe a process known as the Poisson process. This process plays an important role in waiting time models.

The Poisson process is a special case of a more general class of processes that can be described by time dependent probability functions, called continuous parameter Markov chains. Starting with a set of equations that the probability functions must satisfy, it is shown that the functions satisfy a system of differential equations. Using matrix calculus, a method is developed for solving such systems.

8.2 POISSON PROCESS

Consider an experimental situation in which events occur at random times. For example, calls to a mainframe computer may arrive at random times $0 \le t_1 \le t_2 \le \cdots$. If a counter is initiated at time 0, it will increase to 1 at time t_1, increase to 2 at time t_2, and so forth. The outcome in this case is

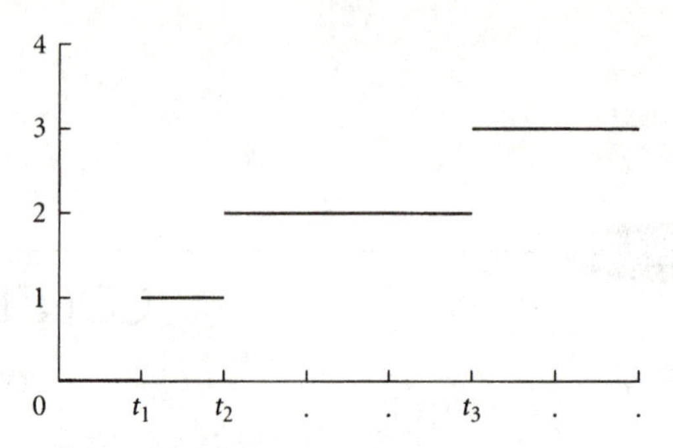

FIGURE 8.1 Counter outcome.

a function $\omega(t)$ on $[0, \infty)$ with graph as depicted in Figure 8.1. A probability space for this type of experimental situation would consist of all such ω. The construction of such a probability space is better left to more advanced texts. A different approach, which avoids such constructions, will be followed here. This approach entails the derivation of some equations based on heuristic arguments.

In the following discussion, $o(h)$ (read "little o of h") will be a generic symbol for a function of h that satisfies the condition

$$\lim_{h \to 0} \frac{o(h)}{h} = 0.$$

We will assume the following properties of the counter process just described. Independently of the number of occurrences of events in the interval $(0, t)$, for small $h > 0$:

(i) The probability that an event will occur in the interval $(t, t + h)$ is $\lambda h + o(h)$,

(ii) The probability that no event will occur in the interval $(t, t + h)$ is $1 - \lambda h + o(h)$, and

(iii) The probability of two or more events occurring in the interval $(t, t + h)$ is $o(h)$,

where λ is a positive constant. These might be reasonable assumptions for the situation described above for periods when saturation is unlikely.

Assuming there is an appropriate probability space for which these assumptions are valid, let $P_n(t)$ be the probability that n events will occur in the time interval $(0, t)$. Consider adjacent time intervals $(0, t)$ and $(t, t + h)$. If $n \geq 1$ events occur in the interval $(0, t + h)$, then one of three things must be true: (1) n of the events occur in $(0, t)$ and none occur in $(t, t + h)$, (2) $n - 1$ events occur in $(0, t)$ and one occurs in $(t, t + h)$, or (3) two or more of the n events occur in $(t, t + h)$. Since these are mutually exclusive possibilities and there is independence between events occurring in $(0, t)$ and $(t, t+h)$,

$$P_n(t+h) = P_n(t)(1 - \lambda h + o(h)) + P_{n-1}(t)(\lambda h + o(h)) + o(h)$$

or

$$P_n(t+h) = P_n(t)(1 - \lambda h) + P_{n-1}(t)\lambda h + o(h)$$

where the $o(h)$ in the first equation has been replaced by an $o(h)$ of the form $P_n(t)o(h) + P_{n-1}(t)o(h)$. Therefore,

$$\frac{P_n(t+h) - P_n(t)}{h} = -\lambda P_n(t) + \lambda P_{n-1}(t) + \frac{o(h)}{h}.$$

Letting $h \to 0+$,

$$P_n'(t) = -\lambda P_n(t) + \lambda P_{n-1}(t), \qquad n \geq 1, t > 0, \tag{8.1}$$

assuming, of course, that the $P_n(t)$ are differentiable. It is necessary to consider the $n = 0$ case separately since only (1) holds in this case. Thus,

$$P_0(t+h) = P_0(t)(1 - \lambda h + o(h)),$$

so that

$$\frac{P_0(t+h) - P_0(t)}{h} = -\lambda P_0(t) + \frac{o(h)}{h}.$$

Letting $h \to 0+$, we obtain the differential equation

$$P_0'(t) = -\lambda P_0(t), \qquad t > 0. \tag{8.2}$$

In deriving Equations 8.1 and 8.2, we let $h \to 0+$ so that the above derivatives are really right derivatives. By considering the intervals $(0, t-h)$ and $(t-h, t)$, these equations can be seen to hold for left derivatives also and therefore for unrestricted derivatives. The $P_n(t)$, $n \geq 0$, must satisfy the initial conditions

$$P_0(0) = 1 \tag{8.3}$$

and

$$P_n(0) = 0, \qquad n \geq 1. \tag{8.4}$$

We will now undertake to solve these differential equations subject to the stated initial conditions. Consider first the differential equation $P_0'(t) = -\lambda P_0(t)$ subject to the initial condition $P_0(0) = 1$. It is easily seen that the solution is

$$P_0(t) = e^{-\lambda t}, \qquad t \geq 0.$$

Putting $n = 1$ in Equation 8.1 and substituting $e^{-\lambda t}$ for $P_0(t)$, we find that $P_1(t)$ satisfies the equation

$$P_1'(t) = -\lambda P_1(t) + \lambda e^{-\lambda t}$$

and the initial condition $P_1(0) = 0$. To solve the equation, write it as

$$P_1'(t) + \lambda P_1(t) = \lambda e^{-\lambda t}.$$

Multiplying both sides by the factor $e^{\lambda t}$, the equation can be written

$$\frac{d}{dt}(e^{\lambda t} P_1(t)) = \lambda.$$

Integrating, $P_1(t) = \lambda t e^{-\lambda t} + c e^{-\lambda t}$. The integration constant c must be zero to satisfy the initial condition $P_1(0) = 0$. Thus,

$$P_1(t) = \lambda t e^{-\lambda t}.$$

Repeating the same steps for the $n = 2$ case, we find

$$P_2(t) = \frac{\lambda^2 t^2}{2} e^{-\lambda t}.$$

By mathematical induction,

$$P_n(t) = \frac{\lambda^n t^n}{n!} e^{-\lambda t}, \qquad n \geq 0, t \geq 0. \tag{8.5}$$

The heuristic description has resulted in a collection of specific functions. If there is any validity to this procedure, we should be able to start with the end product—namely, the $P_n(t)$ functions—and construct a probability model from which the differential equations above can be deduced.

To construct such a process, we would take Ω to consist of all real-valued nondecreasing step functions ω on $[0, \infty)$ with $\omega(0) = 0$ that increase only by unit jumps. The probability function P would be defined as follows. For each $t \geq 0$, let $X_t(\omega) = \omega(t)$. If $0 \leq t_1 \leq t_2 \leq \cdots \leq t_k$ and $n_1 \leq n_2 \leq \cdots \leq n_k$, let

$$\begin{aligned} P(X_{t_1} &= n_1, \ldots, X_{t_k} = n_k) \\ &= P(X_{t_1} - X_0 = n_1, X_{t_2} - X_{t_1} = n_2 - n_1, \ldots, X_{t_k} - X_{t_{k-1}} = n_k - n_{k-1}) \\ &= P_{n_1}(t_1) P_{n_2 - n_1}(t_2 - t_1) \times \cdots \times P_{n_k - n_{k-1}}(t_k - t_{k-1}); \end{aligned}$$

i.e., probabilities are assigned so that the increments

$$X_{t_1} - X_0, X_{t_2} - X_{t_1}, \ldots, X_{t_k} - X_{t_{k-1}}$$

are independent random variables with

$$\begin{aligned} P(X_{t_j} - X_{t_{j-1}} &= n) \\ &= P_n(t_j - t_{j-1}) = \frac{\lambda^n (t_j - t_{j-1})^n}{n!} e^{-\lambda(t_j - t_{j-1})}, \qquad j = 1, \ldots, k. \end{aligned}$$

We can now state the formal definition of a Poisson process on a given probability space $(\Omega, \mathcal{F}, P)$.

Definition 8.1 *The family of random variables* $\{X_t : t \geq 0\}$ *is a* Poisson process *with rate* $\lambda > 0$ *if*

1. $P(X_0 = 0) = 1$,
2. $X_{t_2} - X_{t_1}, \ldots, X_{t_k} - X_{t_{k-1}}$ *are independent random variables whenever* $0 \leq t_1 \leq \cdots \leq t_k$, *and*
3. $P(X_t - X_s = n) = (\lambda^n (t-s)^n / n!)\, e^{-\lambda(t-s)}, n \in N$, *whenever* $0 \leq s \leq t$. ■

Let $\{X_t : t \geq 0\}$ be a Poisson process with parameter $\lambda > 0$. Taking $s = 0$ in (3), X_t has a Poisson density with parameter λt, so that $E[X_t] = \lambda t$ and $\operatorname{var} X_t = \lambda t$ (see Example 4.15). We can consider the time at which the first event occurs by letting W_1 be the first time t that $X_t = 1$. Assuming that W_1 is a random variable, the density of W_1 can be obtained as follows. If $t \geq 0$, then $W_1 \leq t$ if and only if $X_t \geq 1$, so that $F_{W_1}(t) = P(W_1 \leq t) = P(X_t \geq 1) = P(X_t - X_0 \geq 1)$, and therefore

$$F_{W_1}(t) = \sum_{n=1}^{\infty} \frac{\lambda^n t^n}{n!} e^{-\lambda t} = e^{-\lambda t}(e^{\lambda t} - 1) = 1 - e^{-\lambda t}.$$

It follows that the density of W_1 is given by

$$f_{W_1}(t) = \begin{cases} \lambda e^{-\lambda t} & \text{if } t \geq 0 \\ 0 & \text{if } t \leq 0; \end{cases}$$

i.e., W_1 has an exponential density with parameter λ, and it follows from Exercise 7.2.5 that $E[W_1] = 1/\lambda$. The parameter λ is called the rate. The greater the rate of occurrence of events, the smaller the waiting time for an event to occur.

EXERCISES 8.2

1. Let $\{X_t : t \geq 0\}$ be a Poisson process with rate $\lambda > 0$. If $0 < s < t$ and $0 \leq k \leq n$, calculate $P(X_s = k \mid X_t = n)$.

2. Let $\{X_t^{(i)} : t \geq 0\}_{i=1}^n$ be independent Poisson processes with the same rate $\lambda > 0$. Find the density of the waiting time for all n of the processes to have at least one event occur.
3. Let $\{X_t : t \geq 0\}$ be a Poisson process with rate $\lambda > 0$. By writing $X_t = \sum_{k=1}^{[t]}(X_k - X_{k-1}) + (X_t - X_{[t]})$, where $[t]$ denotes the largest integer less than or equal to t, show that $P(\lim_{t\to\infty} X_t = +\infty) = 1$ by considering the events $A_k = (X_k - X_{k-1} = 1)$.
4. Suppose the rate $\lambda = \lambda(t)$ is a nonnegative function on $[0, +\infty)$ that is Riemann integrable on each finite interval and define $P_n(t)$ as before. Then $P_n(t)$ satisfies Equations 8.1, 8.3, and 8.4. Let

$$G(t, s) = \sum_{n=0}^{\infty} P_n(t)s^n, \qquad -1 < s < 1, t \geq 0,$$

be the generating function of the sequence $\{P_n(t)\}_{n=0}^{\infty}$.

(a) Verify that $G(t, s)$ satisfies the equation

$$\frac{\partial G}{\partial t} = -\lambda(t)(1 - s)G, \qquad -1 < s < 1, t > 0,$$

and the initial condition

$$G(0, s) = 1, \qquad -1 < s < 1.$$

(b) Verify that $G(t, s) = e^{-(1-s)\int_0^t \lambda(u)\,du}$ satisfies these conditions.

(c) Verify that

$$P_n(t) = \frac{1}{n!}\left(\int_0^t \lambda(u)\,du\right)^n e^{-\int_0^t \lambda(u)\,du}$$

and that

$$E[X_t] = \int_0^t \lambda(u)\,du.$$

5. Use the results of the previous problem to determine $P_n(t), n \geq 0$ and $E[X_t]$ for $\lambda(t) = 1/(1 + t)$.
6. Use the result of Problem 5 to approximate $P_{10}(100) = P(X_{100} = 10)$ when $\lambda(t) = 1/(1 + t)$.

8.3 BIRTH AND DEATH PROCESSES

The Poisson process is an example of a birth process in which the population size only increases. Realistic models for population growth must not only incorporate deaths but also allow the possibility that birth and death rates depend upon population size.

Assuming that there is a probability space $(\Omega, \mathcal{F}, P)$ and a process $\{X_t : t \geq 0\}$ reflecting population growth, we will assume that for $h > 0$,

1. $P(X_{t+h} - X_t = 1 \mid X_t = n) = \beta_n h + o(h)$,
2. $P(X_{t+h} - X_t = -1 \mid X_t = n) = \delta_n h + o(h)$,
3. $P(X_{t+h} - X_t = 0 \mid X_t = n) = 1 - (\beta_n + \delta_n)h + o(h)$, and
4. $P(|X_{t+h} - X_t| \geq 2) = o(h)$,

where $\beta_n \geq 0$, $\delta_0 = 0$, and $\delta_n \geq 0$ for all $n \geq 1$. We will also assume that $X_0 = n_0$ where $n_0 \geq 1$ is the initial population size, so that X_t represents the population size at time t. Note that there is no mention of independence between population size in $(0, t)$ and changes in $(t, t+h)$; in fact, independence will be lacking because birth and death rates in $(t, t+h)$ can depend upon population size in $(0, t)$. Letting $P_n(t) = P(X_t = n)$, a system of differential equations for the $P_n(t)$ can be derived as follows. Note first of all that $P_{n_0}(0) = P(X_0 = n_0) = 1$ and that $P_n(0) = P(X_0 = n) = 0$ for all $n \neq n_0$. Consider first $P_0(t)$. For $h > 0$,

$$
\begin{aligned}
&P_0(t+h) = P(X_{t+h} = 0) \\
&= P(X_{t+h} = 0, X_t = 0) + P(X_{t+h} = 0, X_t = 1) \\
&+ P(X_{t+h} = 0, X_t \geq 2) \\
&= P(X_{t+h} = 0 \mid X_t = 0)P(X_t = 0) + P(X_{t+h} = 0 \mid X_t = 1)P(X_t = 1) \\
&+ P(X_{t+h} = 0, X_t \geq 2) \\
&= (1 - \beta_0 h + o(h))P_0(t) + (\delta_1 h + o(h))P_1(t) + o(h).
\end{aligned}
$$

Therefore,

$$\frac{P_0(t+h) - P_0(t)}{h} = -\beta_0 P_0(t) + \delta_1 P_1(t) + \frac{o(h)}{h}.$$

Letting $h \to 0+$, we obtain

$$P_0'(t) = -\beta_0 P_0(t) + \delta_1 P_1(t), \qquad t \geq 0.$$

The derivative in this equation is a right derivative, but by replacing t by $t - h$, the same equation holds for the left derivative and therefore for the derivative. Consider now $P_n(t)$ for $n \geq 1$. Proceeding as before,

$$
\begin{aligned}
P_n(t+h) &= \sum_{k=0}^{\infty} P(X_{t+h} = n, X_t = k) \\
&= \sum_{k=0}^{\infty} P(X_{t+h} - X_t = n - k, X_t = k) \\
&= P(X_{t+h} - X_t = -1, X_t = n+1) \\
&+ P(X_{t+h} - X_t = 0, X_t = n) \\
&+ P(X_{t+h} - X_t = 1, X_t = n-1) \\
&+ \sum_{|k-n|\geq 2} P(X_{t+h} - X_t = n - k, X_t = k).
\end{aligned}
$$

Since $\sum_{|k-n|\geq 2} P(X_{t+h} - X_t = n - k, X_t = k) \leq P(|X_{t+h} - X_t| \geq 2) = o(h)$,

$$
\begin{aligned}
P_n(t+h) = (\delta_{n+1}h + o(h))P_{n+1}(t) + (1 - (\beta_n + \delta_n)h + o(h))P_n(t) \\
+ (\beta_{n-1}h + o(h))P_{n-1}(t) + o(h).
\end{aligned}
$$

Therefore,

$$
\frac{P_n(t+h) - P_n(t)}{h} = \delta_{n+1}P_{n+1}(t) - (\beta_n + \delta_n)P_n(t) + \beta_{n-1}P_{n-1}(t) + \frac{o(h)}{h}.
$$

Letting $h \to 0+$ (and also letting $h \to 0+$ after replacing t by $t - h$),

$$
P_n'(t) = \delta_{n+1}P_{n+1}(t) - (\beta_n + \delta_n)P_n(t) + \beta_{n-1}P_{n-1}(t).
$$

The functions $P_0(t), P_1(t), \ldots$ therefore satisfy the system of differential equations

$$
\begin{cases}
P_0'(t) = -\beta_0 P_0(t) + \delta_1 P_1(t), & t \geq 0 \\
P_n'(t) = \delta_{n+1}P_{n+1}(t) - (\beta_n + \delta_n)P_n(t) + \beta_{n-1}P_{n-1}(t)
\end{cases}
\tag{8.6}
$$

subject to the initial conditions

$$
\begin{cases}
P_{n_0}(0) = 1 \\
P_n(0) = 0, & n \neq n_0.
\end{cases}
\tag{8.7}
$$

We will now specialize by considering a pure birth process for which $\delta_n = 0, n \geq 0$. In this case, the system of differential equations becomes

$$
\begin{cases}
P_0'(t) = -\beta_0 P_0(t) \\
P_n'(t) = -\beta_n P_n(t) + \beta_{n-1}P_{n-1}(t), & n \geq 1.
\end{cases}
\tag{8.8}
$$

Suppose $n_0 > 0$. The general solution of the first equation has the form $P_0(t) = c_0 e^{-\beta_0 t}$; to satisfy the initial condition in Equation 8.7, we must have

$c_0 = 0$, in which case $P_0(t) \equiv 0$ and the second equation in 8.8 for $n = 1$ reduces to

$$P_1'(t) = -\beta_1 P_1(t).$$

If $n_0 > 1$, again $P_1(t) \equiv 0$ and the second equation in 8.8 reduces to

$$P_2'(t) = -\beta_2 P_2(t).$$

Continuing in this way, we arrive at the fact that $P_n(t) \equiv 0$ for all $0 \le n < n_0$ and that

$$P_{n_0}(t) = e^{-\beta_{n_0} t}.$$

Consider the second equation in 8.8 for $n > n_0$; after multiplying both sides by $e^{\beta_n t}$, it can be written

$$\frac{d}{dt}(e^{\beta_n t} P_n(t)) = \beta_{n-1} e^{\beta_n t} P_{n-1}(t).$$

Integrating from 0 to t and using the initial condition $P_n(0) = 0$ for $n \ne n_0$, we obtain the recurrence relation

$$P_n(t) = \beta_{n-1} e^{-\beta_n t} \int_0^t e^{\beta_n s} P_{n-1}(s)\,ds. \tag{8.9}$$

Since $P_{n_0}(t)$ is known, this equation can be used to generate the $P_n(t)$ successively. Note that $P_n(t) \ge 0$ for all $n \ge 1, t \ge 0$.

It is conceivable that in certain populations the birth rates might be so great that the population "explodes" or becomes infinite in a finite time interval, and it is of interest to consider the probability

$$P(X_t = +\infty) = 1 - \sum_{n=0}^{\infty} P(X_t = n) = 1 - \sum_{n=0}^{\infty} P_n(t).$$

Thus, $P(X_t = +\infty) > 0$ if and only if $\sum_{n=0}^{\infty} P_n(t) < 1$. An explosion will not occur if $P(X_t = +\infty) = 0$ for all $t > 0$; i.e., if $\sum_{n=0}^{\infty} P_n(t) = 1$ for all $t > 0$. A criterion for the latter is given by the following theorem.

Theorem 8.3.1 *$\sum_{n=0}^{\infty} P_n(t) = 1$ for all $t \ge 0$ if and only if the series $\sum_{n=0}^{\infty} 1/\beta_n$ diverges.*

PROOF: Letting $S_k(t) = \sum_{n=0}^{k} P_n(t)$, $S_k'(t) = \sum_{n=0}^{k} P_n'(t)$. By Equation 8.8,

$$S_k'(t) = -\beta_0 P_0(t) + \sum_{n=1}^{k} (-\beta_n P_n(t) + \beta_{n-1} P_{n-1}(t)) = -\beta_k P_k(t).$$

Integrating from 0 to t and using the initial conditions in 8.7,

$$1 - S_k(t) = \beta_k \int_0^t P_k(s)ds. \tag{8.10}$$

Since the terms defining the sums $S_k(t)$ are nonnegative, for each t the left side decreases monotonically as k increases, and so the right side decreases in the same way. Let

$$\mu(t) = \lim_{k\to\infty} \beta_k \int_0^t P_k(s)\, ds \geq 0.$$

Thus,

$$\int_0^t P_k(s)\, ds \geq \frac{\mu(t)}{\beta_k}.$$

Using the fact that $S_k(t) = \sum_{n=0}^k P_n(t) = \sum_{n=0}^k P(X_t = n) = P(0 \leq X_t \leq k) \leq 1$ and summing $k = 0, \ldots, n$,

$$t \geq \int_0^t S_n(s)\, ds \geq \mu(t) \sum_{k=0}^n \frac{1}{\beta_k}.$$

If the series $\sum_{n=0}^\infty 1/\beta_n$ diverges, then $\mu(t)$ must be zero for all t,

$$\lim_{k\to\infty} (1 - S_k(t)) = \lim_{k\to\infty} \beta_k \int_0 tP_k(s)\, ds = \mu(t) = 0,$$

and therefore $\sum_{n=0}^\infty P_n(t) = \lim_{k\to\infty} S_k(t) = 1$ for all $t \geq 0$. On the other hand, by Equation 8.10,

$$\int_0^t S_k(s)\, ds = \sum_{n=0}^k \int_0^t P_n(s)\, ds = \sum_{n=0}^k \frac{1 - S_n(t)}{\beta_n} \leq \sum_{n=0}^k \frac{1}{\beta_n}.$$

Since $S_k(s)$ increases with k, the integral on the left increases with k and

$$\lim_{k\to\infty} \int_0^t S_k(s)\, ds \leq \sum_{n=0}^\infty \frac{1}{\beta_n}.$$

If the limit can be taken past the integral sign (which is permissible in this case),

$$\int_0^t (\lim_{k\to\infty} S_k(s))\, ds \leq \sum_{n=0}^\infty \frac{1}{\beta_n}.$$

If $\sum_{n=0}^{\infty} P_k(t) = \lim_{k\to\infty} S_k(t) = 1$ for all $t \geq 0$, we would have

$$t \leq \sum_{n=0}^{\infty} \frac{1}{\beta_n}$$

for all $t > 0$. Since t can be arbitrarily large, $\sum_{n=0}^{\infty} 1/\beta_n = +\infty$. Thus, $\sum_{n=0}^{\infty} P_k(t) = 1$ for all $t \geq 0$ implies that the series $\sum_{n=0}^{\infty} 1/\beta_n$ diverges. ■

Generally speaking, the differential equations in 8.6 for a birth and death process are difficult to solve because they must be solved simultaneously, as opposed to the pure birth case where they can be solved sequentially starting with the first equation in 8.8. There are methods of obtaining qualitative information about population growth even if the equations in 8.6 cannot be solved.

EXAMPLE 8.1 Consider a birth and death process with $\beta_n = \beta n$ and $\delta_n = \delta n, n \geq 0$, where $\beta, \delta > 0$. We will assume that $n_0 = 1$, so that $X_0 = 1$. The initial conditions are then

$$\begin{cases} P_1(0) = 1 \\ P_n(0) = 0, \qquad n \neq 1. \end{cases}$$

Consider $M(t) = E[X_t] = \sum_{n=0}^{\infty} nP(X_t = n) = \sum_{n=0}^{\infty} nP_n(t)$. Note that $M(0) = \sum_{n=0}^{\infty} nP_n(0) = 1$ and that $M'(t) = \sum_{n=1}^{\infty} nP'_n(t)$, formally at least. Multiplying both sides of the second equation in 8.6 by n and summing over $n = 1, 2, \ldots,$

$$\begin{aligned} M'(t) &= \delta \sum_{n=1}^{\infty} n(n+1)P_{n+1}(t) - (\beta + \delta)\sum_{n=1}^{\infty} n^2 P_n(t) \\ &\quad + \beta \sum_{n=1}^{\infty} n(n-1)P_{n-1}(t) \\ &= (\beta - \delta)\sum_{n=1}^{\infty} nP_n(t) \\ &= (\beta - \delta)M(t). \end{aligned}$$

The average population size $M(t)$ therefore satisfies the differential equation

$$M'(t) = (\beta - \delta)M(t)$$

subject to the initial condition $M(0) = 1$. Thus,

$$M(t) = \begin{cases} e^{(\beta-\delta)t} & \text{if } \beta \neq \delta \\ 1 & \text{if } \beta = \delta. \end{cases}$$

This function also gives qualitative information about the long-term behavior of population size, since

$$\lim_{t\to+\infty} M(t) = \begin{cases} +\infty & \text{if } \beta > \delta \\ 0 & \text{if } \beta < \delta \\ 1 & \text{if } \beta = \delta. \ \blacksquare \end{cases}$$

EXERCISES 8.3

1. Consider a pure birth process with $\beta_n = \beta n, \delta_n = 0, n \geq 0$, and $X_0 = 1$. Calculate $P_1(t), P_2(t)$, and $P_3(t)$.
2. Consider the pure birth process of the previous problem. Use mathematical induction to prove that $P_n(t) = e^{-\beta t}(1 - e^{-\beta t})^{n-1}, n \geq 2$.
3. Explain what happens to the pure birth equations in 8.8 when $\beta_i > 0, i = 1, \ldots, n-1, \beta_n = 0$.
4. Calculate $E[X_t]$ for a birth and death process with $\beta_n = \alpha + \beta n, \delta_n = \delta n, n \geq 0$, and $X_0 = 1$.
5. Consider the pure birth process for which $\delta_n = 0, \beta_n = \beta n, n \geq 0$. Then the $P_n(t)$ satisfy Equation 8.8 and the initial conditions in 8.7. Let

$$G(t,s) = \sum_{n=n_0}^{\infty} P_n(t)s^n, \qquad -1 < s < 1, t \geq 0$$

be the generating function of the sequence $\{P_n(t)\}_{n=0}^{\infty}$.

(a) Verify that $G(t,s)$ satisfies the equation

$$\frac{\partial G}{\partial t} + \beta s(1-s)\frac{\partial G}{\partial s} = 0$$

and the initial condition $G(0,s) = s^{n_0}$.

(b) Verify that

$$G(t,s) = s^{n_0}\left(\frac{e^{-\beta t}}{1 - s(1 - e^{-\beta t})}\right)^{n_0}$$

satisfies the conditions of the previous problem.

(c) Determine $E[X_t]$.

(d) Determine $P_n(t), n \geq 0$, in the $n_0 = 1$ case.

8.4 MARKOV CHAINS

Let $S = \{s_1, s_2, \ldots\}$ be a finite or countably infinite collection of objects called *states*. We will now describe a model for moving among the states in a random way.

Definition 8.2 *A collection of random variables* $\{X_t : t \geq 0\}$ *is called a* Markov chain *if*

$$P(X_{t_n} = s_{i_n} \mid X_{t_1} = s_{i_1}, \ldots, X_{t_{n-1}} = s_{i_{n-1}}) = P(X_{t_n} = s_{i_n} \mid X_{t_{n-1}} = s_{i_{n-1}})$$

whenever $0 \leq t_1 \leq t_2 \leq \ldots \leq t_n$ *and* $s_{i_1}, \ldots, s_{i_n} \in S$. ∎

This defining property is called the *Markov property*. If we regard t_{n-1} as the present time, this property requires that probability of a future event given the past does not depend upon the remote past or, put another way, the process does not have a memory.

Definition 8.3 *The* transition function *of the Markov chain* $\{X_t : t \geq 0\}$ *is defined by*

$$p_{i,j}(s,t) = P(X_t = s_j \mid X_s = s_i), \qquad 0 \leq s \leq t, s_i, s_j \in S.$$

The chain has stationary transition functions *if the* $p_{i,j}(s,t)$ *depend only upon* $t - s$. ∎

We will consider only Markov chains $\{X_t : t \geq 0\}$ with stationary transition function

$$p_{i,j}(t) = P(X_{t+s} = s_j \mid X_s = s_i), \qquad i,j \geq 1, s,t \geq 0, \quad (8.11)$$

which satisfies the additional continuity condition that

$$\lim_{t \to 0+} p_{i,j}(t) = \delta_{i,j} \qquad (8.12)$$

where $\delta_{i,j} = 1$ if $i = j$ and $\delta_{i,j} = 0$ if $i \neq j$. For fixed $t \geq 0$, the numbers $p_{i,j}(t)$ can be displayed in the following matrix form:

$$P(t) = \begin{bmatrix} p_{1,1}(t) & p_{1,2}(t) & \ldots & \ldots \\ p_{2,1}(t) & p_{2,2}(t) & \ldots & \ldots \\ \vdots & \vdots & & \\ p_{i,1}(t) & p_{i,2}(t) & \ldots & \ldots \\ \vdots & \vdots & \ddots & \end{bmatrix},$$

which is called the *stationary transition matrix*. If S is finite, this matrix is $|S| \times |S|$; if S is countably infinite, it has an infinite number of rows and columns.

EXAMPLE 8.2 Let $\{X_t : t \geq 0\}$ be a Poisson process with rate $\lambda > 0$, let $0 \leq t_1 \leq t_2 \leq \cdots \leq t_k$ and $0 \leq n_1 \leq n_2 \leq \cdots \leq n_k$, and let

$P_n(t) = P(X_t = n)$ as in Section 8.2. Since the process has independent increments,

$$\begin{aligned}
P&(X_{t_k} = n_k \mid X_{t_1} = n_1, \ldots, X_{t_{k-1}} = n_{k-1}) \\
&= \frac{P(X_{t_1} - X_0 = n_1, X_{t_2} - X_{t_1} = n_2 - n_1, \ldots, X_{t_k} - X_{t_{k-1}} = n_k - n_{k-1})}{P(X_{t_1} - X_0 = n_1, \ldots, X_{t_{k-1}} - X_{t_{k-2}} = n_{k-1} - n_{k-2})} \\
&= P(X_{t_k} - X_{t_{k-1}} = n_k - n_{k-1}) \\
&= P_{n_k - n_{k-1}}(t_k - t_{k-1}).
\end{aligned}$$

Calculating $P(X_{t_k} = n_k \mid X_{t_{k-1}} = n_{k-1})$ in the same way, it too is equal to $P_{n_k - n_{k-1}}(t_k - t_{k-1})$. According to Equation 8.5,

$$\begin{aligned}
P&(X_{t_k} = n_k \mid X_{t_1} = n_1, \ldots, X_{t_{k-1}} = n_{k-1}) \\
&= P(X_{t_k} = n_k \mid X_{t_{k-1}} = n_{k-1}) \\
&= P_{n_k - n_{k-1}}(t_k - t_{k-1}) \\
&= \frac{(\lambda(t_k - t_{k-1}))^{n_k - n_{k-1}} e^{-\lambda(t_k - t_{k-1})}}{(n_k - n_{k-1})!},
\end{aligned}$$

so that $\{X_t : t \geq 0\}$ is a Markov chain with stationary transition function

$$p_{m,n}(t) = \begin{cases} (\lambda^{n-m} t^{n-m}/(n-m)!) e^{-\lambda t} & \text{if } n \geq m, t \geq 0 \\ 0 & \text{otherwise.} \end{cases} \qquad \blacksquare$$

The stationary transition functions $p_{i,j}$ inherit an important property from the Markov property. The equation in *(iii)* of the following theorem is known as the *Chapman-Kolmogorov equation.*

Theorem 8.4.1 *The transition functions $p_{i,j}$ have the following properties:*

(i) $p_{i,j}(0) = \delta_{i,j}$ *if* $i, j \geq 1$.

(ii) $\sum_j p_{i,j}(t) = 1$ *for all* $t \geq 0$.

(iii) $p_{i,j}(s+t) = \sum_k p_{i,k}(s) p_{k,j}(t)$ *if* $s, t \geq 0$.

PROOF: We first prove *(iii)*. By the Markov property,

$$\begin{aligned}
p_{i,j}(s+t) &= P(X_{t+s} = s_j \mid X_0 = s_i) \\
&= \sum_k \frac{P(X_{t+s} = s_j, X_s = s_k, X_0 = s_i)}{P(X_0 = s_i)}
\end{aligned}$$

$$= \sum_k P(X_{t+s} = s_j \mid X_0 = s_i, X_s = s_k) P(X_s = s_k \mid X_0 = s_i)$$

$$= \sum_k P(X_{t+s} = s_j \mid X_s = s_k) P(X_s = s_k \mid X_0 = s_i)$$

$$= \sum_k p_{i,k}(s) p_{k,j}(t).$$

Since $P(X_0 = s_j \mid X_0 = s_i) = P(X_0 = s_i, X_0 = s_j)/P(X_0 = s_i)$ is 0 or 1 according to whether $i \neq j$ or $i = j$, respectively, $p_{i,j}(0) = \delta_{i,j}$ and *(i)* holds. Finally, $\sum_j p_{i,j}(t) = \sum_j P(X_t = s_j \mid X_0 = s_i) = P(X_t \in S \mid X_0 = s_i) = 1$ and *(ii)* holds. ■

The three properties listed in this theorem in conjunction with the continuity assumption 8.12 imply a good deal more about the transition functions $p_{i,j}$. Additional properties will be reviewed briefly for the purpose of clarifying applications. Proofs of these facts require a little more mathematical background than is presupposed here. Complete details are available in Chapter 14 of the book by Karlin and Taylor listed at the end of the chapter. The first facts concern differentiability of the transition functions $p_{i,j}$.

Theorem 8.4.2 *(i) For every i,*

$$q_{i,i} = p'_{i,i}(0) = \lim_{t \to 0+} \frac{p_{i,i}(t) - 1}{t}$$

exists but may be $-\infty$.

(ii) For all i, j with $i \neq j$,

$$q_{i,j} = p'_{i,j}(0) = \lim_{t \to 0+} \frac{p_{i,j}(t)}{t}$$

exists and is finite.

Since $\sum_j p_{i,j}(t) = 1$, it might be tempting to take the derivative term by term to show that

$$\sum_j q_{i,j} = 0.$$

This is certainly a valid step if the sum $\sum_j p_{i,j}(t)$ is a finite sum, as it is in many applications, but in general the most that can be proved is that for each i,

$$-q_{i,i} \geq \sum_{j \neq i} q_{i,j}.$$

We will assume in the remainder of this section that

$$+\infty > -q_{i,i} = \sum_{j \neq i} q_{i,j}. \tag{8.13}$$

This requirement means that all of the entries in the following matrix, called the *q-matrix*, are finite and that each row sum is zero:

$$Q = \begin{bmatrix} q_{1,1} & q_{1,2} & \cdots & \cdots \\ q_{2,1} & q_{2,2} & \cdots & \cdots \\ \vdots & \vdots & & \\ q_{i,1} & q_{i,2} & \cdots & \cdots \\ \vdots & \vdots & \ddots & \end{bmatrix}.$$

In matrix notation, $Q\mathbf{1} = 0$ where $\mathbf{1}$ is a column vector all of whose entries are 1.

Not only are the $p_{i,j}(t)$ differentiable at 0, but $p'_{i,j}(t)$ is defined for all $t > 0$. Consider the equations

$$\begin{aligned} p_{i,j}(t+s) - p_{i,j}(t) &= \sum_k p_{i,k}(s) p_{k,j}(t) - p_{i,j}(t) \\ &= \sum_{k \neq i} p_{i,k}(s) p_{k,j}(t) + (p_{i,i}(s) - 1) p_{i,j}(t). \end{aligned}$$

Proceeding formally by dividing by s and letting $s \to 0+$, we obtain the following equations, which are known as the *Kolmogorov backward equations:*

$$p'_{i,j}(t) = \sum_{k \neq i} q_{i,k} p_{k,j}(t) + q_{i,i} p_{i,j}(t), \qquad t > 0. \tag{8.14}$$

On the other hand, we can write

$$\begin{aligned} p_{i,j}(t+s) - p_{i,j}(s) &= \sum_k p_{i,k}(s) p_{k,j}(t) - \sum_k p_{i,k}(s) p_{k,j}(0) \\ &= \sum_k p_{i,k}(s) p_{k,j}(t) - \sum_k p_{i,k}(s) \delta_{k,j} \\ &= \sum_k p_{i,k}(s)(p_{k,j}(t) - \delta_{k,j}). \end{aligned}$$

Operating formally again by dividing by t and letting $t \to 0+$, we obtain the following equations, which are known as the *Kolmogorov forward equations:*

$$p'_{i,j}(t) = \sum_{k \neq j} p_{i,k}(t) q_{k,j} + q_{j,j} p_{i,j}(t), \qquad t > 0. \tag{8.15}$$

If the state space S is infinite, then both sets of forward and backward equations represent an infinite system of differential equations that must be solved simultaneously. Both sets of equations take on deceptively simple forms if expressed in matrix notation. Letting $P'(t) = [p'_{i,j}(t)]$ be the matrix with entries $p'_{i,j}(t)$, the backward equations take on the form

$$P'(t) = QP(t), \qquad t > 0, \tag{8.16}$$

and the forward equations take on the form

$$P'(t) = P(t)Q, \qquad t > 0, \tag{8.17}$$

with both equations subject to the initial continuity condition 8.12.

EXAMPLE 8.3 Consider a Poisson process $\{X_t : t \geq 0\}$ with rate $\lambda > 0$. In this case, the transition functions $p_{i,j}$ are given by

$$p_{i,j}(t) = \frac{\lambda^{j-i} t^{j-i}}{(j-i)!} e^{-\lambda t}, \qquad j \geq i \geq 0,$$

and it is easily seen that $q_{i,i} = p'_{i,i}(0) = -\lambda$, $q_{i,i+1} = p'_{i,i+1}(0) = \lambda$, and $q_{i,j} = 0$ for $j \neq \{i, i+1\}$ and all $i \geq 0$. The q-matrix is given by

$$Q = \begin{bmatrix} -\lambda & \lambda & 0 & \cdots \\ 0 & -\lambda & \lambda & \cdots \\ 0 & 0 & -\lambda & \cdots \\ \vdots & \vdots & \vdots & \ddots \end{bmatrix}. \blacksquare$$

Application of these results lies in the choice of the q-matrix. From the definition of the $q_{i,j}$, for all i,

$$p_{i,i}(h) = 1 + q_{i,i}h + o(h),$$

and for $i \neq j$,

$$p_{i,j}(h) = q_{i,j}h + o(h).$$

These equations frequently suggest how the $q_{i,j}$ should be chosen. It is then a matter of solving the Kolmogorov backward or forward equations for the $p_{i,j}$.

EXAMPLE 8.4 Consider a system consisting of a single unit that has a failure rate of μ and a repair rate of λ; i.e., if the unit is in operation at time t, the probability that it will fail in the interval $(t, t+h)$ is $\mu h + o(h)$; if not in operation at time t, the probability that it will be repaired and put back into operation in the interval $(t, t+h)$ is $\lambda h + o(h)$. This system can be in one of the two states 0, 1 with the q-matrix

$$Q = \begin{bmatrix} -\lambda & \lambda \\ \mu & -\mu \end{bmatrix};$$

e.g., the probability of going from state 1 to state 0 in a time interval of length h is approximately $q_{1,0}h = \mu h$. In this case, there are four transition functions to be determined: $p_{0,0}, p_{0,1}, p_{1,0}, p_{1,1}$. The Kolmogorov forward equations are

$$\begin{cases} p'_{i,0}(t) = p_{i,1}(t)\mu - \lambda p_{i,0}(t), & i = 0, 1 \\ p'_{i,1}(t) = p_{i,0}(t)\lambda - \mu p_{i,1}(t), & i = 0, 1. \end{cases}$$

Since $p_{i,1}(t) = 1 - p_{i,0}(t)$ and $p_{i,0}(t) = 1 - p_{i,1}(t)$, the equations can be written

$$\begin{cases} p'_{i,0}(t) + (\lambda + \mu)p_{i,0}(t) = \mu \\ p'_{i,1}(t) + (\lambda + \mu)p_{i,1}(t) = \lambda. \end{cases}$$

Multiplying both sides of the first equation by $e^{(\lambda+\mu)t}$, it can be written

$$\frac{d}{dt}(e^{(\lambda+\mu)t}p_{i,0}(t)) = \mu e^{(\lambda+\mu)t}.$$

Integrating and then multiplying both sides by $e^{-(\lambda+\mu)t}$,

$$p_{i,0}(t) = \frac{\mu}{\lambda+\mu} + c_{i,0}e^{-(\lambda+\mu)t}$$

where the integration constants must be chosen to satisfy the continuity condition 8.12. After so choosing the constants and applying the same procedure to the $p_{i,1}(t)$, we obtain

$$p_{0,0}(t) = \frac{\mu + \lambda e^{-(\lambda+\mu)t}}{\lambda+\mu} \qquad p_{0,1}(t) = \frac{\lambda - \lambda e^{-(\lambda+\mu)t}}{\lambda+\mu}$$

$$p_{1,0}(t) = \frac{\mu - \mu e^{-(\lambda+\mu)t}}{\lambda+\mu} \qquad p_{1,1}(t) = \frac{\lambda + \mu e^{-(\lambda+\mu)t}}{\lambda+\mu}. \quad \blacksquare$$

EXERCISES 8.4

1. Consider the q-matrix

$$Q = \begin{bmatrix} -\lambda & \lambda & 0 & \cdots \\ 0 & -\lambda & \lambda & \cdots \\ 0 & 0 & -\lambda & \cdots \\ \vdots & \vdots & \vdots & \ddots \end{bmatrix}$$

and the corresponding Kolmogorov forward equations for $P(t) = [p_{i,j}(t)]$.
(a) Use the forward equations to show that $p_{i,0}(t) \equiv 0$ for all $i \geq 1$.
(b) For each $i \geq 1$, use an induction argument to show that $p_{i,j}(t) \equiv 0$ for $j \leq i - 1$.
(c) Calculate $p_{i,j}(t)$ for $j \geq i$.

2. A system consists of m components, some of which are in operation and some of which are not at any given time t. If there are k components in operation at time t, the probability that one of them will fail in the interval $(t, t + h)$ is $\mu kh + o(h)$, the probability that one will be put back into operation in the interval is $\lambda h + o(h)$, and the probability that two or more changes will occur is $o(h)$. If the state of the system is the number of components in operation, what is the q-matrix for the system?
3. A system consists of two components that are connected in parallel with one online and the other on standby. The one in operation at time t will fail in the interval $(t, t + h)$ with probability $\lambda h + o(h)$. A component cannot fail while on standby. A component in failed condition at time t will be repaired in the interval $(t, t+h)$ with probability $\mu h + o(h)$. The probability of two or more changes taking place in an interval of length h is $o(h)$. If the state of the system is the number of failed components, what is the q-matrix for this system?
4. Suppose the system of the previous problem is modified so that a component on standby at time t will fail in the interval $(t, t + h)$ with probability $\lambda_s h + o(h)$. If the number of failed components is the state, what is the q-matrix for the system?

8.5 MATRIX CALCULUS

In the previous sections, we have seen that the differential equation $p'(t) = ap(t)$ has the solution $p(t) = ce^{at}$. We have seen also that the transition matrix $P(t) = [p_{i,j}(t)]$ satisfies the matrix equation $P'(t) = QP(t)$, where Q, the q-matrix, is a constant matrix. We could try to solve the equation $P'(t) = QP(t)$ by means of the function $P(t) = e^{tQ}$ if only we knew what e^{tQ} represents. Since

$$e^{at} = 1 + at + \frac{a^2t^2}{2!} + \frac{a^3t^3}{3!} + \cdots,$$

we might try replacing a by Q to obtain

$$e^{tQ} = I + tQ + \frac{t^2Q^2}{2!} + \frac{t^3Q^3}{3!} + \cdots,$$

where it is necessary to replace the number 1 by the matrix $I = [\delta_{i,j}]$. The individual terms involving powers of Q on the right side make sense. Except for the fact that the right side is a sum of an infinite series all of whose terms are matrices, this equation can be taken as the definition of e^{tQ}. Since sums of infinite series are defined in terms of limits, we must digress to discuss sequences of matrices and limits of such sequences.

In keeping within the bounds of introductory material, we will limit ourselves to matrices with a finite number of rows and columns. We will deal with $r \times s$ matrices $A = [a_{i,j}]$ with r rows and s columns. Either $\{A^{(n)}\}$ or $A^{(1)}, A^{(2)}, \ldots$ will denote an infinite sequence of such matrices, all of size $r \times s$. The dependence upon n is specified by putting the n in parentheses, since the notation A^n stands for the nth power of A when A is a square matrix. Thus, $A^{(n)} = [a_{i,j}^{(n)}]$. If $\lim_{n\to\infty} a_{i,j}^{(n)}$ exists and is equal to $a_{i,j}$ for each i and j, we say that the sequence $\{A^{(n)}\}$ converges to $A = [a_{i,j}]$ and write

$$\lim_{n\to\infty} A^{(n)} = A.$$

Alternatively,

$$\lim_{n\to\infty} [a_{i,j}^{(n)}] = [\lim_{n\to\infty} a_{i,j}^{(n)}] = [a_{i,j}].$$

EXAMPLE 8.5

$$\lim_{n\to\infty} \begin{bmatrix} 1-(1/n) & (1+(1/n))^n \\ 2^{1/n} & (n^2+1)/(n^3+n^2+1) \end{bmatrix} = \begin{bmatrix} 1 & e \\ 1 & 0 \end{bmatrix}. \blacksquare$$

Let $\{A^{(n)}\}$ be a convergent sequence of $r \times s$ matrices and let $\{B^{(n)}\}$ be a convergent sequence of $s \times t$ matrices with $\lim_{n\to\infty} A^{(n)} = A$ and $\lim_{n\to\infty} B^{(n)} = B$, respectively. Then the sequence $\{A^{(n)}B^{(n)}\}$ converges and $\lim_{n\to\infty} A^{(n)}B^{(n)} = AB$. This follows from the fact that the entry in the ith row and jth column of $A^{(n)}B^{(n)}$ is $\sum_{k=1}^{s} a_{i,k}^{(n)} b_{k,j}^{(n)}$ with limit $\sum_{k=1}^{s} a_{i,k} b_{k,j}$, which is the corresponding entry in AB.

Given a sequence of matrices $\{A^{(n)}\}$, all of the same size, we can form the expression $\sum_{n=1}^{\infty} A^{(n)}$, since matrices of the same size can be added. The definition of the sum of the infinite series $\sum_{n=1}^{\infty} A^{(n)}$ mimics the calculus definition. For $n \geq 1$, let

$$S^{(n)} = [s_{i,j}^{(n)}] = \sum_{k=1}^{n} A^{(k)} = A^{(1)} + A^{(2)} + \cdots + A^{(n)}$$

where $s_{i,j}^{(n)} = \sum_{k=1}^{n} a_{i,j}^{(k)}$. If the sequence $\{S^{(n)}\}$ has a limit $S = [s_{i,j}]$, we say that the series $\sum_{n=1}^{\infty} A^{(n)}$ converges and has sum S. Note that

$$s_{i,j} = \lim_{n \to \infty} s_{i,j}^{(n)} = \sum_{k=1}^{\infty} a_{i,j}^{(k)}.$$

In what follows, an infinite series will begin with a zero term as in $\sum_{n=0}^{\infty} A^{(n)}$. If the nth term is A^n for some matrix A, A^0 by convention will be the identity matrix $I = [\delta_{i,j}]$. We will also adopt the notation $A^n = [a_{i,j}^{(n)}]$.

Definition 8.4 *If A is an $r \times r$ matrix, we define $e^A = \sum_{n=0}^{\infty} A^n/n!$ provided each of the series*

$$\sum_{n=0}^{\infty} \frac{a_{i,j}^{(n)}}{n!}$$

converges absolutely, $i, j = 1, \ldots, r$. ■

Lemma 8.5.1 *Let $A = [a_{i,j}]$ be an $r \times r$ matrix and let $A^n = [a_{i,j}^{(n)}]$. If there is a constant M such that $|a_{i,j}^{(n)}| \le M^n$ for $1 \le i, j \le r$ and $n \ge 1$, then e^A is defined. Moreover, if $a_{i,j} \ge 0$ for $1 \le i, j \le r$, then $a_{i,j}^{(n)} \ge 0$ for $1 \le i, j \le r, n \ge 1$, and $\sum_{n=0}^{\infty} a_{i,j}^{(n)}/n! \ge 0$.*

PROOF: Since the series $\sum_{n=0}^{\infty} M^n/n!$ converges, the series $\sum_{n=0}^{\infty} a_{i,j}^{(n)}/n!$ converges absolutely by the comparison test, and e^A is defined. Since $a_{i,j}^{(2)} = \sum_{k=1}^{r} a_{i,k}a_{k,j} \ge 0$, $a_{i,j}^{(2)} \ge 0$ for $1 \le i, j \le r$. It is easily seen that $a_{i,j}^{(n)} \ge 0$ for $1 \le i, j \le r, n \ge 1$, using mathematical induction. ■

If a and b are real numbers, then $ab = ba$ and $e^a e^b = e^{a+b}$; but if A and B are $r \times r$ matrices, it need not be true that $AB = BA$ nor that $e^A e^B = e^{A+B}$ when the latter are defined. For example, if

$$A = \begin{bmatrix} 1 & 1 \\ 0 & 1 \end{bmatrix} \quad \text{and} \quad B = \begin{bmatrix} 0 & 1 \\ 1 & 1 \end{bmatrix},$$

then

$$AB = \begin{bmatrix} 1 & 2 \\ 1 & 1 \end{bmatrix} \neq \begin{bmatrix} 0 & 1 \\ 1 & 2 \end{bmatrix} = BA.$$

Granted even that $AB = BA$, we are confronted immediately with the binomial theorem in dealing with e^{A+B}.

Lemma 8.5.2 *Let A and B be $r \times r$ matrices such that $AB = BA$. Then $(A + B)^n = \sum_{k=0}^{n} \binom{n}{k} A^k B^{n-k}$.*

PROOF: Since $\sum_{k=0}^{1} A^k B^{1-k} = B + A = A + B$, the assertion is true for $n = 1$. Assume it is true for $n - 1$. Then

$$(A+B)^n = (A+B)(A+B)^{n-1} = (A+B)\sum_{k=0}^{n-1}\binom{n-1}{k}A^k B^{n-1-k}.$$

Since $B\,A^k B^{n-1-k} = B\,A\,A^{k-1}B^{n-1-k} = A\,B\,A^{k-1}B^{n-1-k} = \cdots = A^k B B^{n-1-k} = A^k B^{n-k}$,

$$\begin{aligned}(A+B)^n &= \sum_{k=0}^{n-1}\binom{n-1}{k}A^{k+1}B^{n-1-k} + \sum_{k=0}^{n-1}\binom{n-1}{k}A^k B^{n-k}\\ &= \sum_{j=1}^{n}\binom{n-1}{j-1}A^j B^{n-j} + \sum_{j=0}^{n-1}\binom{n-1}{j}A^j B^{n-j}\\ &= A^n + \sum_{j=1}^{n-1}\left(\binom{n-1}{j-1} + \binom{n-1}{j}\right)A^j B^{n-j} + B^n.\end{aligned}$$

Since the sum of the binomial coefficients is $\binom{n}{j}$ (see Exercise 1.3.2),

$$A + B = A^n + \sum_{j=1}^{n-1}\binom{n}{j}A^j B^{n-j} + B^n = \sum_{j=0}^{n}\binom{n}{j}A^j B^{n-j}$$

and the assertion is true for n. By the principle of mathematical induction, the assertion is true for all $n \geq 1$. ■

For the proof of the following lemma, the entry in the ith row and jth column of the matrix A will be denoted by $(A)_{i,j}$.

Lemma 8.5.3 *Let A and B be $r \times r$ matrices for which e^A and e^B are defined and $AB = BA$. Then e^{A+B} is defined and $e^{A+B} = e^A e^B$.*

PROOF: Since

$$e^A = \left[\sum_{n=0}^{\infty}\frac{a_{i,j}^{(n)}}{n!}\right] \quad \text{and} \quad e^B = \left[\sum_{m=0}^{\infty}\frac{b_{i,j}^{(m)}}{m!}\right],$$

$$\left(e^A e^B\right)_{i,j} = \sum_{k=1}^{r}\left(\sum_{n=0}^{\infty}\frac{a_{i,k}^{(n)}}{n!}\right)\left(\sum_{m=0}^{\infty}\frac{b_{k,j}^{(m)}}{m!}\right).$$

Since the indicated series are absolutely convergent, the product of their sums can be written $\sum_{n=0}^{\infty} c_{i,j,k}^{(n)}$ where

$$c_{i,j,k}^{(n)} = \sum_{\ell=0}^{n} \frac{a_{i,k}^{(\ell)}}{\ell!} \frac{b_{k,j}^{(n-\ell)}}{(n-\ell)!}.$$

By Lemma 8.5.2,

$$\begin{aligned}
\left(e^{A}e^{B}\right)_{i,j} &= \sum_{k=1}^{r}\sum_{n=0}^{\infty}\sum_{\ell=0}^{n} \frac{a_{i,k}^{(\ell)}}{\ell!} \frac{b_{k,j}^{(n-\ell)}}{(n-\ell)!} \\
&= \sum_{n=0}^{\infty}\sum_{\ell=0}^{n}\left(\sum_{k=1}^{r} \frac{a_{i,k}^{(\ell)}}{\ell!} \frac{b_{k,j}^{(n-\ell)}}{(n-\ell)!}\right) \\
&= \sum_{n=0}^{\infty}\sum_{\ell=0}^{n} \frac{1}{n!}\binom{n}{\ell}\sum_{k=1}^{r} a_{i,k}^{(\ell)} b_{k,j}^{(n-\ell)} \\
&= \sum_{n=0}^{\infty} \frac{1}{n!}\sum_{\ell=0}^{n}\binom{n}{\ell}(A^{\ell}B^{n-\ell})_{i,j} \\
&= \sum_{n=0}^{\infty} \frac{1}{n!}((A+B)^{n})_{i,j} = (e^{A+B})_{i,j}. \quad \blacksquare
\end{aligned}$$

Theorem 8.5.4 *Let $Q = [q_{i,j}]$ be an $r \times r$ matrix for which $q_{i,j} \geq 0$ if $i \neq j$ and $\sum_{j=1}^{r} q_{i,j} = 0, i = 1, \ldots, r$. Then $P(t) = [p_{i,j}(t)] = e^{tQ}$ is defined, and the $p_{i,j}(t)$ have the following properties:*

(i) $\lim_{t\to 0+} p_{i,j}(t) = \delta_{i,j}$.

(ii) $p_{i,j}(t) \geq 0, t > 0$.

(iii) $\sum_{j=1}^{r} p_{i,j}(t) = 1, t > 0$.

(iv) $p_{i,j}(s+t) = \sum_{k=1}^{r} p_{i,k}(s)p_{k,j}(t)$.

(v) $p_{i,j}'(t) = \sum_{k=1}^{r} q_{i,k}p_{k,j}(t)$.

PROOF: We first show that e^{tQ} is defined by showing that there is a constant M such that

$$|q_{i,j}^{(n)}| \leq M^{n}, \qquad 1 \leq i,j \leq r, n \geq 1.$$

To see this, let $M = \max_{1\leq i\leq r}(\sum_{j=1}^{r}|q_{i,j}|)$. Since $|q_{i,j}| \leq M$, the assertion is true for $n = 1$. Assume it is true for $n-1$; i.e., $|q_{i,j}^{(n-1)}| \leq M^{n-1}$ for $1 \leq i,j \leq r$. Since $q_{i,j}^{(n)} = \sum_{k=1}^{r} q_{i,k}q_{k,j}^{(n-1)}$,

$$|q_{i,j}^{(n)}| \leq \sum_{k=1}^{r} |q_{i,k}||q_{k,j}^{(n-1)}| \leq M^{n-1}\sum_{k=1}^{r}|q_{i,k}| \leq M^{n},$$

and the assertion is true for n whenever it is true for $n - 1$. It is therefore true for all $n \geq 1$ by mathematical induction. Since

$$\left| \frac{t^n q_{i,j}^{(n)}}{n!} \right| \leq \frac{t^n M^n}{n!}$$

and the series $\sum_{n=0}^{\infty} (Mt)^n/n!$ converges, the series

$$\sum_{n=0}^{\infty} \frac{t^n q_{i,j}^{(n)}}{n!}$$

converges absolutely for $1 \leq i,j \leq r$. Thus, e^{tQ} is defined for all $t > 0$. In fact, this argument shows that this power series in t has $(-\infty, \infty)$ as its interval of convergence. Since

$$p_{i,j}(t) = \delta_{i,j} + \sum_{n=1}^{\infty} \frac{q_{i,j}^{(n)}}{n!} t^n$$

and the sum of a power series is continuous on its interval of convergence, $\lim_{t \to 0+} p_{i,j}(t) = \delta_{i,j}, 1 \leq i,j \leq r$. This proves *(i)*. Consider the matrix $Q + \alpha I$. The off-diagonal elements of $Q + \alpha I$ are the same as those of Q. The diagonal elements of Q are negative, but an $\alpha > 0$ can be chosen so that $Q + \alpha I$ has nonnegative elements. Writing $e^{tQ} = e^{-\alpha t I} e^{t(Q+\alpha I)}$, note that the entries of the matrix $e^{t(Q+\alpha I)}$ are all nonnegative by Lemma 8.5.1. Since it is easily checked that $e^{-\alpha t I} = e^{-\alpha t} I$, the same is true for the entries of $e^{-\alpha t I}$. Thus, the entries of $P(t) = [p_{i,j}(t)] = e^{tQ}$ are nonnegative. This proves *(ii)*. Since

$$\begin{aligned} p_{i,j}(t) &= \delta_{i,j} + t q_{i,j} + \sum_{n=2}^{\infty} \frac{t^n q_{i,j}^{(n)}}{n!} \\ &= \delta_{i,j} + t q_{i,j} + \sum_{n=2}^{\infty} \frac{t^n}{n!} \sum_{k=1}^{r} q_{i,k}^{(n-1)} q_{k,j}, \end{aligned}$$

$$\sum_{j=1}^{r} p_{i,j}(t) = \sum_{j=1}^{r} \delta_{i,j} + t \sum_{j=1}^{r} q_{i,j} + \sum_{n=2}^{\infty} \frac{t^n}{n!} \sum_{k=1}^{r} q_{i,k}^{(n-1)} \sum_{j=1}^{r} q_{k,j} = \delta_{i,i} = 1$$

and *(iii)* is proved. Since e^{sQ} and e^{tQ} are defined, so is $e^{(s+t)Q}$, and $e^{(s+t)Q} = e^{sQ} e^{tQ}$; i.e.,

$$p_{i,j}(s+t) = (e^{(s+t)Q})_{i,j} = \sum_{k=1}^{r} (e^{sQ})_{i,k} (e^{tQ})_{k,j} = \sum_{k=1}^{r} p_{i,k}(s) p_{k,j}(t)$$

and *(iv)* is satisfied. It remains only to show that $P'(t) = [p_{i,j}(t)] = QP(t)$. Since a power series can be differentiated term by term within its interval of convergence,

$$\begin{aligned}
p'_{i,j}(t) &= \frac{d}{dt}\left(\sum_{n=0}^{\infty} \frac{t^n q_{i,j}^{(n)}}{n!}\right) \\
&= \sum_{n=1}^{\infty} \frac{t^{n-1}}{(n-1)!} q_{i,j}^{(n)} \\
&= \sum_{n=1}^{\infty} \frac{t^{n-1}}{(n-1)!} \sum_{k=1}^{r} q_{i,k} q_{k,j}^{(n-1)} \\
&= \sum_{k=1}^{r} q_{i,k} \sum_{n=1}^{\infty} \frac{t^{n-1} q_{k,j}^{(n-1)}}{(n-1)!} \\
&= \sum_{k=1}^{r} q_{i,k} p_{k,j}(t)
\end{aligned}$$

and *(v)* is true. ■

Theorem 8.5.4 provides the answer to whether or not a q-matrix determines a Markov chain. Direct application of the theorem is impractical except possibly for the simplest cases, as will be seen in the following example, and useful information about the resulting Markov chain must be obtained indirectly, as in the next section.

EXAMPLE 8.6 Consider the q-matrix

$$Q = \begin{bmatrix} -1 & 1 & 0 \\ 0 & -1 & 1 \\ 1 & 0 & -1 \end{bmatrix}.$$

Then

$$Q^2 = \begin{bmatrix} 1 & -2 & 1 \\ 1 & 1 & -2 \\ -2 & 1 & 1 \end{bmatrix}, Q^3 = \begin{bmatrix} 0 & 3 & -3 \\ -3 & 0 & 3 \\ 3 & -3 & 0 \end{bmatrix},$$

$$Q^4 = \begin{bmatrix} -3 & -3 & 6 \\ 6 & -3 & -3 \\ -3 & 6 & -3 \end{bmatrix}, Q^5 = \begin{bmatrix} 9 & 0 & -9 \\ -9 & 9 & 0 \\ 0 & -9 & 9 \end{bmatrix},$$

$$Q^6 = \begin{bmatrix} -18 & 9 & 9 \\ 9 & -18 & 9 \\ 9 & 9 & -18 \end{bmatrix}.$$

Upon calculating Q^7, it is seen that $Q^7 = (-27)Q$, and from this it follows that

$$Q^{k+6n} = (-27)^n Q^k, \qquad 1 \leq k \leq 6, n \geq 0.$$

Thus,

$$\begin{aligned} e^{tQ} &= \sum_{m=0}^{\infty} \frac{t^m Q^m}{m!} \\ &= \sum_{k=1}^{6} \sum_{n=0}^{\infty} \frac{(-27)^n t^{k+6n}}{(k+6n)!} Q^k \\ &= \sum_{n=0}^{\infty} (-27)^n t^{6n} \sum_{k=1}^{6} \frac{t^k}{(k+6n)!} Q^k. \end{aligned}$$

In particular,

$$\begin{aligned} p_{1,3}(t) = \sum_{n=0}^{\infty} (-27)^n t^{6n} \bigg(&\frac{t^2}{(2+6n)!} - 3\frac{t^3}{(3+6n)!} + 6\frac{t^4}{(4+6n)!} \\ &- 9\frac{t^5}{(5+6n)!} + 9\frac{t^6}{(6+6n)!} \bigg). \quad \blacksquare \end{aligned}$$

EXERCISES 8.5

1. Show that $e^{\alpha t I} = e^{\alpha t} I$.
2. If $\lambda, \mu > 0$ and

$$Q = \begin{bmatrix} -\lambda & \lambda \\ \mu & -\mu \end{bmatrix},$$

solve the matrix equation $P'(t) = QP(t), t \geq 0$, using the methods of this section.

3. Solve the matrix equation $P'(t) = QP(t), t \geq 0$, using the methods of this section, where

$$Q = \begin{bmatrix} -1 & 1 & 0 & 0 \\ 1 & -1 & 0 & 0 \\ 0 & 0 & -1 & 1 \\ 0 & 0 & 1 & -1 \end{bmatrix}.$$

8.6 STATIONARY DISTRIBUTIONS

Let $Q = [q_{i,j}]$ be an $r \times r$ q-matrix and let $P(t) = [p_{i,j}(t)]$ be the associated matrix of transition functions. One of the problems related to the $P(t)$ is the long-range behavior of the $p_{i,j}(t)$ as $t \to \infty$. In general, the long-range behavior can be complicated, and we will deal only with the simplest case.

Letting $Q^n = [q_{i,j}^{(n)}]$, by definition of matrix multiplication,

$$q_{i,j}^{(2)} = \sum_{k=1}^{r} q_{i,k} q_{k,j}.$$

Writing $Q^3 = QQ^2$,

$$q_{i,j}^{(3)} = \sum_{k=1}^{r} q_{i,k} q_{k,j}^{(2)} = \sum_{k=1}^{r} \sum_{\ell=1}^{r} q_{i,k} q_{k,\ell} q_{\ell,j}.$$

More generally,

$$q_{i,j}^{(n)} = \sum_{i_1=1}^{r} \sum_{i_2=1}^{r} \cdots \sum_{i_{n-1}=1}^{r} q_{i,i_1} q_{i_1,i_2} \cdots q_{i_{n-1},j}. \tag{8.18}$$

Definition 8.5 *The q-matrix $Q = [q_{i,j}]$ is irreducible if for every pair $i, j \in \{1, \ldots, r\}$ with $i \neq j$ either $q_{i,j} > 0$ or there is a finite sequence $i_1, \ldots, i_m$ such that*

$$q_{i,i_1} q_{i_1,i_2} \times \cdots \times q_{i_m,j} \neq 0. \quad \blacksquare$$

We can assume that the $i, i_1, \ldots, i_m, j$ in this definition are distinct, because if $i_p = i_q$ for some $p < q$, then the factor

$$q_{i_p,i_{p+1}} \times \cdots \times q_{i_{q-1},i_q}$$

can be deleted since it is preceded by q_{i_{p-1},i_p} and followed by $q_{i_p,i_{q+1}}$. The resulting product will remain nonzero. Since the $q_{k,\ell} \geq 0$ whenever $k \neq \ell$, we can assume that

$$q_{i,i_1} \times \cdots \times q_{i_n,j} > 0$$

in addition to the $i, i_1, \ldots, i_m, j$ being distinct.

EXAMPLE 8.7 The q-matrix

$$Q = \begin{bmatrix} -1 & 1 & 0 \\ 0 & -1 & 1 \\ 1 & 0 & -1 \end{bmatrix}$$

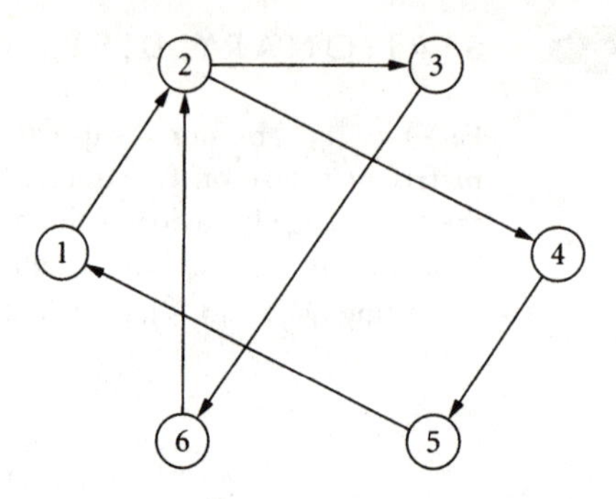

FIGURE 8.2 State diagram.

is irreducible since $q_{1,2} = 1 > 0, q_{1,2}q_{2,3} = 1 > 0, q_{2,3}q_{3,1} = 1 > 0, q_{2,3} = 1 > 0, q_{3,1} = 1 > 0$, and $q_{3,1}q_{1,2} = 1 > 0$. ■

There is an easy way to determine if an $r \times r$ q-matrix is irreducible by drawing a diagram as in Figure 8.2. The numbers in the circles represent the states $1, 2, \ldots, 6$. The arrow connecting 2 to 4 signifies that $q_{2,4} > 0$. If it is possible to find a path connecting all the states by following the arrows, then Q is irreducible. In this case, the matrix Q is irreducible since there is a path connecting all the states.

Theorem 8.6.1 *If the q-matrix $Q = [q_{i,j}]$ is irreducible, then $p_{i,j}(t) > 0$ for all $t > 0, 1 \le i,j \le r$.*

PROOF: We first show that $p_{i,i}(t) > 0$ for $t > 0$. Since $\lim_{t\to 0+} p_{i,i}(t) = 1$, there is a $\delta > 0$ such that $p_{i,i}(t) > 0$ for all $0 < t \le \delta$. Consider any $t > 0$ and choose k such that $k\delta \le t < (k+1)\delta$. Since $P(t) = P(k\delta)P(t-k\delta) = P(\delta)P(\delta)\times\cdots\times P(\delta)P(t-k\delta)$, $p_{i,j}(t)$ is a sum of terms as in Equation 8.18, with the $q_{k,\ell}$ replaced by $p_{k,\ell}$, and

$$p_{i,i}(t) \ge p_{i,i}(\delta)p_{i,i}(\delta)\times\cdots\times p_{i,i}(\delta)p_{i,i}(t-k\delta);$$

since $t - k\delta < \delta$, all the factors on the right are positive and therefore $p_{i,i}(t) > 0$ for all $t > 0$. Suppose now that $i \ne j$. Let m be the smallest integer for which there is a finite sequence $i_1, \ldots, i_m$ such that

$$q_{i,i_1}q_{i_1,i_2}\times\cdots\times q_{i_m,j} > 0.$$

Then $q_{i,j}^{(n)} = 0$ for all $n < m$. Consider

$$q_{i,j}^{(m)} = \sum_{i_1=1}^{r}\cdots\sum_{i_m=1}^{r} q_{i,i_1}q_{i_1,i_2}\times\cdots\times q_{i_m,j}.$$

Each term on the right must be nonnegative, because if some term were strictly negative then some factor $q_{i_j,i_{j+1}}$ would be strictly negative, and this can only happen if $i_j = i_{j+1}$, contradicting the minimality of m. Because all terms are nonnegative and at least one is positive, $q_{i,j}^{(m)} > 0$. But $q_{i,j}^{(n)} = 0$ for all $n < m$ implies that

$$p_{i,j}(t) = \sum_{n=m}^{\infty} \frac{t^n}{n!} q_{i,j}^{(n)}.$$

Since

$$\frac{p_{i,j}(t)}{t^m} - \frac{q_{i,j}^{(m)}}{m!} = \sum_{n=m+1}^{\infty} \frac{t^{n-m}}{n!} q_{i,j}^{(n)} \to 0 \text{ as } t \to 0+,$$

$p_{i,j}(t)$ has the same sign as $q_{i,j}^{(m)}$ for all small t, and therefore $p_{i,j}(t) > 0$ for all small t. Choosing $\delta > 0$ such that $p_{i,j}(t) > 0$ for $0 < t < \delta$, $p_{i,j}(t+s) \geq p_{i,j}(t)p_{j,j}(s) > 0$ for $0 < t < \delta$ and all $s > 0$. Therefore, $p_{i,j}(t) > 0$ for all $t > 0$. ■

To study the long-range behavior of the $p_{i,j}(t)$, we will recall the discrete parameter version first. Let $P = [p_{i,j}]$ be an $r \times r$ stochastic matrix; i.e., $p_{i,j} \geq 0$ for $1 \leq i,j \leq r$ and $\sum_{j=1}^{r} p_{i,j} = 1, 1 \leq i \leq r$. Moreover, let $P^n = [p_{i,j}^{(n)}]$ be the matrix of n-step transition probabilities. According to Theorem 5.2.2, if there is a positive integer N such that $p_{i,j}^{(N)} > 0$ for $1 \leq i,j \leq r$, then $\lim_{n\to\infty} p_{i,j}^{(n)}$ exists and is independent of i. It should be noted that the conclusion of the following theorem does not require that the q-matrix be irreducible, but the limit may depend upon i.

Theorem 8.6.2 *Let $Q = [q_{i,j}]$ be an irreducible $r \times r$ q-matrix and let $P(t) = [p_{i,j}(t)]$ be the associated matrix of Markov transition functions. Then there is a probability density $\pi = \{\pi_1, \ldots, \pi_r\}$ such that*

$$\lim_{t\to\infty} p_{i,j}(t) = \pi_j, \qquad j = 1, \ldots, r,$$

independently of i.

PROOF: Fix $j \in \{1, \ldots, r\}$ and let $\varepsilon > 0$. Since $\lim_{h\to 0+} p_{i,j}(h) = \delta_{i,j}, 1 \leq i,j \leq r$,

$$\lim_{h\to 0+} \sum_{i=1}^{r} |p_{i,j}(h) - \delta_{i,j}| = 0$$

and there is an $h_0 > 0$ such that

$$\sum_{i=1}^{r} |p_{i,j}(h) - \delta_{i,j}| < \frac{\varepsilon}{2}$$

whenever $0 < h < h_0$. Fixing h with $0 < h < h_0$, we will now show that $\pi_j = \lim_{n\to\infty} p_{i,j}(nh)$ exists independently of i. To see this, note that $P(h) = [p_{i,j}(h)]$ is a stochastic matrix and that $p_{i,j}(h) > 0$ by Theorem 8.6.1, since Q is irreducible. Moreover, $[p_{i,j}(nh)] = P(nh) = [P(h)]^n = [p_{i,j}^{(n)}(h)]$ and $\pi_j = \lim_{n\to\infty} p_{i,j}(nh)$ exists and is independent of i, since this is true of the $p_{i,j}^{(n)}(h)$ by Theorem 5.2.2. Any $t > 0$ can be written $t = n(t)h + r(t)$ where $n(t)$ is a nonnegative integer and $0 \le r(t) < h$. From the inequalities

$$|p_{i,j}(t) - \pi_j| \le |p_{i,j}(t) - p_{i,j}(n(t)h)| + |p_{i,j}(n(t)h) - \pi_j|$$

and

$$|p_{i,j}(t) - p_{i,j}(n(t)h)| = \left|\sum_{k=1}^{r} p_{i,k}(n(t)h)p_{k,j}(r(t)) - \sum_{k=1}^{r} p_{i,k}(n(t)h)p_{k,j}(0)\right|$$
$$\le \sum_{k=1}^{r} |p_{k,j}(r(t)) - \delta_{k,j}|,$$

$$|p_{i,j}(t) - \pi_j| \le \frac{\varepsilon}{2} + |p_{i,j}(n(t)h) - \pi_j|.$$

Since $n(t) \to \infty$ as $t \to \infty$, the second term on the right can be made less than $\varepsilon/2$ for large t; i.e., $\lim_{t\to\infty} p_{i,j}(t) = \pi_j$. Clearly, $\pi_j \ge 0, 1 \le j \le r$. Since $\sum_{j=1}^{r} p_{i,j}(t) = 1$ and there are only a finite number of terms in the sum, $\sum_{j=1}^{r} \pi_j = 1$. ■

Except for relatively simple q-matrices, using limits to calculate π is difficult. Fortunately, there is a simple algebraic method for finding π, assuming that the q-matrix is irreducible.

Definition 8.6 *The probability density π on $\{1, \ldots, r\}$ is a* stationary density *for the Markov transition functions $p_{i,j}(t)$ if*

$$\pi_j = \sum_{k=1}^{r} \pi_k p_{k,j}(t), \qquad j = 1, \ldots, r. \ ■$$

Theorem 8.6.3 *Let $Q = [q_{i,j}]$ be an $r \times r$ irreducible q-matrix and let $P(t) = [p_{i,j}(t)]$ be the associated matrix of transition functions. Then $P(t)$ has a unique stationary* *density π that satisfies the equation*

$$\sum_{i=1}^{r} \pi_i q_{i,j} = 0, \qquad j = 1, \ldots, r. \tag{8.19}$$

PROOF: Let $\pi_j = \lim_{t\to\infty} p_{i,j}(t)$. Since

$$p_{i,j}(s+t) = \sum_{k=1}^{r} p_{i,k}(s) p_{k,j}(t),$$

letting $s \to \infty$ we obtain $\pi_j = \sum_{k=1}^{r} \pi_k p_{k,j}(t)$, and π_j is a stationary density for $P(t)$. Let σ be a second stationary density. Since

$$\sigma_j = \sum_{k=1}^{r} \sigma_k p_{k,j}(t), \qquad t > 0,$$

letting $t \to \infty$,

$$\sigma_j = \sum_{k=1}^{r} \sigma_k \pi_j = \pi_j \sum_{k=1}^{r} \sigma_k = \pi_j, \qquad j = 1, \ldots, r,$$

and $\sigma = \pi$. Thus π is unique. Since

$$\pi_j = \sum_{k=1}^{r} \pi_k p_{k,j}(t) = \sum_{k=1}^{r} \pi_k \sum_{n=0}^{\infty} \frac{t^n}{n!} q_{k,j}^{(n)} = \sum_{n=0}^{\infty} \frac{t^n}{n!} \left(\sum_{k=1}^{r} \pi_k q_{k,j}^{(n)} \right),$$

the sum of the power series in t on the right is a constant, and therefore all the coefficients of $t, t^2, \ldots$ must be zero; in particular

$$\sum_{k=1}^{r} \pi_k q_{k,j}^{(1)} = \sum_{k=1}^{r} \pi_k q_{k,j} = 0. \; \blacksquare$$

Equation 8.19 alone does not determine the stationary density π uniquely. The equation $\sum_{j=1}^{r} \pi_j = 1$ must be used in conjunction with Equation 8.19.

EXAMPLE 8.8 Consider the q-matrix

$$Q = \begin{bmatrix} -4 & 1 & 2 & 1 \\ 1 & -4 & 2 & 1 \\ 2 & 1 & -6 & 3 \\ 0 & 0 & 1 & -1 \end{bmatrix}.$$

It is easily seen that Q is irreducible. In this case, Equation 8.19 becomes

$$\begin{aligned} -4\pi_1 + \pi_2 + 2\pi_3 \quad &= 0 \\ \pi_1 - 4\pi_2 + \pi_3 \quad &= 0 \\ 2\pi_1 + 2\pi_2 - 6\pi_3 + \pi_4 &= 0 \\ \pi_1 + \pi_2 + 3\pi_3 - \pi_4 &= 0. \end{aligned}$$

Applying the usual row and column operations, these equations are easily seen to be linearly dependent, and the equation

$$\pi_1 + \pi_2 + \pi_3 + \pi_4 = 1$$

must be used to find the stationary density π. In this case, $\pi_1 = 1/10$, $\pi_2 = 1/15$, $\pi_3 = 1/6$, $\pi_4 = 2/3$. If the $p_{i,j}(t)$ are the transition functions corresponding to the above q-matrix, then $\lim_{t\to\infty} p_{i,1}(t) = 1/10$, $\lim_{t\to\infty} p_{i,2}(t) = 1/15$, $\lim_{t\to\infty} p_{i,3}(t) = 1/6$, $\lim_{t\to\infty} p_{i,4}(t) = 2/3$ for $i = 1, 2, 3, 4$. ■

EXERCISES 8.6

1. Which of the following q-matrices are irreducible?

$$Q_1 = \begin{bmatrix} -1 & 1 & 0 & 0 & 0 \\ 0 & -1 & 0 & 1 & 1 \\ 0 & 1 & -1 & 0 & 0 \\ 0 & 0 & 1 & -1 & 0 \\ 1 & 0 & 0 & 0 & -1 \end{bmatrix}$$

$$Q_2 = \begin{bmatrix} -1 & 1 & 0 & 0 & 0 \\ 1 & -1 & 0 & 0 & 0 \\ 1 & 0 & -2 & 1 & 0 \\ 0 & 0 & 0 & -1 & 1 \\ 0 & 0 & 1 & 0 & -1 \end{bmatrix}.$$

2. Consider the system of Exercise 8.4.3. If the failure rate is $\lambda = .01$ and the repair rate is $\mu = 2$, what is the limiting distribution of the system?
3. Determine the limiting distribution of the transition functions $p_{i,j}(t)$ corresponding to the following q-matrix:

$$Q = \begin{bmatrix} -3 & 0 & 1 & 1 & 1 \\ 1 & -3 & 0 & 1 & 1 \\ 1 & 1 & -4 & 1 & 1 \\ 1 & 0 & 1 & -2 & 0 \\ 0 & 0 & 0 & 1 & -1 \end{bmatrix}.$$

The following problems require mathematical software such as Mathematica or Maple V.

4. Consider the Markov chain $P(t)$ determined by the q-matrix

$$Q = \begin{bmatrix} -1 & 1/2 & 1/2 \\ 1/3 & -1 & 2/3 \\ 1/4 & 3/4 & -1 \end{bmatrix}.$$

Approximate with three-place accuracy the Markov transition function $P(t) = \{p_{i,j}(t)\}$ when $t = 2$, and determine $\pi_j = \lim_{t\to\infty} p_{i,j}(t)$, $1 \leq j \leq 3$.

5. Approximate with three-place accuracy the limiting distribution of the transition function $P(t) = [p_{i,j}(t)]$ corresponding to the q-matrix

$$Q = \begin{bmatrix} -1 & .10 & .25 & .45 & .20 \\ .05 & -1 & .15 & .65 & .15 \\ .10 & .10 & -1 & .50 & .30 \\ .25 & .25 & .25 & -1 & .25 \\ .50 & .10 & .20 & .20 & -1 \end{bmatrix}.$$

SUPPLEMENTAL READING LIST

S. Karlin and H. M. Taylor (1981). *A Second Course in Stochastic Processes.* New York: Academic Press.

SOLUTIONS TO EXERCISES

CHAPTER 1

Exercises 1.2

1. $P(A) = 3/8$.
2. $|\Omega| = 2^n$.
3. $P(A) = 3/8$.
4. $P(A) = 1/4$.
5. 21 configurations.
6. Let $P_1(n)$ be the statement that $1 + 2 + \cdots + n = n(n+1)/2$. Since $1 = 1(1+1)/2$, $P_1(1)$ is true. Assume $P_1(n)$ is true. Then

$$1 + 2 + \cdots + n + (n+1) = \frac{n(n+1)}{2} + (n+1) = \frac{(n+1)(n+2)}{2},$$

which is just the statement $P_1(n+1)$. Therefore, $P_1(n+1)$ is true whenever $P_1(n)$ is true, and $P_1(n)$ is true for all $n \geq 1$ by the principle of mathematical induction. Now let $P_2(n)$ be the statement $1^2 + 2^2 + \cdots + n^2 = n(n+1)(2n+1)/6$. Since $1^2 = 1(1+1)(2+1)/6$, $P_2(1)$ is true. Assume $P_2(n)$ is true. Then

$$\begin{aligned} 1^2 + 2^2 + \cdots + n^2 + (n+1)^2 &= \frac{n(n+1)(2n+1)}{6} + (n+1)^2 \\ &= (n+1)\left(\frac{2n^2+n}{6} + n + 1\right) \end{aligned}$$

$$= (n+1)\left(\frac{2n^2+n+6n+6}{6}\right)$$
$$= \frac{(n+1)(n+2)(2n+3)}{6},$$

which is just the statement $P_2(n+1)$. Therefore, $P_2(n+1)$ is true whenever $P_2(n)$ is true, and $P_2(n)$ is true for all $n \geq 1$ by the principle of mathematical induction.

Exercises 1.3

1. (a) is obvious. (b) Since the number of ways of selecting n individuals out of m for inclusion in a sample is the same as the number of ways of selecting $m - n$ to not be included, the two binomial coefficients are equal.
2. Express $C(n-1, r)$ and $C(n-1, r-1)$ in terms of factorials, take the sum, and simplify the resulting equation.
3. (a) $f(0) = 1$ and for $k \leq n, f^{(k)}(0) = n(n-1)\times\cdots\times(n-k+1) = (n)_k, f^{(k)}(0)/k! = (n)_k/k! = \binom{n}{k}$; for $k > n$, $f^{(k)}(0) = 0$, which is also equal to $\binom{n}{k}$ since $k > n$. Therefore,
$$\sum_{k=0}^{\infty} \frac{f^{(k)}(0)}{k!}t^k = \sum_{k=0}^{\infty}\binom{n}{k}t^k = \sum_{k=0}^{n}\binom{n}{k}t^k.$$
4. The kth derivative of $f(t)$ at $t = 0$ is $\alpha(\alpha-1)\times\cdots\times(\alpha-k+1) = (\alpha)_k$.
5. Let $a = 1$ and $b = t$ in Equation 1.9, differentiate with respect to t, and set $t = 1$.
6. Put $a = 1$ and $b = t$ in Equation 1.9, differentiate twice with respect to t, and put $t = 1$.
7. Let A be the collection of outcomes not having a 1. Then A is an ordered sample of size n with replacement from the population {2,3,4,5,6} of size 5. The number of such samples is 5^n. Thus, $P(A) = 5^n/6^n$.
8. Let A be the collection of outcomes for which the number of heads and tails are equal. The total number of outcomes is 2^{2n}. To count the number of outcomes in A, select n out of the $2n$ positions in a label to be filled with H's, which can be done in $\binom{2n}{n}$ ways, and fill the remaining positions with T's. Thus, $P(A) = \binom{2n}{n}/2^{2n}$.
9. (a) $\binom{3n-1}{n}$. (b) Let A be the collection of outcomes for which boxes $1, 2, \ldots, n$ are empty. The n particles are then distributed among the

remaining boxes numbered $n+1,\ldots,2n$ in $\binom{2n-1}{n}$ ways. Thus,

$$P(A) = \binom{2n-1}{n} \Big/ \binom{3n-1}{n}.$$

10.
$$\binom{-x}{k} = \frac{(-x)_k}{k!} = \frac{(-x)(-x-1)\times\cdots\times(-x-k+1)}{k!}$$
$$= (-1)^k \frac{(x+k-1)\times\cdots\times(x)}{k!}$$
$$= (-1)^k \frac{(x+k-1)_k}{k!}$$
$$= (-1)^k \binom{x+k-1}{k}.$$

11. .176197.
12. 53,130.

Exercises 1.4

1. $p(2) = 1/16$, $p(3) = 1/8$, $p(4) = 3/16$, $p(5) = 1/4$, $p(6) = 3/16$, $p(7) = 1/8$, $p(8) = 1/16$.
2. $p(3) = 1/64$, $p(4) = 3/64$, $p(5) = 6/64$, $p(6) = 10/64$, $p(7) = 12/64$, $p(8) = 12/64$, $p(9) = 10/64$, $p(10) = 6/64$, $p(11) = 3/64$, $p(12) = 1/64$.
3. $p(3) = 1/216$, $p(4) = 3/216$, $p(5) = 6/216$, $p(6) = 10/216$, $p(7) = 15/216$, $p(8) = 21/216$, $p(9) = 25/216$, $p(10) = 27/216$, $p(11) = 27/216$, $p(12) = 25/216$, $p(13) = 21/216$, $p(14) = 15/216$, $p(15) = 10/216$, $p(16) = 6/216$, $p(17) = 3/216$, $p(18) = 1/216$.
4. $1/\binom{54}{6} = 3.8719 \times 10^{-8}$; or one chance in 25,827,165.
5. $1/(48)_6 \approx .11318 \times 10^{-9}$.
6.
$$\frac{\binom{50}{2}\binom{950}{8}}{\binom{1000}{10}}$$
or, what is the same,
$$\frac{\binom{10}{2}\binom{990}{48}}{\binom{1000}{50}}.$$
7. $(12)_4/12^4 = .573$, accurate to three decimal places.

8. $$\frac{13 \times 12 \times \binom{4}{3}\binom{4}{2}}{\binom{52}{5}}.$$

9. $10 \cdot 4/\binom{52}{5}$.

10. $$\frac{\binom{100}{5}\binom{n-100}{95}}{\binom{n}{100}}.$$

One would guess 2000.

Exercises 1.5

1. 8/7.
2. $1/(15 \cdot 16^3)$.
3. 6/11.
4. 5/12.
5. An outcome is an ordered sample of size 10 from a population of size 2 with replacement. (a)$P(A_1) = P(A_2) = 2^9/2^{10} = 1/2, P(A_1 \cap A_2) = 2^8/2^{10} = 1/4, P(A_2 \mid A_1) = 1/2$. (b) $P(A_1 \cap A_2) = P(A_1)P(A_2)$. (c) $P(A_2 \mid A_1) = P(A_2)$. (d) If 10 is replaced by 20, there is no change in any of these probabilities.
6. 1/2.
7. 22/63.
8. From Figure 1.2, $P(A \cap B) = 1/6, P(B) = 1/2$, so that $P(A \mid B) = 1/3$. Since $P(A) = 1/3$, the partial information does not change the probability of A.
9. What does random mean in this exercise? If all n keys are placed in a row and tried one at a time, then we are dealing with an ordered sample of size n without replacement from a population of size n, and there are $n!$ outcomes. The number of outcomes with the good key in the rth place is $(n-1)!$. The required probability is $(n-1)!/n! = 1/n$.
10. (a) $0 \leq p(k) \leq 1$. (b) Since

$$\sum_{k=1}^{\infty} p(k) = \sum_{k=1}^{\infty}\left(\frac{1}{k} - \frac{1}{k+1}\right) = \lim_{n\to\infty}\sum_{k=1}^{n}\left(\frac{1}{k} - \frac{1}{k+1}\right)$$
$$= \lim_{n\to\infty}\left(1 - \frac{1}{n+1}\right) = 1,$$

the $p(k)$ can be used as weights in a probability model.

CHAPTER 2

Exercises 2.2

1. (a) Correct. (b) Incorrect since 2 is not a set. (c) Incorrect since $\{1, 2, 3\}$ has no sets as members. (d) Correct.
2. The proposition is "$(i,j) \in \Omega$ and $i > j$." $A = \{(i,j) : (i,j) \in \Omega$ and $i > j\} = \{(2, 1), (3, 1), (4, 1), (5, 1), (6, 1), (3, 2), (4, 2), (5, 2), (6, 2), (4, 3), (5, 3), (6, 3), (5, 4), (6, 4), (6, 5)\}$.
3. Since $2^{1/n} > 1$, $[0, 1] \subset A_n$ for all $n \geq 1$. Thus, $[0, 1] \subset \cap A_n$. Since $\lim_{n\to\infty} 2^{1/n} = 1$, there is no $x > 1$ in $\cap A_n$. Thus, $\cap A_n = [0, 1]$.
4. Examine the graph of the equation $y = x^n$. $\cap A_n = \{(x, y) : 0 \leq x < 1, y = 0\}$.
5. $\cap A_n = \{(x, y) : 0 \leq x < 1, y = 0\} \cup \{(x, y) : x = 1, 0 \leq y \leq 1\}$.
6. If $\omega \in A$, then ω is not in A^c; i.e., $\omega \in (A^c)^c$. If $\omega \in (A^c)^c$, then ω is not in A^c; i.e., $\omega \in A$. Thus, $(A^c)^c \subset A$ and $A = (A^c)^c$.
7. Assume $A \subset B$. If $\omega \in B^c$, then ω cannot be in A since if it were it would be in B also, which is impossible. Thus, $\omega \notin A$ and $B^c \subset A^c$. Assume now that $B^c \subset A^c$. If $\omega \in A$, then $\omega \notin B^c$ since if it were then it would be in A^c, which is impossible; thus, $\omega \in (B^c)^c = B$, and so $A \subset B$.
8. Not true in general. Let $U = \{1, 2, 3, 4\}, X = \{1, 2, 3\}, Y = \{3, 4\}, Z = \{2, 3, 4\}$. Then $Y \cup Z = \{2, 3, 4\}, X \cap (Y \cup Z) = \{2, 3\}$, whereas $X \cap Y = \{3\}, (X \cap Y) \cup Z = \{2, 3, 4\}$. Clearly, $X \cap (Y \cup Z) \neq (X \cap Y) \cup Z$.
9. By de Morgan's laws, the distributive laws, and the facts that $X \cap X^c = \varnothing, Y \cap Y^c = \varnothing$, $(X \cup Y) \cap (X \cap Y)^c = (X \cup Y) \cap (X^c \cup Y^c) = ((X \cup Y) \cap X^c) \cup ((X \cup Y) \cap Y^c) = (X \cap X^c) \cup (Y \cap X^c) \cup (X \cap Y^c) \cup (Y \cap Y^c) = (Y \cap X^c) \cup (X \cap Y^c)$.
10. If n is odd, $A_n = [0, 1]$; if n is even, $A_n = [0, 0] = \{0\}$. For each $n \geq 1$, $\cap_{k\geq n} A_k = \{0\}, \cup_{n=1}^{\infty}(\cap_{k\geq n} A_k) = \cup_{n=1}^{\infty}\{0\} = \{0\}$. Also, $\cup_{k\geq n} A_k = [0, 1], \cap_{n=1}^{\infty}(\cup_{k\geq n} A_k) = [0, 1]$.

Exercises 2.3

1. The domain is $[-1, 1]$. The range is $[0, 1]$.
2. $f = \{(x, y) : x \in R, y \in R, y = \sqrt{1 - x^4}, -1 \leq x \leq 1\}$. The domain is $[-1, 1]$. The range is $[0, 1]$.
3. $f = \{(x, y) : x \in R, y \in R, y = 1/\sqrt{1 - x^2}, -1 < x < 1\}$. The domain is $(-1, 1)$, and the range is $[1, \infty)$.
4. Define $\alpha : N \to X$ by putting $\alpha(p) = p/q$. Then X is the range of α, and X is countable.
5. For $i = 1, 2, \ldots, m$, let $X_i = \{x_{i1}, x_{i2}, \ldots\}$. Since each X_i is countably infinite, there is a mapping $\alpha_i : N \to X_i$ having X_i as its range. Consider

the array

$$\begin{matrix} x_{11} & x_{12} & x_{13} & \dots \\ x_{21} & x_{22} & x_{23} & \dots \\ x_{31} & x_{32} & x_{33} & \dots \\ \vdots & \vdots & \vdots & \ddots \\ x_{m1} & x_{m2} & x_{m3} & \dots \end{matrix}$$

The elements of this array can be arranged in a sequence $\{\alpha(n)\}_{n=1}^{\infty}$ by going down the first column, then down the second column, and so forth. To identify $\alpha(n)$ with some $\alpha_i(m)$, let p be the number of columns to the left of the nth element $\alpha(n)$. The $n - pm$ element of the $(p+1)$st column is then the nth element $\alpha(n)$. More precisely, each $n \in N$ has a unique representation $n = pm + q$ where $p \in N \cup \{0\}$ and $q \in \{1, 2, \dots, m\}$. Letting $X = \cup X_i$, define $\alpha : N \to X$ as follows. If $n = pm + q$ as above, put

$$\alpha(n) = \alpha_q(p+1).$$

Then X is the range of α.

6. Assume X is countable; i.e., X is the range of an infinite sequence $\{x_n\}_{n=1}^{\infty}$ where each x_n is an infinite sequence of 0's and 1's. Let $x_n = \{x_{n,k}\}_{k=1}^{\infty}$. Consider the infinite sequence $y = \{y_k\}_{k=1}^{\infty}$ where $y_k = 1 - x_{k,k}$. Since $y_n \neq x_{n,n}$ for each $n \geq 1$, $y \neq x_n$ for all $n \geq 1$. This is a contradiction since $y \in X$ but is not in the range of $\{x_n\}_{n=1}^{\infty}$. Therefore, X is not countable.
7. $N \times N = \cup_{k=2}^{\infty} A_k$. Since each A_k is countable, $N \times N$ is countable by Theorem 2.3.1.
8. We can assume that A and B are ranges of infinite sequences $\{a_m\}_{m=1}^{\infty}$ and $\{b_n\}_{n=1}^{\infty}$, respectively. Then $A \times B = \cup_{k=2}^{\infty}\{(a_m, b_n) : m + n = k\}$.
9. All three are countable.
10. Suppose $X_i = \{x_{i1}, x_{i2}, \dots\}, i \geq 1$. Fix $n \in N$ and consider the set $A = \{m : m \in N, ((m+1)(m+2))/2 \geq n\}$. $A \subset N$. Since

$$\frac{(n+1)(n+2)}{2} \geq \frac{3n}{2} \geq n,$$

$n \in A$ and therefore $A \neq \emptyset$. It follows from the well-ordering property that A has a least element $m(n)$ for which

$$\frac{(m(n)+1)(m(n)+2)}{2} \geq n.$$

Since $m(n)$ is the smallest integer with this property, $m(n) - 1$ does not have this property, and so

$$\frac{m(n)(m(n)+1)}{2} < n$$

and $m(n)$ is the largest element of N with this property. Now define $l(n)$ so that

$$n = \frac{m(n)(m(n)+1)}{2} + l(n)$$

where $1 \leq l(n) \leq m(n) + 1$. Define $\alpha : N \to X = \cup_{n=1}^{\infty} X_n$ by putting

$$\alpha(n) = x_{m(n)+2-l(n),l(n)}.$$

Exercises 2.4

1. Consider the statement

$$P(n) : A_1, \ldots, A_n \in \mathcal{A} \quad \text{implies} \quad \bigcup_{j=1}^{n} A_j \in \mathcal{A}.$$

When $n = 1$, the statement simply says that $A_1 \in \mathcal{A}$ implies $A_1 \in \mathcal{A}$, which is trivially true. Assume that $P(n)$ is true and consider $P(n+1)$. Let $A_1, \ldots, A_n, A_{n+1} \in \mathcal{A}$. By the induction hypothesis that $P(n)$ is true, $A_1 \cup \cdots \cup A_n \in \mathcal{A}$. Since $\mathcal{A}$ is an algebra, $A_1 \cup \cdots A_n \cup A_{n+1} = (A_1 \cup \cdots A_n) \cup A_{n+1} \in \mathcal{A}$. Therefore, $P(n+1)$ is true. It follows from the principle of mathematical induction that $P(n)$ is true for all positive integers n.

2. Suppose $A_1, \ldots, A_n \in \mathcal{A}$. Since $\mathcal{A}$ is closed under complementation, $A_1^c, \ldots, A_n^c \in \mathcal{A}$. Since $\mathcal{A}$ is closed under finite unions by Problem 1, $\cup_{j=1}^{n} A_j^c \in \mathcal{A}$. Since $\mathcal{A}$ is closed under complementation, $(\cup_{j=1}^{n} A_j^c)^c \in \mathcal{A}$. By de Morgan's laws,

$$\bigcap_{j=1}^{n} A_j = \left(\bigcup_{j=1}^{n} A_j^c \right)^c \in \mathcal{A}.$$

3. Let $\{A_j\}$ be a finite or infinite sequence in the σ-algebra $\mathcal{F}$. If finite, the intersection is in $\mathcal{F}$ by Problem 2 since $\mathcal{F}$ is an algebra. Assume $\{A_j\}$ is an infinite sequence. Since $\mathcal{F}$ is closed under complementation, the sequence $\{A_j^c\}$ is in $\mathcal{F}$ and $\cup A_j^c$ is in $\mathcal{F}$. Thus, $\cap A_j = (\cup A_j^c)^c \in \mathcal{F}$.

4. $\Omega \in \mathcal{F}$ since $\Omega^c = \varnothing$ is countable. Suppose $A \in \mathcal{F}$. If A is countable, then the complement of A^c is countable and $A^c \in \mathcal{F}$; if A is not countable, then A^c is countable and $A^c \in \mathcal{F}$. In either case, $A^c \in \mathcal{F}$. Let $\{A_j\}$ be a sequence in $\mathcal{F}$. If every A_j is countable, then $\cup A_j$ is countable and belongs to $\mathcal{F}$; if some A_{j_0} is not countable, then $A_{j_0}^c$ is countable; $(\cup A_j)^c = (\cap A_j^c)$ is countable since it is a subset of the countable A_{j_0}, and so $\cup A_j \in \mathcal{F}$. In either case, $\cup A_j \in \mathcal{F}$.
5. For each $n \geq 1$, let A_{2n} be the event "success occurs for the first time on trial numbered $2n$." Let $A = \cup_{n=1}^{\infty} A_{2n}$. Since the A_{2n} are disjoint events, $P(A) = \sum_{n=1}^{\infty} P(A_{2n})$. But $P(A_{2n}) = q^{2n-1}p$ and

$$P(A) = \sum_{n=1}^{\infty} q^{2n-1}p = \frac{p}{q}\sum_{n=1}^{\infty}(q^2)^n = pq\sum_{n=1}^{\infty}(q^2)^{n-1} = \frac{pq}{1-q^2}$$
$$= \frac{q}{1+q}.$$

6. $(5/36) + (31/36)^2(5/36) + (31/36)^4(5/36) + (31/36)^6(5/36) + \cdots = .5373$.
7. Let W_n, R_n, and B_n be the events that a white chip, a red chip, and a black chip are chosen on the nth drawing, respectively. The event A of interest can be decomposed into disjoint events according to the trial at which a white chip appears for the first time while being preceded by red chips; i.e., $A = W_1 \cup (R_1 \cap W_2) \cup (R_1 \cap R_2 \cap W_3) \cup \cdots$ and $P(A) = P(W_1) + P(R_1 \cap W_2) + P(R_1 \cap R_2 \cap W_3) + \cdots = w/(w+b)$.
8. Define the sequence $\{B_j\}_{j=1}^{\infty}$ by $B_1 = A_1$ and $B_j = A_j \cap A_{j-1}^c$ for $j \geq 2$. The B_j are disjoint. Since each $B_j \subset A_j, j \geq 1, \cup B_j \subset \cup A_j$. Suppose $\omega \in \cup A_j$. Then there is a largest $k \geq 1$ such that $\omega \in A_k$ and $\omega \notin \cup_{j=1}^{k-1} A_j$, so that $\omega \in A_k \cap A_{k-1}^c = B_k \subset \cup B_j$. Thus, $\cup A_j \subset \cup B_j$ and the two are equal.
9. Fix $n \geq 1$. Since $A = \cap_{j=1}^{\infty} A_j \subset A_n, A_n = A \cup (A_n \cap A^c)$. Let $B_n = A_n \cap A^c$. Since $A_{n+1} \subset A_n, B_{n+1} = A_{n+1} \cap A^c \subset A_n \cap A^c = B_n$ and $\{B_j\}_{j=1}^{\infty}$ is a decreasing sequence. Now $\cap_{j=1}^{\infty} B_j = \cap_{j=1}^{\infty}(A_j \cap A^c)$. It is easy to check that the last set is equal to $A^c \cap \cap_{j=1}^{\infty} A_j = A^c \cap A = \varnothing$.

Exercises 2.5

1. Let A be the event "outer diameter of the sleeve ω is too large" and let B be the event "inner diameter of the sleeve ω is too large." Then $P(A) = .05$ and $P(B) = .03$. Since we know nothing about $P(A \cap B)$, $P(A \cup B) \leq P(A) + P(B) = .08$. Since $.05 = P(A) \leq P(A \cup B)$ and $.03 = P(B) \leq P(A \cup B), .05 \leq P(A \cup B) \leq P(A) + P(B) = .08$.
2. Let A, B, and C be three arbitrary events. Then, by Equation 2.11,

$$P(A \cup B \cup C) = P((A \cup B) \cup C) = P(A \cup B) + P(C) - P((A \cup B) \cap C).$$

But

$$P(A \cup B) = P(A) + P(B) - P(A \cap B)$$

and

$$\begin{aligned} P((A \cup B) \cap C) &= P((A \cap C) \cup (B \cap C)) \\ &= P(A \cap C) + P(B \cap C) - P((A \cap C) \cap (B \cap C)) \\ &= P(A \cap C) + P(B \cap C) - P(A \cap B \cap C). \end{aligned}$$

Thus,

$$\begin{aligned} P(A \cup B \cup C) &= P(A) + P(B) - P(A \cap B) + P(C) \\ &\quad - P(A \cap C) - P(B \cap C) + P(A \cap B \cap C). \end{aligned}$$

3. 65/96.
4. (a) $P = 4 \cdot (1000/10^4) - 6 \cdot (100/10^4) + 4 \cdot (10/10^4) - (1/10^4) = .3439$ (b) $P = 1 - (9/10)^4 = .3439$.
5. The required probability is $P((A \cap B^c) \cup (B \cap A^c)) = P(A \cap B^c) + P(B \cap A^c) - P((A \cap B^c) \cap (B \cap A^c)) = P(A) - P(A \cap B) + P(B) - P(A \cap B) = P(A) + P(B) - 2P(A \cap B)$.
6. No. The probability of getting at least one ace with the throw of four dice is

$$1 - \left(\frac{5}{6}\right)^4 \approx .51775,$$

whereas the probability of getting a double ace in 24 throws of a pair of dice is

$$1 - \left(\frac{35}{36}\right)^{24} \approx .49140.$$

7. Suppose $\omega_0 = \{\delta_j\}_{j=1}^{\infty}$ where each δ_i is a 1 or a 0. For each $n \geq 1$, let $A_n = \{\omega : \omega = \{x_j\}_{j=1}^{\infty}, x_1 = \delta_1, \ldots, x_n = \delta_n\}$. Then $\{\omega_0\} = \bigcap_{j=1}^{\infty} A_j \subset A_n$ for all $n \geq 1$. Let $\alpha = \max(p, q) < 1$. Since

$$\begin{aligned} 0 \leq P(A_n) &= p^{\sum_{i=1}^n \delta_i} q^{(n - \sum_{i=1}^n \delta_i)} \\ &\leq \alpha^{\sum_{i=1}^n \delta_i} \alpha^{(n - \sum_{i=1}^n \delta_i)} \\ &= \alpha^n \end{aligned}$$

and $\lim_{n \to \infty} \alpha^n = 0$, $\lim_{n \to \infty} P(A_n) = 0$, and so $P(\omega_0) = 0$.

8. Since $A \cap B \subset A$ and $A \cap B \subset B$, $P(A \cap B) \leq P(A)$ and $P(A \cap B) \leq P(B)$, so that $P(A \cap B) \leq \min(P(A), P(B))$. On the other hand, $P(A \cap B) = P(A) + P(B) - P(A \cup B) \geq P(A) + P(B) - 1$ since $P(A \cup B) \leq 1$.

9. The inequality is trivially true when $n = 1$. Assume the inequality is true for n. By Problem 8,

$$\begin{aligned} &P(A_1 \cap \cdots \cap A_n \cap A_{n+1}) \\ &\geq P(A_1 \cap \cdots A_n) + P(A_{n+1}) - 1 \\ &\geq (P(A_1) + \cdots + P(A_n) - (n-1)) + P(A_{n-1}) - 1 \\ &= P(A_1) + \cdots + P(A_{n+1}) - n, \end{aligned}$$

and the inequality is true for $n + 1$ whenever it is true for n. By the principle of mathematical induction, it is true for all $n \geq 1$.

Exercises 2.6

1. By mutual independence, the pairs A and B, A and C, B and C are independent. By Theorem 2.6.1, the pairs A and B^c, B^c and C are independent. Thus, the pairs A and B^c, A and C, B^c and C are independent. We need only show that $P(A \cap B^c \cap C) = P(A)P(B^c)P(C)$. By Equation 2.8, $P(A \cap B^c \cap C) = P((A \cap C) \cap B^c) = P(A \cap C) - P(A \cap C \cap B) = P(A)P(C) - P(A)P(B)P(C) = P(A)P(C)(1 - P(B)) = P(A)P(B^c)P(C)$. Thus, A, B^c, and C are mutually independent.

2. Let $A_1, \ldots, A_n$ be mutually independent. This means that if $1 \leq i_1 < \cdots i_k \leq n$, then

$$P(A_{i_1} \cap \cdots \cap A_{i_k}) = \prod_{j=1}^{k} P(A_{i_j}).$$

If we can show that this equation remains valid whenever some A_{i_j} is replaced by its complement, then this procedure can be repeated as often as necessary to arrive at the B_{i_j}. Just by interchanging the positions of the A_{i_j}, we can assume that we want to replace A_{i_1} by its complement. By Equation 2.8,

$$\begin{aligned} P(A_{i_1}^c \cap A_{i_2} \cap \cdots \cap A_{i_k}) &= P(A_{i_2} \cap \cdots \cap A_{i_k}) - P(A_{i_1} \cap \cdots \cap A_{i_k}) \\ &= \prod_{j=2}^{k} P(A_{i_j}) - \prod_{j=1}^{k} P(A_{i_j}) \\ &= \left(\prod_{j=2}^{k} P(A_{i_j})\right)\left(1 - P(A_{i_1})\right) \\ &= P(A_{i_1}^c)P(A_{i_2}) \times \cdots \times P(A_{i_k}). \end{aligned}$$

3. (a) $P(A \cap B \cap C) = P(A \cap C)B \cap C)) = P(A \mid B \cap C)P(C) = P(A \mid B \cap C)P(B \mid C)P(C)$. (b) $P(A_1 \cap \cdots \cap A_n) = P(A_n \mid A_1 \cap \cdots \cap A_{n-1})P(A_{n-1} \mid A_1 \cap \cdots \cap A_{n-2}) \times \cdots \times P(A_2 \mid A_1)P(A_1)$, provided the conditional probabilities are defined.

4. The only outcomes (i, j, k) with positive probabilities are $(11, 12, 13)$, $(11, 12, 11)$, $(11, 10, 11)$, $(11, 10, 9)$, $(9, 10, 11)$, $(9, 10, 9)$, $(9, 8, 9)$, $(9, 8, 7)$. Let R_i be the event "the ith ball selected is red," $i = 1, 2, 3$. Then

$$\begin{aligned} P(11, 12, 13) &= P(R_1^c \cap R_2^c \cap R_3^c) \\ &= P(R_3^c \mid R_1^c \cap R_2^c)P(R_2^c \mid R_1^c)P(R_1^c) \\ &= \frac{8}{20} \cdot \frac{9}{20} \cdot \frac{1}{2} \\ &= \frac{9}{100}. \end{aligned}$$

Similarly, $P(11, 12, 11) = 27/200$, $P(11, 10, 11) = 11/80$, $P(11, 10, 9) = 11/80$, $P(9, 10, 11) = 11/80$, $P(9, 10, 9) = 11/80$, $P(9, 8, 9) = 27/200$, $P(9, 8, 7) = 9/100$.

5. We are given $P(A) = P(B) = P(C) = 1/2$, $P(A \cap B) = P(A \cap C) = P(B \cap C) = P(A \cap B \cap C) = 1/4$. (a) Since $P(A \cap B) = P(A)P(B)$, $P(A \cap C) = P(A)P(C)$, and $P(B \cap C) = P(B)P(C)$, A, B, and C are pairwise independent. (b) Since $P(A \cap B \cap C) = 1/4 \neq 1/8 = P(A)P(B)P(C)$, the three events are not mutually independent.

6. Let A be the event "1 is sent" and let B be the event "1 is received." By Bayes' rule,
(a) $P(A \mid B) = .9896$.
(b) $P(A^c \mid B^c) = .9519$.

7. For $i = 1, 2, 3$, let A_i be the event "the ith digit sent is 1" and let B_i be the event "the ith digit received is 1." By independence and the previous problem,

$$\begin{aligned} &P(A_1 \cap A_2 \cap A_3 \mid B_1 \cap B_2^c \cap B_3) \\ &= \frac{P((A_1 \cap B_1) \cap (A_2 \cap B_2^c) \cap (A_3 \cap B_3))}{P(B_1 \cap B_2^c \cap B_3)} \\ &= \frac{P(A_1 \cap B_1)P(A_2 \cap B_2^c)P(A_3 \cap B_3)}{P(B_1)P(B_2^c)P(B_3)} \\ &= P(A_1 \mid B_1)P(A_2 \mid B_2^c)P(A_3 \mid B_3) \\ &= (.9896)^2(.0481) = .0471. \end{aligned}$$

8. Let C_i be the event "Chest i is selected" and let G be the event "Gold coin is observed." The given facts are $P(C_1) = P(C_2) = P(C_3) = 1/3$, $P(G \mid C_1) = 1$, $P(G \mid C_2) = 1/2$, $P(G \mid C_3) = 0$. Given that the observed coin is gold, the only way that the other coin can be gold is for

the outcome to be in Chest 1, which means that we are required to find $P(C_1 \mid G)$. Apply Bayes' rule. $P(C_1 \mid G) = 2/3$.

9. $P((A \cup B) \cap (C \cap D)) = P((A \cap C \cap D) \cup (B \cap C \cap D)) = P(A \cap C \cap D) + P(B \cap C \cap D) - P(A \cap B \cap C \cap D) = P(A)P(C)P(D) + P(B)P(C)P(D) - P(A)P(B)P(C)P(D) = P(C)P(D)\{P(A) + P(B) - P(A)P(B)\} = P(C \cap D)\{P(A) + P(B) - P(A \cap B)\} = P(C \cap D)P(A \cup B)$.
10. Since the events $A \cap A_j, j \geq 1$, are disjoint, $P(A \cap (\cup A_j)) = P(\cup(A \cap A_j)) = \sum P(A \cap A_j) = \sum P(A)P(A_j) = P(A)P(\cup A_j)$.
11. (a) .9007. (b) .9021. Not much is gained by adding B_3.
12. Three of the B_i and two of the C_j at a cost of \$460.

Exercises 2.7

1. $28{,}319/44{,}800 \approx .6321$.
2. (a) For equalization to occur on the $2n$th trial, there must be n heads and n tails. (b) Using the fact that $p(1-p) < 1/4$ and the limit ratio test, the infinite series

$$\sum_{n=1}^{\infty} P(A_{2n}) = \sum_{n=1}^{\infty} \binom{2n}{n} p^n (1-p)^n$$

converges, and so $P(A_{2n}\, i.o.) = 0$.

3. Let $B_{n,r}$ be the event consisting of those outcomes for which there is a run of length r beginning on the nth trial. Since $P(B_{n,r}) = 1/2^{r+1}$, the events $B_{r+2,r}, B_{2r+4,r}, \ldots, B_{n(r+2),r}, \ldots$ are independent events, and

$$\sum_{n=1}^{\infty} P(B_{n(r+2),r}) = \sum_{n=1}^{\infty} \frac{1}{2^{r+1}} = +\infty,$$

$P(B_{n(r+2),r}\; i.o.) = 1$. Since $(B_{n(r+2),r}\; i.o.) \subset (B_{n,r}\; i.o.) \subset (A_{n,r}\; i.o.)$, $P(A_{n,r}\; i.o.) = 1$.

4. Since

$$\sum_{n=1}^{\infty} P(A_{n,r_n}) = \sum_{n=1}^{\infty} \frac{1}{2^{(1+\delta)\log_2 n}} = \sum_{n=1}^{\infty} \frac{1}{n^{1+\delta}} < \infty,$$

$P(A_{n,r_n}\; i.o.) = 0$.

5. For Ω, take all sequences $\omega = \{x_j\}_{j=1}^{\infty}$ where each $x_i \in \{2, 3, \ldots, 12\}$. For $n \geq 1, 1 \leq i_1 < i_2 < \cdots < i_n$, and $\delta_1, \ldots, \delta_n \in \{2, 3, \ldots, 12\}$, let

$$A_{i_1,\ldots,i_n}(\delta_1, \ldots, \delta_n) = \{\omega : \omega = \{x_i\}_{i=1}^{\infty}, x_{i_1} = \delta_1, \ldots, x_{i_n} = \delta_n\}$$

and let $\mathcal{F}$ be the smallest σ-algebra containing all such sets. Define

$$\begin{aligned} P(A_{i_1,\ldots,i_n}(\delta_1,\ldots,\delta_n)) &= P(A_{i_1}(\delta_1)\cap\cdots\cap A_{i_n}(\delta_n)) \\ &= \prod_{j=1}^{n} P(A_{i_j}(\delta_i)) \\ &= \prod_{j=1}^{n} p(\delta_i) \end{aligned}$$

where p is the weight function given in Figure 1.3.

6. $\sum_{n=1}^{\infty}(5/36)(25/36)^{n-1}(5/36) = 25/(11\cdot 36)$.
7. $6/36 + 2/36 + 2\{1/36 + 2/45 + 25/(11)(36)\} \approx .4929$.
8. The probability of losing is $1/36 + 2/36 + 1/36 + 2\{1/18 + 1/15 + 5/66\} \approx .507$. There is probability 1 that the game will terminate, since the sum of the probabilities of winning and losing is 1.
9. .63210558.
10. Let T be the number of purchases required to obtain a complete set of collectibles. Then

$$P(T > n) = \sum_{r=1}^{8}(-1)^{r-1}\binom{8}{r}\left(.3+.7\left(1-\frac{r}{8}\right)\right)^n.$$

Since $P(T > 55) > .05$, $P(T > 56) < .05$, 56 purchases are required.

CHAPTER 3

Exercises 3.2

1. .4050.
2. 47.
3. .2202.
4.

$$f_X(x) = \begin{cases} \frac{2(-x+n+1)}{n(n+1)} & \text{if } x = 1,2,\ldots n \\ 0 & \text{otherwise.} \end{cases}$$

$$f_Y(y) = \begin{cases} \frac{2(-y+n+1)}{n(n+1)} & \text{if } y = 1,2,\ldots n \\ 0 & \text{otherwise.} \end{cases}$$

5. Let $z \in \{1, 2, \ldots, 6\}$. By Equation 3.6,

$$f_Z(z) = \sum_{x=1}^{6} f_{X,Z}(x, z) = \sum_{x=1}^{z-1} \frac{1}{36} + \sum_{x=z}^{z} \frac{x}{36} = \frac{z-1}{36} + \frac{z}{36} = \frac{2z-1}{36}.$$

6. The recursion formula is

$$b(k; n, p) = \frac{(n-k+1)p}{kq} b(k-1; n, p).$$

The ratio is

$$\frac{b(k; n, p)}{b(k-1; n, p)} = 1 + \frac{(n+1)p - k}{kq}, \qquad k = 1, 2, \ldots, n.$$

Let $m = [(n+1)p]$, the greatest integer less than or equal to $(n+1)p$. If $(n+1)p$ is not an integer, the $b(k; n, p)$ are strictly increasing for $k \leq m$ and then are strictly decreasing for $k \geq m$, so that the maximum value is $b(m; n, p)$. If $(n+1)p$ is an integer, then both $b(m; n, p)$ and $b(m-1; n, p)$ are maximum values.

7. f_X is a Poisson density with parameter α, and f_Y is a Poisson density with parameter β.

8. $f_X(x_i) = cf(x_i)$ where $c = \sum_{j=1}^{\infty} g(y_j)$, and $f_Y(y_j) = dg(y_j)$ where $d = \sum_{i=1}^{\infty} f(x_i)$.

9. $f_X(x) = .25$ for $x = 1, 2, 3, 4$. $f_Y(1) = .12, f_Y(2) = .27, f_Y(3) = .20, f_Y(4) = .25, f_Y(5) = .16$. $P(Y \geq X) = .70$

10. Think of Ω as points (i, j) in the plane with integer coordinates $1 \leq i, j \leq 50$, except that points on the diagonal $i = j$ are not included. Points on the two diagonals adjacent to the diagonal are not included in the event.

$$P(|X - Y| \geq 2) = \frac{2352}{2450} = .96.$$

11. .9341.

12. 372.

Exercises 3.3

1. For all real t, $(1+t)^{a+b} = (1+t)^a(1+t)^b$. By the generalized binomial theorem,

$$\sum_{z=0}^{\infty} \binom{a+b}{z} t^z = \Big(\sum_{x=0}^{\infty} \binom{a}{x} t^x\Big)\Big(\sum_{y=0}^{\infty} \binom{b}{y} t^y\Big)$$

$$= \sum_{z=0}^{\infty} c_z t^z$$

where

$$c_z = \sum_{x=0}^{z} \binom{a}{x}\binom{b}{z-x}.$$

If two power series agree on an open interval about 0, then their coefficients must be equal. Thus,

$$\binom{a+b}{z} = \sum_{x=0}^{z} \binom{a}{x}\binom{b}{z-x}.$$

2. For $z = 0, 1, 2, \ldots,$

$$\begin{aligned} f_Z(z) &= \sum_{x=0}^{z} \binom{-r}{x} p^r (-q)^x \binom{-s}{z-x} p^s (-q)^{z-x} \\ &= p^{r+s}(-q)^z \sum_{x=0}^{z} \binom{-r}{x}\binom{-s}{z-x} \\ &= \binom{-(r+s)}{z} p^{r+s}(-q)^z. \end{aligned}$$

Thus, Z has a negative binomial density with parameters $r + s$ and p.

3. $P(X \geq Y) = ((n+1)/2n)$, $P(X = Y) = 1/n$.

4.
$$\begin{aligned} f_{X,Y}(x,y) &= P(X = x, Y = y) = P(X = x, Y = y, N = x + y) \\ &= P(X = x, Y = y \mid N = x + y)P(N = x + y) \\ &= \binom{x+y}{x} p^x q^y \frac{\lambda^{x+y} e^{-\lambda}}{(x+y)!} \\ &= \frac{(\lambda p)^x}{x!} e^{-\lambda p} \frac{(\lambda q)^y}{y!} e^{-\lambda q}. \end{aligned}$$

Since $X \leq N$,

$$\begin{aligned} f_X(x) = P(X = x) &= \sum_{n=x}^{\infty} P(X = x \mid N = n)P(N = n) \\ &= \sum_{n=x}^{\infty} \binom{n}{x} p^x q^{n-x} \frac{\lambda^n e^{-\lambda}}{n!} \\ &= \frac{(\lambda p)^x}{x!} e^{-\lambda} \sum_{n=x}^{\infty} \frac{(\lambda q)^{n-x}}{(n-x)!} \end{aligned}$$

$$= \frac{(\lambda p)^x}{x!} e^{-\lambda} e^{\lambda q}$$
$$= \frac{(\lambda p)^x}{x!} e^{-\lambda p}.$$

Similarly,

$$P(Y = y) = \frac{(\lambda q)^y}{y!} e^{-\lambda q}.$$

Thus, $f_{X,Y}(x, y) = f_X(x) f_Y(y)$, and X and Y are independent random variables.

5. $P(X \geq Y) = 1/(2-p), P(X = Y) = p/(2-p)$.
6. For $z = 2, 3, \ldots,$

$$f_Z(z) = \sum_{x=0}^{z} f_X(x) f_Y(y) = \sum_{x=1}^{z-1} pq^{x-1} pq^{z-x-1} = (z-1)p^2 q^{z-2}.$$

7.
$$f_Z(z) = \begin{cases} \frac{z-1}{n^2} & \text{if } 2 \leq z \leq n \\ \frac{-z+2n+1}{n^2} & \text{if } n < z \leq 2n \\ 0 & \text{otherwise.} \end{cases}$$

8. Suppose the ranges of X and Y are $\{x_1, x_2, \ldots\}$ and $\{y_1, y_2, \ldots\}$, respectively. Suppose ϕ_i and ψ_j are in the ranges of $\phi(X)$ and $\psi(Y)$, respectively. Let $\{x_{i_1}, \ldots, x_{i_\alpha}\}$ be the set of values of X such that $\phi(x_{i_m}) = \phi_i$ and let $\{y_{j_1}, \ldots, y_{j_\beta}\}$ be the set of values of Y such that $\psi(y_{j_n}) = \psi_j$. Then $(\phi(X) = \phi_i, \psi(Y) = \psi_j) = \cup_{m,n}(X = x_{i_m}, Y = y_{j_n})$. Since the latter sets are disjoint,

$$P(\phi(X) = \phi_i, \psi(Y) = \psi_j) = \sum_{m,n} P(X = x_{i_m}) P(Y = y_{j_m})$$
$$= \left(\sum_m P(X = x_{i_m})\right)\left(\sum_n P(Y = y_{j_n})\right)$$
$$= P(\cup_m (X = x_{i_m})) P(\cup_n (Y = y_{j_n}))$$
$$= P(\phi(X) = \phi_i) P(\psi(Y) = \psi_j).$$

9. $f_Z(1) = f_Z(13) = .000, f_Z(2) = f_Z(12) = .004, f_Z(3) = f_Z(11) = .018, f_Z(4) = f_Z(10) = .057, f_Z(5) = f_Z(9) = .122, f_Z(6) = f_Z(8) = .189, f_Z(7) = .219$.

Exercises 3.4

1. $\hat{f}_X(t) = t(1-t^n)/n(1-t)$.

2. Since $A(t) = 1-(1-t^2)^{1/2} = 1-\sum_{j=0}^{\infty}\binom{1/2}{j}(-t^2)^j$, $a_0 = 0$, $a_{2j+1} = 0$ for all $j \geq 0$, and $a_{2j} = (-1)^{j+1}\binom{1/2}{j}$ for all $j \geq 1$.

3. (a) Geometric density with $p = 1/3$. (b) Poisson density with $\lambda = 1/4$. (c) Negative binomial density with $r = 5$ and $p = 1/8$.

4. $X = X_1+\cdots+X_N$ where $X_1, X_2, \ldots$ is an infinite sequence of independent random variables with $\hat{f}_{X_j}(t) = (1/2)+(1/2)t$, N has generating function $\hat{f}_N(t) = (t/6)((1-t^6)/(1-t))$, and the random variables $N, X_1, X_2, \ldots$ are independent. Thus,

$$\hat{f}_X(t) = \hat{f}_N(\hat{f}_{X_1}(t)) = \frac{.5+.5t}{6} \times \frac{1-(.5+.5t)^6}{1-(.5+.5t)}.$$

5. $\hat{f}(t)$ is the generating function of a random number of random variables $S_N = X_1+\cdots+X_N$ where the X_j's are Bernoulli random variables with $p = 1/3$ and N has a uniform density on $\{1, 2, \ldots, 6\}$. This could arise from tossing a die to determine how many times a basic S or F trial with $p = 1/3$ should be performed.

6. $f_X(2x) = (2^x e^{-2})/x!$ and $f_X(2x+1) = 0$ for $x = 0, 1, \ldots$.

7. Letting $m = [n/2]$, m is the largest integer such that $2m \leq n$. Write $E_n = (\sum_{j=1}^n X_j \in \{0, 2. \ldots, 2m\})$. Stratifying E_n using the values of X_1,

$$\begin{aligned}
E_n &= (X_1 = 0, \sum_{j=1}^n X_j \in \{0, 2, \ldots, 2m\}) \\
&\cup (X_1 = 1, \sum_{j=1}^n X_j \in \{0, 2, \ldots, 2m\}) \\
&= (X_1 = 0, \sum_{j=2}^n X_j \in \{0, 2, \ldots, 2m\}) \\
&\cup \left(X_1 = 1, \left(\sum_{j=2}^n X_j \in \{0, 2, \ldots, 2m\}\right)^c\right),
\end{aligned}$$

and so

$$\begin{aligned}
p_n = P(E_n) &= P(X_1 = 0, \sum_{j=2}^n X_j \in \{0, 2, \ldots, 2m\}) \\
&+ P\left(X_1 = 1, \left(\sum_{j=2}^n X_j \in \{0, 2, \ldots, 2m\}\right)^c\right).
\end{aligned}$$

Since X_1 and $\sum_{j=2}^{n} X_j$ are independent by Lemma 3.3.3,

$$p_n = P(X_1 = 0)P\left(\sum_{j=2}^{n} X_j \in \{0, 2, \ldots, 2m\}\right) + P(X_1 = 1)\left(1 - P\left(\sum_{j=2}^{n} X_j \in \{0, 2, \ldots, 2m\}\right)\right).$$

Since $P(\sum_{j=2}^{n} X_j \in \{0, 2, \ldots, 2m\}) = P(\sum_{j=1}^{n-1} X_j \in \{0, 2, \ldots, 2m\}) = p_{n-1}$,

$$p_n = qp_{n-1} + p(1 - p_{n-1}).$$

8. The difference equation is $q_n = (1/2)q_{n-1} + (1/4)q_{n-2}$, $n \geq 2$, subject to the initial conditions $q_0 = q_1 = 1$. The solution is

$$q_n = \frac{1+\sqrt{5}}{5-\sqrt{5}} \frac{(-1)^n}{(1-\sqrt{5})^n} + \frac{1-\sqrt{5}}{5+\sqrt{5}} \frac{(-1)^n}{(1+\sqrt{5})^n}, \qquad n \geq 0.$$

9. .03262.
10. $f_Z(1) = 1/192$, $f_Z(2) = 1/32$, $f_Z(3) = 1/12$, $f_Z(4) = 13/96$, $f_Z(5) = 31/192)$, $f_Z(6) = 1/6$, $f_Z(7) = 31/192$, $f_Z(8) = 13/96$, $f_Z(9) = 1/12$, $f_Z(10) = 1/32$, $f_Z(11) = 1/192$.
11. $f_X(0) = .00790$, $f_X(1) = .04972$, $f_X(2) = .13998$, $f_X(3) = ..23204$, $f_X(4) = .25083$, $f_X(5) = .18474$, $f_X(6) = .09390$, $f_X(7) = .03252$, $f_X(8) = .00735$, $f_X(9) = .00098$, $f_X(10) = .00006$.

Exercises 3.5

1. Add the right sides of Equations 3.14 and 3.16 to verify that $p_x + q_x = 1$ in the $p \neq q$ case and the right sides of Equations 3.15 and 3.17 to verify that $p_x + q_x = 1$ in the $p = q = 1/2$ case.
2. $\lim_{a \to \infty} q_x = 1$ is the probability of eventual ruin against an infinitely rich adversary in the unfair situation $q > p$.
3. If $Y_2 + \cdots + Y_j = y$ for some $j \geq 2$, then there is a smallest integer for which this is true; i.e.,

$$(Y_2 + \cdots + Y_j = y) =$$

$$(Y_2 = y) \cup \bigcup_{k=0}^{\infty} (Y_2 \neq y, \ldots, Y_2 + \cdots + Y_{2+k} \neq y, Y_2 + \cdots + Y_{3+k} = y).$$

Since the events on the right side are disjoint and each is independent of $(Y_1 = 1)$, the events $(Y_1 = 1)$ and $(Y_2 + \cdots Y_j = y$ for some $j \geq 2)$ are independent (see Exercises 2.6.9).

4. The difference equation for the probability of ruin q_x is

$$q_x = \alpha q_{x+1} + \beta q_x + \gamma q_{x-1}, \qquad 1 < x < a - 1,$$

subject to the boundary conditions

$$q_0 = 1, q_a = 0.$$

The equation is the same as the difference equation

$$q_x = \frac{\alpha}{1-\beta} q_{x+1} + \frac{\gamma}{1-\beta} q_{x-1}.$$

The solution to this equation is obtained by replacing p by $\alpha/(1-\beta)$ and q by $\gamma/(1-\beta)$ in Equations 3.14 and 3.15 to obtain

$$q_x = \frac{(\gamma/\alpha)^a - (\gamma/\alpha)^x}{(\gamma/\alpha)^a - 1}, \qquad 1 \leq x \leq a - 1,$$

in the $\alpha \neq \beta$ case and

$$q_x = 1 - \frac{x}{a}, \qquad 1 \leq x \leq a - 1,$$

in the $\alpha = \beta$ case.

CHAPTER 4

Exercises 4.2

1. $E[X^2] = (1+q)/p^2$.
2. $E[X^2] = \lambda^2 + \lambda$.
3. $E[X] = 2$.
4. $E[1/(X+1)] = (1 - e^{-\lambda})/\lambda$.
5. $E[1/(X+1)] = p/(q(r-1))$.
6. $E[X] = \hat{g}_X(1) = \sum_{x=0}^{\infty} g_X(x) = \sum_{x=0}^{\infty} P(X > x) = \sum_{x=0}^{\infty} P(X \geq x+1) = \sum_{x=1}^{\infty} P(X \geq x)$.
7. The generating function of S_N is given by $\hat{f}_{S_N}(t) = \hat{f}_N(\hat{f}_{X_1}(t))$ and $E[S_N] = \hat{f}'_{S_N}(1)$. Since $\hat{f}'_{S_N}(t) = \hat{f}'_N(\hat{f}_{X_1}(t))\hat{f}'_{X_1}(t), E[N] = \hat{f}'_N(1)$,

$E[X_1] = \hat{f}'_{X_1}(1)$, and $\hat{f}_{X_1}(1) = 1$,

$$\begin{aligned} E[S_N] &= \hat{f}'_N(\hat{f}_{X_1}(1))\hat{f}'_{X_1}(1) \\ &= \hat{f}'_N(1)\hat{f}'_{X_1}(1) \\ &= E[N]E[X_1]. \end{aligned}$$

8. Let $T = n$ if the 10-digit combination occurs on the nth trial for the first time. Then

$$\hat{g}_T(t) = \frac{\sum_{j=0}^{9}(t/2)^j}{2 - \sum_{j=1}^{10}(t/2)^j},$$

and $E[T] = \hat{g}_T(1) = 2046$.

9. $P(T \leq 11) = 1 - P(T > 11) = 1 - g_T(11) = 279/1024$.

Exercises 4.3

1. $E[XZ] = 154/9$.
2. $E[U] = 91/36$.
3. $\text{var}\, X = q/p^2$.
4. $\text{var}\, X = 32$.
5. $\text{var}\, X = r(q/p^2)$.
6. $d = 377$.
7. $n = 750$.
8. $E[V] = V_1 r(n_1/n) + \cdots + V_s r(n_s/n)$.
9. 106.5 pounds.
10. $\text{var}\, X_j = r(n_j/n)((n-r)(n-n_j)/n(n-1))$.
11. Let $\{x_1, x_2, \ldots\}$ be the range of X. By hypothesis, $\sum_j (x_j - \mu)^2 f_X(x_j) = 0$. For any j with $x_j - \mu \neq 0, f_X(x_j) = 0$. Since $\sum_j f_X(x_j) = 1$, there is some x_i in the range of X with $f_X(x_i) > 0$ that cannot be different from μ. Therefore, $x_i = \mu, f_X(\mu) > 0$ and $f_X(\mu) = 1$; i.e., $P(X = \mu) = 1$.
12. $\text{var}\, X = 2\hat{g}'(1) + \hat{g}(1) - [\hat{g}(1)]^2$.
13. $\sigma_T = 27.09$.

Exercises 4.4

1. $\mu_X = .5, \sigma_X^2 = 1.05, \mu_Y = 2.6, \sigma_Y^2 = .94, E[XY] = 1.7, \rho(X,Y) \approx .40$.
2. $\text{var}(X + 2Y) = 6 + 3\sqrt{2}$.
3. Note that $X + Y = 2$. Since $\text{cov}(X,2) = 0$ and $\text{cov}(X,X) = \sigma_X^2$, $\text{cov}(X,Y) = \text{cov}(X, 2-X) = \text{cov}(X,2) - \text{cov}(X,X) = -\sigma_X^2$. Since $\sigma_{2-X}^2 = \sigma_X^2, \rho(X,Y) = -\sigma_X^2/\sigma_X\sigma_{2-X} = -1$.

4. $\rho(X_1 + 2X_2 - X_3, 3X_1 - X_2 + X_3) = 5/(2\sqrt{170})$.
5. $\rho(X, Y) = -1/2$.
6. (a) $E[X_i] = (r-1)^n/r^n = (1-(1/r))^n$.
   (b) $E[X_iX_j] = (r-2)^n/r^n = (1-(2/r))^n$.
   (c) $E[S_r] = r((r-1)^n/r^n) = r(1-(1/r))^n$.
   (d) Since $X_i^2 = X_i$, $\operatorname{var} X_i = (1-(1/r))^n - (1-(1/r))^{2n}$. For $i \neq j$, $\operatorname{cov}(X_iX_j) = (1-(2/r))^n - (1-(1/r))^{2n}$. Thus,

$$\begin{aligned}\operatorname{var} S_r &= \sum_{i=1}^{r} \operatorname{var} X_i + 2 \sum_{1\le i<j\le r} \operatorname{cov}(X_i, X_j) \\ &= r\left(\left(1-\frac{1}{r}\right)^n - \left(1-\frac{1}{r}\right)^{2n}\right) \\ &\quad + r(r-1)\left(\left(1-\frac{2}{r}\right)^n - \left(1-\frac{1}{r}\right)^{2n}\right).\end{aligned}$$

7. (a) $E[I_{i,k}I_{j,k}] = 0$ for $i \neq j$.
   (b) $E[I_{i,k}I_{j,\ell}] = p_ip_j$ for $k \neq \ell$.
   (c) $E[Y_i] = np_i$ and $E[Y_iY_j] = n(n-1)p_ip_j$ for $i \neq j$.
   (d) $\operatorname{var} Y_i = np_i(1-p_i)$.
   (e) $\operatorname{cov}(Y_i, Y_j) = -np_ip_j$ and $\rho(Y_i, Y_j) = -\sqrt{p_ip_j/(1-p_i)(1-p_j)}$ for $i \neq j$.
8. Suppose the range of X is $\{a, b\}$ and the range of y is $\{c, d\}$. If E is any event, let I_E be the indicator function of E; i.e., $I_E(\omega) = 1$ or 0 according to whether ω is in E or not. Let $A = \{\omega : X(\omega) = a\}$, $C = \{\omega : Y(\omega) = c\}$. Then $X = aI_A + bI_{A^c}$, $Y = cI_C + dI_{C^c}$, $XY = acI_{A\cap C} + adI_{A\cap C^c} + bcI_{A^c\cap C} + bdI_{A^c\cap C^c}$. Thus,

$$\begin{aligned}E[XY] &= (a-b)(c-d)P(A\cap C) + d(a-b)P(A) \\ &\quad + b(c-d)P(C) + bd\end{aligned}$$

and

$$\begin{aligned}E[X]E[Y] &= (a-b)(c-d)P(A)P(C) + d(a-b)P(A) \\ &\quad + b(c-d)P(C) + bd.\end{aligned}$$

Since $\operatorname{cov}(X, Y) = E[XY] - E[X]E[Y] = 0$, $P(A\cap C) = P(A)P(C)$. Therefore, the pair $\{A, C\}$ are independent events, as are the pairs $\{A^c, C\}$, $\{A, C^c\}$, and $\{A^c, C^c\}$. Thus, X and Y are independent.

9. Minimize the function of two variables $g(c, d) = E[(Y - cX - d)^2]$. Then

$$a = \frac{\operatorname{cov}(X, Y)}{\sigma_X^2}$$

$$b = E[Y] - \frac{\operatorname{cov}(X, Y)}{\sigma_X^2} E[X].$$

Exercises 4.5

1. $E[Y] = 25.75$, $\operatorname{var} Y = 490.1875$.
2. $E[Y] = 25$, $\operatorname{var} Y = 500$.
3. Since $X_1 + X_2$ has a Poisson density $p(\cdot; \lambda_1 + \lambda_2)$ by Theorem 3.3.4, if n is any positive integer,

$$\begin{aligned} P(X_1 = x \mid X_1 + X_2 = n) &= \frac{P(X_1 = x, X_1 + X_2 = n)}{P(X_1 + X_2 = n)} \\ &= \frac{P(X_1 = x)P(X_2 = n - x)}{P(X_1 + X_2 = n)} \\ &= \frac{p(x; \lambda_1)p(n - x; \lambda_2)}{p(n; \lambda_1 + \lambda_2)} \\ &= \binom{n}{x}\left(\frac{\lambda_1}{\lambda_1 + \lambda_2}\right)^x \left(\frac{\lambda_2}{\lambda_1 + \lambda_2}\right)^{n-x}. \end{aligned}$$

4. By Theorem 3.3.4, $X + Y$ has a binomial density $b(\cdot; m + n, p)$. For $z = 0, \ldots, m + n$ and $x = 0, \ldots, z$,

$$\begin{aligned} f_{X|X+Y}(x \mid z) &= \frac{P(X = x, X + Y = z)}{P(X + Y = z)} \\ &= \frac{P(X = x)P(Y = z - x)}{P(X + Y = z)} \\ &= \frac{b(x; m, p)b(z - x; n, p)}{b(z; m + n, p)} \\ &= \frac{\binom{m}{x}\binom{n}{z - x}}{\binom{m + n}{z}}. \end{aligned}$$

By Equation 1.11,

$$E[X \mid X + Y = z] = \sum_{x=0}^{z} x \frac{\binom{m}{x}\binom{n}{z - x}}{\binom{m + n}{z}}$$

$$= \frac{m}{\binom{m+n}{z}} \binom{m+n-1}{z-1}$$

$$= \frac{m}{m+n} z.$$

5. $E[X] = 50.$
6. Since N and X_n are independent, $f_{X_N|N}(x \mid n) = f_{X_n}(x)$ and

$$E[X_N \mid N = n] = \sum_x x f_{X_N|N}(x \mid n) = \sum_x x f_{X_n}(x) = E[X_n].$$

7. By Problem 6,

$$E[S_N] = \sum_{n=0}^{\infty} E[S_N \mid N = n] f_N(n) = \sum_{n=0}^{\infty} E[S_n] f_N(n)$$

$$= \sum_{n=0}^{\infty} n\mu f_N(n) = \mu E[N].$$

Also,

$$E[S_N^2] = \sum_{n=0}^{\infty} E[S_N^2 \mid N = n] f_N(n) = \sum_{n=0}^{\infty} E[S_n^2] f_N(n)$$

$$= \sum_{n=0}^{\infty} (\operatorname{var} S_n + (E[S_n])^2) f_N(n) = \sum_{n=0}^{\infty} (n\sigma^2 + n^2\mu^2) f_N(n)$$

$$= \sigma^2 E[N] + \mu^2 E[N^2].$$

Therefore, $\operatorname{var} S_N = \sigma^2 E[N] + \mu^2 E[N^2] - \mu^2 (E[N])^2 = \sigma^2 E[N] + \mu^2 \operatorname{var} N.$

8. Let $Z = c$. Then

$$f_{Z|X}(z \mid x) = \frac{P(Z = z, X = x)}{P(X = x)} = \begin{cases} 1 & \text{if } z = c \\ 0 & \text{if } z \neq c \end{cases}$$

whenever $f_X(x) > 0$. Therefore,

$$E[Z \mid X = x] = \sum_z z f_{Z|X}(z \mid x) = c$$

whenever $f_X(x) > 0$.

9. By Inequality 4.8, $\phi(X)Y$ has finite expectation and $E[\phi(X)Y \mid X = x]$ is defined whenever $f_X(x) > 0$. Suppose $f_X(x) > 0$. Letting $Z = \phi(X)Y$, note that

$$E[Z \mid X = x] = \sum_z z f_{Z|X}(z \mid x) = \sum_{z \neq 0} z f_{Z|X}(z \mid x).$$

For $z \neq 0$,

$$f_{Z|X}(z \mid x) = P(\phi(X)Y = z, X = x)/P(X = x).$$

Suppose $\phi(x) \neq 0$. Then

$$\begin{aligned} f_{Z|X}(z \mid x) &= \frac{P(Y = z/\phi(x), X = x)}{P(X = x)} \\ &= f_{Y|X}\left(\frac{z}{\phi(x)} \,\middle|\, x\right) \end{aligned}$$

and

$$\begin{aligned} E[Z \mid X = x] &= \sum_{z \neq 0} z f_{Y|X}\left(\frac{z}{\phi(x)} \,\middle|\, x\right) \\ &= \phi(x) \sum_{z \neq 0} \frac{z}{\phi(x)} f_{Y|X}\left(\frac{z}{\phi(x)} \,\middle|\, x\right) \\ &= \phi(x) E[Y \mid X = x]. \end{aligned}$$

If $\phi(x) = 0$, then $f_{Z|X}(z \mid x) = P(0 = z, X = x)/P(X = x) = 0$, and

$$E[Z \mid X = x] = \sum_{z \neq 0} z f_{Z|X}(z \mid x) = 0 = \phi(x) E[Y \mid X = x]$$

whenever $f_X(x) > 0$.

10. D_x will satisfy the difference equation

$$D_x = \alpha D_{x+1} + \beta D_x + \gamma D_{x-1} + 1 \qquad 1 \leq x \leq a - 1,$$

and the boundary conditions

$$D_0 = 0, D_a = 0.$$

We can assume that $\beta < 1$, since otherwise neither gambler or adversary ever wins or loses. Thus, D_x satisfies

$$D_x = \frac{\alpha}{1 - \beta} D_{x+1} + \frac{\gamma}{1 - \beta} D_{x-1} + \frac{1}{1 - \beta}, \qquad 1 \leq x \leq a - 1,$$

subject to the boundary conditions

$$D_0 = 0, D_a = 0.$$

Letting $\tilde{D}_x = (1-\beta)D_x$, $\tilde{D}_x$ satisfies the equation

$$\tilde{D}_x = \tilde{p}\tilde{D}_{x+1} + \tilde{q}\tilde{D}_{x-1} + 1$$

and the boundary conditions

$$\tilde{D}_0 = 0, \tilde{D}_a = 0$$

where $\tilde{p} = \alpha/(1-\beta)$, $\tilde{q} = \gamma/(1-\beta)$. Thus, $\tilde{D}_x$ satisfies Equations 4.12 and 4.13 with p and q replaced by $\tilde{p}$ and $\tilde{q}$, respectively. $\tilde{D}_x$ is therefore given by Equations 4.16 and 4.17, so that in the $\alpha \neq \gamma$ case,

$$D_x = \frac{1}{1-\beta}\left(\frac{x}{\gamma-\alpha} - \frac{a}{\gamma-\alpha}\frac{1-(\gamma/\alpha)^x}{1-(\gamma/\alpha)^a}\right), \qquad 1 \le x \le a-1,$$

and in the $\alpha = \gamma$ case,

$$D_x = \frac{1}{1-\beta}x(a-x), \qquad 1 \le x \le a-1.$$

Exercises 4.6

1. $H(X) = 15/8$ bits.
2. (a) $H(X) = -\sum_{x=1}^{\infty}(1/2)^x \log(1/2)^x = \sum_{x=1}^{\infty} x(1/2)^x = (1/2)\sum_{x=1}^{\infty} x(1/2)^{x-1} = 2$ (see Example 4.3). (b) 2 bits.
3. 1 bit.
4. Use the equation $H(X,Y) = H(Y\,|\,X) + H(X)$. Since $H(X) = \log n$ and $H(Y\,|\,X = i) = \log i$, $i = 1,\ldots,n$, $H(Y\,|\,X) = \sum_{i=1}^{n}(1/n)\log i = (1/n)\sum_{i=1}^{n}\log i = (\log(n!))/n$, $H(X,Y) = (\log(n!))/n + \log n$.
5. $H(X) \approx 3.2744$.
6. 5.7 bits.
7. 2 bits.
8. By Lemma 4.6.1,

$$\begin{aligned} &H(X\,|\,Y) - H(X)\\ &= -\sum_{i,j} f_{X|Y}(x_i\,|\,y_j)\log f_{X|Y}(x_i\,|\,y_j) f_Y(y_j) + \sum_{i,j} f_{X,Y}(x_i,y_j)\log f_X(x_i)\\ &= \sum_{i,j} f_{X,Y}(x_i,y_j)\log\left(\frac{f_X(x_i)}{f_{X|Y}(x_i\,|\,y_j)}\right) \end{aligned}$$

$$= \frac{1}{\ln 2} \sum_{i,j} f_{X,Y}(x_i, y_j) \ln\left(\frac{f_X(x_i)}{f_{X|Y}(x_i \mid y_j)}\right)$$
$$\leq \frac{1}{\ln 2} \sum_{i,j} f_{X,Y}(x_i, y_j) \left(\frac{f_X(x_i)}{f_{X|Y}(x_i \mid y_j)} - 1\right)$$
$$= 0.$$

9. $p_1 \approx .10307$, $p_2 \approx .12273$, $p_3 \approx .14615$, $p_4 \approx .17403$, $p_5 \approx .20724$, $p_6 \approx .24678$.

CHAPTER 5

Exercises 5.2

1.
$$P = \begin{bmatrix} 1 & 0 & 0 & 0 & \dots & 0 & 0 & 0 \\ q & 0 & p & 0 & \dots & 0 & 0 & 0 \\ 0 & q & 0 & p & \dots & 0 & 0 & 0 \\ \vdots & \vdots & \vdots & \vdots & \ddots & \vdots & \vdots & \vdots \\ 0 & 0 & 0 & 0 & \dots & q & 0 & p \\ 0 & 0 & 0 & 0 & \dots & 0 & 0 & 1 \end{bmatrix}.$$

2. $n = 5$ and $\nu_1 = 4/59$, $\nu_2 = 8/59$, $\nu_3 = 24/59$, $\nu_4 = 23/59$.
3. $\nu_1 = q/(p+q)$, $\nu_2 = p/(p+q)$.
4. $n = 2$ and $\nu_1 = \nu_2 = \nu_3 = 1/3$.
5. The result is true for $n = 1$ by hypothesis. Assume that the result is true for n. Then

$$\sum_{k=1}^{N} \mu_k p_{k,j}(n+1) = \sum_{k=1}^{N} \mu_k \sum_{\ell=1}^{N} p_{k,\ell} p_{\ell,j}(n)$$
$$= \sum_{\ell=1}^{N} p_{\ell,j}(n) \sum_{k=1}^{N} \mu_k p_{k,\ell}$$
$$= \sum_{\ell=1}^{N} p_{\ell,j}(n) \mu_\ell = \mu_j,$$

and the result is true for $n+1$. By the principle of mathematical induction, the result is true for all $n \geq 1$.

6. First show that each $P(n) = [p_{i,j}(n)]$ is doubly stochastic as follows. The result is true for $n = 1$ by hypothesis. Assume that $P(n)$ is doubly stochastic. Then

$$\sum_{i=1}^{N} p_{i,j}(n+1) = \sum_{i=1}^{N}\sum_{k=1}^{N} p_{i,k}p_{k,j}(n)$$
$$= \sum_{k=1}^{N} p_{k,j}(n) \sum_{i=1}^{N} p_{i,k}$$
$$= \sum_{k=1}^{N} p_{k,j}(n) = 1,$$

and the result is true for $n+1$. By the principle of mathematical induction, $P(n)$ is doubly stochastic for every $n \geq 1$. Since $\sum_{i=1}^{N} p_{i,j}(n) = 1$, $n \geq 1$,

$$1 = \lim_{n\to\infty} \sum_{i=1}^{N} p_{i,j}(n) = \sum_{i=1}^{N} \lim_{n\to\infty} p_{i,j}(n) = \sum_{i=1}^{N} \nu_j = N\nu_j,$$

and therefore $\nu_j = 1/N, j = 1, \ldots, N$.

7. Since P is doubly stochastic, $\nu_j = 1/5$, $j = 1, \ldots, 5$, by the previous exercise.

8. By Problem 5, $\mu_j = \sum_{k=1}^{N} \mu_k p_{k,j}(n)$ for all $n \geq 1$. Letting $\nu_j = \lim_{n\to\infty} p_{i,j}(n), j = 1, \ldots, N$,

$$\mu_j = \sum_{k=1}^{N} \mu_k \nu_j = \nu_j, \quad j = 1, \ldots, N.$$

Thus, $\{\mu_j\}_{j=1}^{N}$ is the asymptotic distribution.

9.
$$P(2n-1) = P^{2n-1} = [p_{i,j}(2n-1)] = \begin{bmatrix} 0 & 1/2 & 1/2 \\ 1 & 0 & 0 \\ 1 & 0 & 0 \end{bmatrix}, \qquad n \geq 1.$$

$$P(2n) = P^{2n} = [p_{i,j}(2n)] = \begin{bmatrix} 1 & 0 & 0 \\ 0 & 1/2 & 1/2 \\ 0 & 1/2 & 1/2 \end{bmatrix}, \qquad n \geq 1.$$

The asymptotic distribution is not defined since $\lim_{n\to\infty} p_{i,j}(n)$ does not exist.

10.
$$P = \begin{bmatrix} 0 & 1 & 0 & 0 & 0 & \ldots & 0 & 0 \\ \frac{1}{N^2} & \frac{2(N-1)}{N^2} & \frac{(N-1)^2}{N^2} & 0 & 0 & \ldots & 0 & 0 \\ 0 & \frac{2^2}{N^2} & \frac{2^2(N-2)}{N^2} & \frac{(N-2)^2}{N^2} & 0 & \ldots & 0 & 0 \\ \vdots & \vdots & \vdots & \vdots & \vdots & \ddots & \vdots & \vdots \\ 0 & \ldots & & & & \ldots & 1 & 0 \end{bmatrix}.$$

$$\nu_k = \binom{N}{k}^2 \Big/ \binom{2N}{N}^2, \qquad k = 0, \ldots, N.$$

11. $$P(X_{m+n} = j \mid X_m = i)$$
$$= \sum_{j_1,\ldots,j_{n-1}} P(X_{m+1} = j_1, \ldots, X_{m+n-1} = j_{n-1}, X_{m+n} = j \mid X_m = i)$$
$$= \sum_{j_1,\ldots,j_{n-1}} P(X_m = i, X_{m+1} = j_1, \ldots, X_{m+n} = j) \times \frac{1}{P(X_m = i)}$$

By Equation 5.2,

$$P(X_{m+n} = j \mid X_m = i)$$
$$= \left(\sum_{j_1,\ldots,j_{n-1}} \sum_{i_0,\ldots,i_{m-1}} \pi(i_0) p_{i_0,i_1} \times \cdots \times p_{i_{m-1},i} p_{i,j_1} \times \cdots \times p_{j_{n-1},j} \right)$$
$$\times \frac{1}{P(X_m = i)}$$
$$= \sum_{j_1,\ldots,j_{n-1}} p_{i,j_1} \times \cdots \times p_{i_{n-1},j},$$

and the latter does not depend upon m.

12. $\nu_1 = .2126372201$, $\nu_2 = .2339009422$, $\nu_3 = .2144524159$, $\nu_4 = .2157489844$, $\nu_5 = .1232604374$.

13. $\nu_1 = .1098019693$, $\nu_2 = .0365008905$, $\nu_3 = .1201612072$, $\nu_4 = .2061555430$, $\nu_5 = .2261706688$, $\nu_6 = .3012097212$.

Exercises 5.3

1. $$\binom{2n}{n} = \frac{(2n)(2n-1)(2n-3)\cdots 3\cdot 2\cdot 1}{n!n!}$$
$$= \frac{2^n(2n-1)(2n-3)\cdots 3\cdot 1}{n!}$$
$$= \frac{(-1)^n 2^{2n}(-1/2)(-3/2)\cdots((1/2)-n)}{n!}$$
$$= \frac{(-4)^n(-1/2)(-3/2)\cdots(-(1/2)-n+1)}{n!}$$
$$= (-4)^n \binom{-1/2}{n}.$$

2. $p = 1/3$.

3. q_x satisfies the difference equation

$$q_x = pq_{x+1} + qq_{x-1}, \qquad 2 \le x \le a-1,$$

subject to the boundary conditions $q_a = 0, q_0 - \delta q_1 = 1 - \delta$. If $p \neq q$,

$$q_x = (1-\delta)\left(\left(\frac{q}{p}\right)^a - \left(\frac{q}{p}\right)^x\right) \div \left((1-\delta)\left(\frac{q}{p}\right)^a + \delta\frac{q}{p} - 1\right).$$

If $p = q$,

$$q_x = \frac{1-\delta}{a(1-\delta)+\delta}(a-x).$$

4. D_x satisfies the difference equation

$$D_x = pD_{x+1} + qD_{x-1} + 1, \qquad 1 \leq x \leq a-1,$$

subject to the boundary conditions $D_a = 0, \delta D_1 = D_0$. In the $p \neq q$ case, if we put

$$D = (1-\delta)\left(\frac{q}{p}\right)^a - \left(1 - \frac{\delta q}{p}\right),$$

then $D_x = E[T_x] = A + B(q/p)^x, 0 \leq x \leq a$, where

$$A = \frac{1}{D(q-p)}\left(\delta\left(\frac{q}{p}\right)^a + a\left(1 - \frac{\delta q}{p}\right)\right)$$

$$B = -\frac{1}{D(q-p)}\{a(1-\delta)+\delta\}.$$

Exercises 5.4

1. $q = -2 + \sqrt{5} \approx .236$.
2. $q_3 = .204325$.
3. $q = (1 - |\alpha - \beta|)/2\beta$.
4. Start with equation $\hat{f}_{X_{j+1}}(s) = \hat{f}_{X_j}(\hat{p}(s))$, differentiate twice, and set $s = 1$ to obtain

$$\hat{f}''_{X_{j+1}}(1) = \hat{f}'_{X_j}(1)\hat{p}''(1) + \hat{f}''_{X_j}(1)(\hat{p}'(1))^2.$$

Use the facts that $\hat{p}'(1) = E[X_1] = \mu, \hat{p}''(1) = \operatorname{var} X_1 - \mu + \mu^2, \hat{f}'_{X_j}(1) = E[X_j] = \mu^j, \hat{f}''_{X_j}(1) = \operatorname{var} X_j - \hat{f}'_{X_j}(1) + (\hat{f}'_{X_j}(1))^2 = \operatorname{var} X_j - \mu^j + \mu^{2j}$ to obtain

$$\hat{f}''_{X_{j+1}}(1) = \mu^j(\sigma^2 - \mu + \mu^2) + \mu^2(\operatorname{var} X_j - \mu^j + \mu^{2j}).$$

Then

$$\operatorname{var} X_{j+1} = \hat{f}''_{X_{j+1}}(1) - \hat{f}'_{X_{j+1}}(1) + (\hat{f}'_{X_{j+1}}(1))^2 = \mu^2 \operatorname{var} X_j + \sigma^2 \mu^j.$$

5. For $j \geq 1$, let $P(j)$ be the proposition

$$\operatorname{var} X_j = \sigma^2(\mu^{2j-2} + \mu^{2j-3} + \cdots + \mu^{j-1}).$$

Since $\operatorname{var} X_1 = \sigma^2$ and $P(1)$ is the proposition $\operatorname{var} X_1 = \sigma^2(\mu^0) = \sigma^2$, $P(1)$ is true. Assume $P(j-1)$ is true. By Problem 3,

$$\begin{aligned}\operatorname{var} X_j &= \mu^2 \operatorname{var} X_{j-1} + \mu^{j-1}\sigma^2 \\ &= \mu^2\sigma^2(\mu^{2j-4} + \mu^{2j-5} + \cdots + \mu^{j-2}) + \mu^{j-1}\sigma^2 \\ &= \sigma^2(\mu^{2j-2} + \mu^{2j-3} + \cdots + \mu^j + \mu^{j-1}),\end{aligned}$$

and it follows that $P(j)$ is true. Therefore, $P(j)$ is true for all $j \geq 1$ by the principle of mathematical induction.

6. $q = .706420$.
7. $q = .203188$.
8. $q = .552719$.
9. $q_1 = .125$, $q_2 = .177979$, $q_3 = .204325$, $q_4 = .218344$, $q_5 = .226058$, $q_6 = .230379$, $q_7 = .232824$, $q_8 = .234214$, $q_9 = .235007$, $q_{10} = .235461$.

Exercises 5.5

1. $R(0) = 3/8$, $R(\pm 1) = 1/4$, $R(\pm 2) = 1/16$, $R(n) = 0$ otherwise. $\hat{X}_n^* = (6/5)X_{n-1} - (9/10)X_{n-2} + (2/5)X_{n-3}$ and $\sigma_3^2 = 21/160 \approx .131$.
2. The independence of $X_n = (1/4)Y_n + (1/2)Y_{n-1} + (1/4)Y_{n-2}$ and $X_{n-4} = (1/4)Y_{n-4} + (1/2)Y_{n-5} + (1/4)Y_{n-6}$ would seem to indicate that nothing would be gained by including X_{n-4} in $\hat{X}_n^*$. There is, however, a reduction in the mean square error by including X_{n-4}. $\sigma_4^2 \approx .117$, and

$$\hat{X}_n^* = \frac{4}{3}X_{n-1} - \frac{6}{5}X_{n-2} + \frac{4}{5}X_{n-3} - \frac{1}{3}X_{n-4}.$$

3. Letting $\sigma_X^2 = \operatorname{var} X_n$, $\sigma_X^2 = \alpha^2\sigma_X^2 + \sigma_\varepsilon^2$ or $\sigma_X^2(1-\alpha^2) = \sigma^2 > 0$. Thus, $\alpha^2 < 1$, and so $|\alpha| < 1$.
4. $a_1 = (\rho_1(1-\rho_2))/(1-\rho_1^2)$, $a_2 = (\rho_2 - \rho_1^2)/(1-\rho_1^2)$.
5. Predicted value $= 2.946$.

CHAPTER 6

Exercises 6.2

1.
$$F(x) = \begin{cases} 0 & \text{if } x < 0 \\ x^2 & \text{if } 0 \le x < 1 \\ 1 & \text{if } x \ge 1. \end{cases}$$

2.
$$F_X(x) = \begin{cases} 0 & \text{if } x < 0 \\ x^3 & \text{if } 0 \le x < 1 \\ 1 & \text{if } x \ge 1. \end{cases}$$

3. 2/5.

4. 3/7.

5.
$$F(x) = \begin{cases} 0 & \text{if } x \le -1 \\ (1/2)(x+1)^2 & \text{if } -1 \le x \le 0 \\ 1-(1/2)(1-x)^2 & \text{if } 0 \le x \le 1 \\ 1 & \text{if } x \ge 1. \end{cases}$$

6. $c = 1/2$ and $P(-1 \le X \le 1) = 1 - 1/e$.

7.
$$F(x) = \begin{cases} 0 & \text{if } x < -1 \\ (1/2)(x+1) & \text{if } -1 \le x < 1 \\ 1 & \text{if } x \ge 1. \end{cases}$$

$$F_Y(y) = \begin{cases} 0 & \text{if } y < 0 \\ \sqrt{y} & \text{if } 0 < y < 1 \\ 1 & \text{if } y \ge 1. \end{cases}$$

8.
$$F_Y(y) = \begin{cases} 0 & \text{if } y > 0 \\ \sqrt[3]{y} & \text{if } 0 \le y < 1 \\ 1 & \text{if } y > 1. \end{cases}$$

9.
$$f(x) = \begin{cases} 0 & \text{if } x \le 0 \\ 2x & \text{if } 0 \le x \le 1/2 \\ 1 & \text{if } 1/2 \le x \le 5/4 \\ 0 & \text{if } x \ge 5/4. \end{cases}$$

10. (a) $\Rightarrow$ (b) For any $x \in R, (X < x) = \cup_{r_j \in Q, r_j < x}(X \le r_j) \in \mathscr{F}$ since each $(X \le r_j) \in \mathscr{F}$ by (a).

(b) $\Rightarrow$ (c) For any $x \in R, (X \ge x) = (X < x)^c \in \mathscr{F}$ since $(X < x) \in \mathscr{F}$ by (b).

(c) $\Rightarrow$ (d) For any $x \in R, (X > x) = \cup_{r_j \in Q, r_j > x}(X \ge r_j) \in \mathscr{F}$ since each $(X \ge x) \in \mathscr{F}$ by (c).

(d) $\Rightarrow$ (a). For any $x \in R, (X \le x) = (X > x)^c \in \mathscr{F}$ since $(X > x) \in \mathscr{F}$ by (d).

Exercises 6.3

1.
$$F'(x) = \begin{cases} 1/4 & \text{if } 1 < x < 2 \\ 1/2 & \text{if } 4 < x < 5 \\ 0 & \text{otherwise.} \end{cases}$$

$$\int_{-\infty}^{x} F'(t)\,dt = \begin{cases} 0 & \text{if } x < 1 \\ (1/4)(x-1) & \text{if } 1 < x < 2 \\ 1/4 & \text{if } 2 < x < 4 \\ (1/4) + (1/2)(x-4) & \text{if } 4 < x < 5 \\ 3/4 & \text{if } x > 5 \end{cases}$$

Since $F(5) = 1 \neq 3/4 = \int_{-\infty}^{5} F'(t)\,dt$, F' is not a density for F.

2. (a) $P(0 \leq X < 1) = 1/4$.

   (b) $P(0 \leq X \leq 1) = 1/2$.

   (c) $P(X = 1) = 1/4$.

   (d) $P(1/2 \leq X \leq 5/2) = 11/16$.

3. Let $G(y) = P(Y \leq y) = P(\sin X \leq y)$. If $y < -1, G(y) = 0$ and if $y > 1, G(y) = 1$. Suppose $-1 < y < 1$. Then $G(y) = P(\sin X \leq y) = P(-1 \leq \sin X \leq y) = P(\arcsin(-1) \leq X \leq \arcsin y) = F(\arcsin y) - F_X(-\pi/2)$. Since

$$F_X(x) = \begin{cases} 0 & \text{if } x < -\pi/2 \\ (1/\pi)(x + (\pi/2)) & \text{if } -\pi/2 \leq x < \pi/2 \\ 1 & \text{if } x \geq \pi/2, \end{cases}$$

$$G(y) = \begin{cases} 0 & \text{if } y < -1 \\ (1/\pi)(\arcsin y + (\pi/2)) & \text{if } -1 \leq y < 1 \\ 1 & \text{if } y \geq 1 \end{cases}$$

and

$$g(y) = \begin{cases} 1/(\pi\sqrt{1-y^2}) & \text{if } -1 < y < 1 \\ 0 & \text{otherwise.} \end{cases}$$

4. Since Y takes on values in $[0, +\infty]$ with probability 1, $G(y) = P(Y \leq y) = 0$ whenever $y < 0$ and $G(y) = P(Y \leq y) = 1$ whenever $y \geq M$. Suppose $0 < y < M$. Then $G(y) = P(\min(X, M) \leq y) = 1 - P(\min(X, M) > y) = 1 - P(X > y, M > y) = 1 - P(X > y) = P(X \leq y) = 1 - e^{-y}$. Thus,

$$G(y) = \begin{cases} 0 & \text{if } y < 0 \\ 1 - e^{-y} & \text{if } 0 \leq y < M \\ 1 & \text{if } y \geq M. \end{cases}$$

Since the graph of G has a jump at M, G is not continuous and does not have a density function.

5.
$$f_Y(y) = \begin{cases} 0 & \text{if } y < 0 \\ 2ye^{-y^2} & \text{if } y \geq 0. \end{cases}$$

6. $f_Y(y) = e^y e^{-e^y}$.

7.
$$F_X(x) = \begin{cases} 0 & \text{if } x < 0 \\ 1 - (2/\pi)\arccos(x/100) & \text{if } 0 \leq x < 100 \\ 1 & \text{if } x \geq 100. \end{cases}$$

$$f_X(x) = \begin{cases} (2/\pi)(1/\sqrt{100^2 - x^2}) & \text{if } 0 < x < 100 \\ 0 & \text{otherwise.} \end{cases}$$

Exercises 6.4

1. $P(Y \leq X) = 1/2$.
2. $P(Y \leq X) = 3/5$.
3. $c = 8/\pi$, $P(X \geq 1/2) = (2/3) - (3\sqrt{3}/4\pi)$.
4. $f_X(x) = \begin{cases} e^{-x} & \text{if } x \geq 0 \\ 0 & \text{if } x < 0 \end{cases}$ $\qquad f_Y(y) = \begin{cases} 1/(1+y)^2 & \text{if } y \geq 0 \\ 0 & \text{if } y < 0. \end{cases}$

5.
$$f_Z(z) = \begin{cases} 0 & \text{if } z < 0 \\ 1 - e^{-z} & \text{if } 0 \leq z < 1 \\ e^{-z}(e-1) & \text{if } z \geq 1 \end{cases}$$

6.
$$f_Z(z) = \begin{cases} 5e^{-5z} & \text{if } z > 0 \\ 0 & \text{if } z < 0. \end{cases}$$

7.
$$f_Z(z) = \begin{cases} 6(e^{-2z} - e^{-3z}) & \text{if } z > 0 \\ 0 & \text{if } z < 0. \end{cases}$$

8.
$$F_Z(z) = \begin{cases} 0 & \text{if } z < 0 \\ z^2/2 & \text{if } 0 \leq z < 1 \\ -1 + 2z - (z^2/2) & \text{if } 1 \leq z < 2 \\ 1 & \text{if } z > 2. \end{cases}$$

$$f_Z(z) = \begin{cases} z & \text{if } 0 < z < 1 \\ 2 - z & \text{if } 1 < z < 2 \\ 0 & \text{otherwise.} \end{cases}$$

Exercises 6.5

1. Since $X = \sigma Z + \mu$ where Z has a standard normal density, $Y = aX + b = a\sigma Z + a\mu + b$, and Y has a $n(a\mu + b, a^2\sigma^2)$ density.
2. Since $P(X \leq 100) = \Phi((100 - \mu)/\sigma) = .9938$ and $P(X \leq 60) = \Phi((60 - \mu)/\sigma) = .9332$, using the table of the normal distribution function,

$$\frac{100 - \mu}{\sigma} = 2.5 \qquad \frac{60 - \mu}{\sigma} = 1.5.$$

   Solving for μ and σ, $\mu = 0, \sigma = 40$.
3. According to Example 6.19, X^2 and Y^2 both have $\Gamma(1/2, 1/2)$ densities. By Theorem 6.5.2, $Z = X^2 + Y^2$ has a $\Gamma(1, 1/2)$ density.
4. 1/24. Integrate by transforming to polar coordinates.
5. By Equation 6.9, X^2 and Y^2 both have $\Gamma(1/2, 1/2\sigma^2)$ densities. By Theorem 6.5.2, $W = X^2 + Y^2$ has a $\Gamma(1, 1/2\sigma^2)$ density. For $z < 0, F_Z(z) = 0$. For $z > 0, F_Z(z) = P(Z \leq z) = P(\sqrt{W} \leq z) = P(W \leq z^2) = F_W(z^2)$. Thus, $f_Z(z) = 2zf_W(z^2)$, and

$$f_Z(z) = \begin{cases} 0 & \text{if } z < 0 \\ (z/\sigma^2)e^{-(z^2/2\sigma^2)} & \text{if } z > 0. \end{cases}$$

6. The function $\phi(s) = (x - a)/(b - a)$ maps the interval (a, b) onto the interval $(0, 1)$. Let $Y = \phi(X)$. Since X takes on values between a and b, Y takes on values between 0 and 1. Thus, $f_Y(y) = 0$ if $y < 0$ or $y > 1$. If $0 < y < 1$, then $F_Y(y) = P(Y \leq y) = P((X - a)/(b - a) \leq y) = P(X \leq y(b - a) + a) = F_X(y(b - a) + a)$, and therefore $f_Y(y) = F'_X(y(b - a) + a)(b - a) = 1$. Thus,

$$f_Y(y) = \begin{cases} 1 & \text{if } 0 < y < 1 \\ 0 & \text{otherwise.} \end{cases}$$

7. Let $Y = \Phi(X)$. Since Φ takes on values between 0 and 1, the same is true of Y, and so $f_Y(y) = 0$ if $y < 0$ or $y > 1$. For $0 < y < 1, F_Y(y) = P(\Phi(X) \leq y) = P(X \leq \Phi^{-1}(y)) = \Phi(\Phi^{-1}(y)) = y$. Thus, $f_Y(y) = 1$ for $0 < y < 1$ and Y has a uniform density on $(0, 1)$.

Exercises 6.6

1. Since the exponential density is the same as the $\Gamma(1, \lambda)$ density, $Z = X_1 + \cdots + X_n$ has a $\Gamma(n, \lambda)$ density by Theorem 6.6.1.
2. Since each X_i has a standard normal density, each X_i^2 has a $\Gamma(1/2, 1/2)$ density. By Theorem 6.6.1, $W = X_1^2 + \cdots + X_n^2$ has a $\Gamma(n/2, 1/2)$ density.

Since $Z = \sqrt{W}, f_Z(z) = 0$ if $z < 0$. For $z > 0, f_Z(z) = 2zf_W(z^2)$. Therefore,

$$f_Z(z) = \begin{cases} (2/2^{n/2})\Gamma(n/2)z^{n-1}e^{-z^2/2} & \text{if } z > 0 \\ 0 & \text{if } z < 0. \end{cases}$$

3. (a) Proof of Theorem 6.6.1. Consider the proposition

$$P(n) : X_1 + \cdots + X_n \text{ has a } \Gamma(\alpha_1 + \cdots + \alpha_n, \lambda) \text{ density.}$$

$P(1)$ is trivially true since X_1 has a $\Gamma(\alpha_1, \lambda)$ density by hypothesis. Assume that $P(n-1)$ is true. Then $X_1 + \cdots + X_{n-1}$ has a $\Gamma(\alpha_1 + \cdots + \alpha_{n-1}, \lambda)$ density. By Theorem 6.5.2, $(X_1 + \cdots + X_{n-1}) + X_n$ has a $\Gamma(\alpha_1 + \cdots + \alpha_n, \lambda)$ density. Therefore, $P(n)$ is true whenever $P(n-1)$ is true. By the principle of mathematical induction, $P(n)$ is true for all $n \geq 1$.
(b) Proof of Theorem 6.6.2. Consider the proposition

$$P(n) : X_1 + \cdots + X_n \text{ has a } n(\mu_1 + \cdots + \mu_n, \sigma_1^2 + \cdots + \sigma_n^2) \text{ density.}$$

$P(1)$ is trivially true since X_1 has a $n(\mu_1, \sigma_1^2)$ density by hypothesis. Assume $P(n-1)$ is true. Then $X_1 + \cdots + X_{n-1}$ has a $n(\mu_1 + \cdots + \mu_{n-1}, \sigma_1^2 + \cdots + \sigma_{n-1}^2)$ density. By Theorem 6.5.1, $(X_1 + \cdots + X_{n-1}) + X_n$ has a $n(\mu_1 + \cdots + \mu_n, \sigma_1^2 + \cdots + \sigma_n^2)$ density. Therefore, $P(n)$ is true whenever $P(n-1)$ is true. By the principle of mathematical induction, $P(n)$ is true for all $n \geq 1$.

4. $$f_Y(y) = \begin{cases} (1/2)\arcsin\sqrt{1-y^2} & \text{if } -1 < y < 1 \\ 0 & \text{otherwise.} \end{cases}$$

5. $$P(X_1 \leq 2X_2 \leq 3X_3) = \int_0^{+\infty}\left(\int_{x_1/2}^{+\infty}\left(\int_{(2/3)x_2}^{+\infty} e^{-(x_1+x_2+x_3)}\,dx_3\right)dx_2\right)dx_1 = \frac{18}{55}.$$

6. Use a mathematical induction argument to show that

$$f_{X_n}(x) = \begin{cases} (1/n!)(-\log x)^n & \text{if } 0 < x < 1 \\ 0 & \text{otherwise.} \end{cases}$$

7. $$F_{U,V}(u,v) = \begin{cases} v^n - (v-u)^n & \text{if } 0 \leq u \leq v \leq 1 \\ 0 & \text{otherwise.} \end{cases}$$

For $0 \le u \le v \le 1$,

$$\int_0^u \int_x^v n(n-1)(y-x)^{n-2}\,dy\,dx = \int_0^u n(v-x)^{n-1}\,dx$$
$$= v^n - (v-u)^n.$$

8.
$$f_X(x) = \begin{cases} 1/(1+x)^2 & \text{if } x \ge 0 \\ 0 & \text{if } x < 0. \end{cases}$$

$$f_Y(y) = \begin{cases} 1/(1+y)^2 & \text{if } y \ge 0 \\ 0 & \text{if } y < 0. \end{cases}$$

$$f_Z(z) = \begin{cases} e^{-z} & \text{if } z \ge 0 \\ 0 & \text{if } z < 0. \end{cases}$$

$$f_{X,Y}(x,y) = \begin{cases} 2/(1+x+y)^3 & \text{if } x \ge 0, y \ge 0 \\ 0 & \text{otherwise.} \end{cases}$$

$$f_{X,Y\,|\,Z}(x,y\,|\,z) = \begin{cases} z^2 e^{-z(x+y)} & \text{if } x, y, z \ge 0 \\ 0 & \text{otherwise.} \end{cases}$$

9. For $t \ge 0$,

$$\begin{aligned} F_T(t) = P(T \le t) &= P(\max(T_1, T_2) \le t) \\ &= P(T_1 \le t, T_2 \le t) \\ &= P(T_1 \le t)P(T_2 \le t) \\ &= F_{T_1}(t)F_{T_2}(t). \end{aligned}$$

$$\begin{aligned} f_T(t) &= F_{T_1}(t)f_{T_2}(t) + f_{T_1}(t)F_{T_2}(t) \\ &= \left(1 - e^{\int_0^t \beta_1(s)\,ds}\right)\beta_2(t)e^{-\int_0^t \beta_2(s)\,ds} \\ &+ \beta_1(t)e^{-\int_0^t \beta_1(s)\,ds}\left(1 - e^{-\int_0^t \beta_2(s)\,ds}\right). \end{aligned}$$

CHAPTER 7

Exercises 7.2

1. $E[\sin X] = 2/\pi$.
2. $E[|X|] = 2/\sqrt{2\pi}$.
3. $E[\min(X, 1/2)] = 3/8, E[\max(X, 1/2)] = 5/8$.
4. $E[U] = 1/(n+1)$, $E[U^2] = 2/(n+1)(n+2)$, $E[V] = n/(n+1)$, $E[V^2] = n/(n+2)$.
5. $E[X] = 1/\lambda, E[X^2] = 2/\lambda^2$.

6. $E[X^r] = ((\alpha + r - 1)(\alpha + r - 2) \times \cdots \times \alpha)/\lambda^r$

7.
$$E[X^r] = \int_0^1 x^r \frac{\Gamma(\alpha_1 + \alpha_2)}{\Gamma(\alpha_1)\Gamma(\alpha_2)} x^{\alpha_1 - 1}(1 - x)^{\alpha_2 - 1}\,dx$$
$$= \frac{\Gamma(\alpha_1 + \alpha_2)}{\Gamma(\alpha_1)\Gamma(\alpha_2)} \frac{\Gamma(\alpha_1 + r)\Gamma(\alpha_2)}{\Gamma(\alpha_1 + \alpha_2 + r)}$$
$$\times \int_0^1 \frac{\Gamma(\alpha_1 + \alpha_2 + r)}{\Gamma(\alpha_1 + r)\Gamma(\alpha_2)} x^{\alpha_1 + r - 1}(1 - x)^{\alpha_2 - 1}dx$$
$$= \frac{(\alpha_1 + r - 1)(\alpha_1 + r - 2) \times \cdots \times \alpha_1}{(\alpha_1 + \alpha_2 + r - 1)(\alpha_1 + \alpha_2 + r - 2) \times \cdots \times (\alpha_1 + \alpha_2)},$$

since the second integral is equal to 1.

8. Choose $a < b$ so that $a < c_i < b, i = 1, \ldots, m$. Then $F(x) = 0$ for $x < a$ and $F(x) = 1$ for $x > b$ and $\int_{-\infty}^{+\infty} \phi(t)\,dF(t) = \int_a^b \phi(t)\,dF(t)$. Since ϕ is continuous at each c_i, given $\varepsilon > 0$, there is a $\delta_i > 0$ such that $|\phi(x) - \phi(c_i)| < \varepsilon$ whenever $|x - c_i| < \delta_i$. Let

$$\delta = \min\{\delta_1, \ldots, \delta_m, c_2 - c_1, \ldots, c_m - c_{m-1}\}.$$

Then

$$|\phi(x) - \phi(c_i)| < \varepsilon \text{ whenever } |x - c_i| < \delta, i = 1, \ldots, m.$$

Let π_ε be any partition of $[a, b]$ such that $|\pi_\varepsilon| < \delta$. Let $\pi = \{x_0, \ldots, x_n\}$ be any partition of $[a, b]$ finer that π_ε. Then each c_i belongs to one and only one interval $(x_{j_i - 1}, x_{j_i}], i = 1, \ldots, m$. Note that $F(x_k) - F(x_{k-1}) = 0$ if k is not one of the j_i's. Let ξ_k be any point of $(x_{k-1}, x_k], k = 1, \ldots, n$. Then

$$\left|\sum_{k=1}^{n} \phi(\xi_k)(F(x_k) - F(x_{k-1})) - \sum_{i=1}^{m} \phi(c_i)(F(c_i) - F(c_i -))\right|$$
$$= \left|\sum_{i=1}^{m} \phi(\xi_{j_i})(F(x_{j_i}) - F(x_{j_i - 1})) - \sum_{i=1}^{m} \phi(c_i)(F(c_i) - F(c_i -))\right|$$
$$= \left|\sum_{i=1}^{m} \phi(\xi_{j_i})(F(c_i) - F(c_i -)) - \sum_{i=1}^{m} \phi(c_i)(F(c_i) - F(c_i -))\right|$$
$$\leq \sum_{i=1}^{m} |\phi(\xi_{j_i}) - \phi(c_i)|(F(c_i) - F(c_i -))$$
$$< \varepsilon \sum_{i=1}^{m} (F(c_i) - F(c_i -)) = \varepsilon.$$

This shows that $\int_{-\infty}^{+\infty} \phi(t)\,dF(t) = \int_a^b \phi(t)\,dF(t) = \lim_{|\pi| \to 0} \sum_{k=1}^{n} \phi(\xi_k)\Delta F_k = \sum_{i=1}^{m} c_i(F(c_i) - F(c_i -))$.

9. $E[X] = \int_0^{+\infty} x f_X(x)\,dx = \int_0^{+\infty}(\int_0^x f_X(x)\,dy)\,dx$. Interchanging the order of integration,

$$E[X] = \int_0^{+\infty}\left(\int_y^{+\infty} f_X(x)\,dx\right)dy = \int_0^{+\infty} P(X > y)\,dy$$
$$= \int_0^{+\infty}(1 - F_X(y))\,dy.$$

10. Since $Y \geq X, (X > x) \subset (Y > y)$ and $1 - F_X(x) \leq 1 - F_Y(x)$. By Problem 5, $E[X] = \int_0^{+\infty}(1 - F_X(x))dx \leq \int_0^{+\infty}(1 - F_Y(x))dx = E[Y]$.

Exercises 7.3

1. $E[\max(X, Y)] = 3/2$.
2. $E[U \mid X = x, Y = y] = x/2$.
3. $E[\Lambda \mid X_1 = x_1, \ldots, X_n = x_n] = (n + 1)/(1 + x_1 + \cdots + x_n)$.
4. $$E[Y \mid X = x] = \begin{cases} (2/3)x & \text{if } 0 < x < 1 \\ 0 & \text{otherwise.} \end{cases}$$

$$E[X \mid Y = y] = \begin{cases} (2/3)(y^2 + y + 1)/(y + 1) & \text{if } 0 < y < 1 \\ 0 & \text{otherwise.} \end{cases}$$

5. The density of X is

$$f_X(x) = \begin{cases} (\alpha\beta^\alpha)/(x + \beta)^{\alpha+1} & \text{if } x > 0 \\ 0 & \text{if } x \leq 0. \end{cases}$$

The conditional density of Λ given that $X = x$ is a $\Gamma(\alpha + 1, x + \beta)$ density.

6. $E[R] = (n - 1)/(n + 1)$, $\operatorname{var} R = 2(n - 1)/((n + 1)^2(n + 2))$.
7. $$f_{U|V}(u \mid v) = \frac{(n - 1)(v - u)^{n-2}}{v^{n-1}}.$$

$$E[U \mid V = v] = \begin{cases} v/n & \text{if } 0 < v < 1 \\ 0 & \text{otherwise.} \end{cases}$$

8. $$f_{Z|X,Y}(z \mid x, y) = \frac{1}{2}(1 + x + y)^3 z^2 e^{-z(1+x+y)} \qquad \text{if } x, y, z \geq 0.$$

$$E[Z \mid X = x, Y = y] = \frac{3}{1 + x + y}.$$

9.

$$E[Y \mid X = x] = \begin{cases} (1+x)/2 & \text{if } -1 < x < 0 \\ (1-x)/2 & \text{if } 0 < x < 1 \\ 0 & \text{otherwise.} \end{cases}$$

10.

$$E[X \mid Y = y] = \begin{cases} (y-1)/(\ln y) & \text{if } 0 < y < 1 \\ 0 & \text{otherwise.} \end{cases}$$

Exercises 7.4

1. .7745.
2. .5819.
3. $P(S_{360} \geq 59000) = .0622$.
4. $n = 757$.
5. $P(|S_{1000}| \leq 30) = .6574$.
6. $P(S_{1000} \geq -50) = .056$.
7. $\lambda = 2.575\sqrt{n/12}$.
8. Let $n = 10^6$. Since $\mu = E[X_j] = 0$ and $\sigma^2 = 10^{-m}/12, P(|S_n| \leq 5 \cdot 10^{-m+2}) \leq P(|S_n^*| \leq \sqrt{3}) \approx 2\Phi(\sqrt{3}) - 1 \approx .9168$.

Exercises 7.5

The identities

$$\cos^2 \alpha = \frac{1 + \cos 2\alpha}{2}$$

and

$$\cos \alpha \cos \beta = \frac{1}{2}(\cos(\alpha + \beta) + \cos(\alpha - \beta))$$

are used in these exercises.

1. $E[X_n] = \sum_{j=1}^{m} a_j E[\cos(nb_j + Z_j)] = \sum_{j=1}^{m} a_j (1/2\pi) \int_0^{2\pi} \cos(nb_j + z)\, dz = 0$.

$$\begin{aligned} E[X_n^2] &= E\left[\left(\sum_{i=1}^{m} a_i \cos(nb_i + Z_i)\right)\left(\sum_{j=1}^{m} a_j \cos(nb_j + Z_j)\right)\right] \\ &= \sum_{i=1}^{m} a_i^2 E[\cos^2(nb_i + Z_i)] \\ &\quad + \sum_{i \neq j} a_i a_j E[\cos(nb_i + Z_i)] E[\cos(nb_j + Z_j)] \end{aligned}$$

$$= \sum_{i=1}^{m} a_i^2 \frac{1}{2\pi} \int_0^{2\pi} \cos^2(nb_i + z)\, dz$$

$$= \sum_{i=1}^{m} a_i^2 \frac{1}{2\pi} \int_0^{2\pi} \frac{1 + \cos(2(nb_i + z))}{2}\, dz$$

$$= \frac{1}{2} \sum_{i=1}^{m} a_i^2.$$

For $\nu \geq 1$,

$$\begin{aligned}
r(\nu) &= E[X_n X_{n+\nu}] \\
&= E\left[\left(\sum_{i=1}^{m} a_i \cos(nb_i + Z_i)\right)\left(\sum_{j=1}^{m} a_j \cos((n+\nu)b_j + Z_j)\right)\right] \\
&= \sum_{i=1}^{m} a_i^2 E[\cos(nb_i + Z_i)\cos((n+\nu)b_i + Z_i)] \\
&+ \sum_{i \neq j} a_i a_j E[\cos(nb_i + Z_i)] E[\cos((n+\nu)b_j + Z_j)] \\
&= \sum_{i=1}^{m} \frac{a_i^2}{2} E[\cos((2n+\nu)b_i + 2Z_i) + \cos(\nu b_i)] \\
&= \sum_{i=1}^{m} \frac{a_i^2}{2} \cos \nu b_i + \sum_{i=1}^{m} \frac{a_i^2}{2} \frac{1}{2\pi} \int_0^{2\pi} \cos((2n+\nu)b_i + 2z)\, dz \\
&= \sum_{i=1}^{m} \frac{a_i^2}{2} \cos \nu b_i.
\end{aligned}$$

Since $r(\nu) = r(-\nu) = r(|\nu|)$,

$$\rho(\nu) = \frac{\sum_{i=1}^{m} a_i^2 \cos \nu b_i}{\sum_{i=1}^{m} a_i^2}.$$

2. Let $\mu = E[X_0]$ and $\hat{r}(\nu) = \operatorname{cov}(X_0, X_\nu)$. Then $E[Y_n] = \mu \sum_{j=1}^{m} a_j$ and

$$\begin{aligned}
&E\left[\left(Y_n - \mu \sum_{j=1}^{m} a_j\right)^2\right] \\
&= E\left[\left(\sum_{i=1}^{m} a_i (X_{n-i+1} - \mu)\right)\left(\sum_{j=1}^{m} a_j (X_{n-j+1} - \mu)\right)\right] \\
&= \sum_{i=1}^{m} \sum_{j=1}^{m} a_i a_j E[(X_{n-i+1} - \mu)(X_{n-j+1} - \mu)]
\end{aligned}$$

$$= \sum_{i=1}^{m}\sum_{j=1}^{m} a_i a_j \hat{r}(i-j),$$

which is independent of n. For $\nu \geq 1$,

$$E\left[\left(Y_n - \mu\sum_{i=1}^{m} a_i\right)\left(Y_{n+\nu} - \mu\sum_{j=1}^{m} a_j\right)\right]$$

$$= \sum_{i=1}^{m}\sum_{j=1}^{m} a_i a_j E[(X_{n-i+1} - \mu)(X_{n+\nu-j+1} - \mu)]$$

$$= \sum_{i=1}^{m}\sum_{j=1}^{m} a_i a_j \hat{r}(i-j+\nu),$$

which is independent of n. Since $r(\nu) = r(-\nu) = r(|\nu|)$,

$$\rho(\nu) = \frac{\sum_{i=1}^{m}\sum_{j=1}^{m} a_i a_j \hat{r}(i-j+|\nu|)}{\sum_{i=1}^{m}\sum_{j=1}^{m} a_i a_j \hat{r}(i-j)}, \qquad \nu \in Z.$$

3. $$f(\lambda) = \frac{1+\alpha^2+2\cos\alpha\lambda}{2\pi(1+\alpha^2)}, \qquad -\pi < \lambda < \pi.$$

4. $$f(\lambda) = \frac{1+\alpha^2+\beta^2+2(\alpha+\alpha\beta)\cos\lambda+2\beta\cos 2\lambda}{2\pi(1+\alpha^2+\beta^2)}, \qquad -\pi < \lambda < \pi.$$

5. $E[X_t] = \sum_{j=1}^{m} p_j (1/2\pi)\int_0^{2\pi}\cos(\lambda_j t + \theta)\,d\theta = 0$, and

$$E[X_t^2] = \sum_{j=1}^{m} p_j \frac{1}{2\pi}\int_0^{2\pi}\cos^2(\lambda_j t+\theta)\,d\theta$$

$$= \sum_{j=1}^{m} p_j \frac{1}{2\pi}\int_0^{2\pi}\frac{1+\cos(2\lambda_j t + 2\theta)}{2}\,d\theta$$

$$= \frac{1}{2}\sum_{j=1}^{m} p_j = \frac{1}{2}.$$

For $h > 0$,

$$\operatorname{cov}(X_t, X_{t+h}) = E[\cos(\Lambda t+\Theta)\cos(\Lambda(t+h)+\Theta)]$$

$$= \frac{1}{2}E[\cos(\Lambda(2t+h)+2\Theta)+\cos\Lambda h]$$

$$= \frac{1}{2}\sum_{j=1}^{m} p_j \frac{1}{2\pi}\int_0^{2\pi}(\cos(\lambda_j(2t+h)+2\theta)+\cos\lambda_j h)\,d\theta$$

$$= \frac{1}{2}\sum_{j=1}^{m} p_j \cos\lambda_j h.$$

Since $r(h) = r(-h) = r(|h|), r(h) = (1/2)\sum_{j=1}^{m} p_j \cos \lambda_j h$, and therefore $\rho(h) = \sum_{j=1}^{m} p_j \cos \lambda_j h$.

CHAPTER 8

Exercises 8.2

1. $P(X_s = k \mid X_t = n) = \binom{n}{k}(s/t)^k(1-(s/t))^{n-k}$.
2. For $i = 1,\ldots,n$, let $W_1^{(i)}$ be the first time that $X_t^{(i)} = 1$ and let $W = \max(W_1^1,\ldots,W_1^{(n)})$. Then $P(W \le t) = (1-e^{-\lambda t})^n$ and $f_W(t) = n\lambda e^{-\lambda t}(1-e^{-\lambda t})^{n-1}$ for $t \ge 0$ and $= 0$ for $t \le 0$.
3. Note that $P(X_t - X_{[t]} < 0) = 0$. Thus, $P(X_t \ge \sum_{k=1}^{[t]}(X_k - X_{k-1})) = 1$, and it suffices to show that $P(\lim_{t\to\infty} X_{[t]} = +\infty) = 1$. Since the events $A_1, A_2, \ldots$ are independent events and $P(A_k) = \lambda e^{-\lambda}, \sum_{k=1}^{\infty} P(A_k) = \infty$. By the Borel-Cantelli lemma, $P(A_n\, i.o.) = 1$ and $P(\sum_{k=1}^{\infty}(X_k - X_{k-1}) = +\infty) = 1$. Since $\lim_{t\to\infty} X_{[t]} = \sum_{k=1}^{\infty}(X_k - X_{k-1}), P(\lim_{t\to\infty} X_{[t]} = \infty) = 1$.

5. $$P_n(t) = \frac{1}{n!}(\ln(1+t))^n \frac{1}{1+t} \qquad E[X_t] = \ln(1+t).$$

6. $P_{10}(100) \approx .012$.

Exercises 8.3

1. $P_1(t) = e^{-\beta t}, P_2(t) = e^{-\beta t} - e^{-2\beta t}, P_3(t) = e^{-\beta t} - 2e^{-2\beta t} + e^{-3\beta t}$.
2. By Problem 1, $P_1(t) = e^{-\beta t}$ and the assertion is true for $n = 1$. Assume that the assertion is true for $n-1$; i.e., $P_{n-1}(t) = e^{-\beta t}(1-e^{-\beta t})^{n-2}$. By Equation 8.9,

$$\begin{aligned}
P_n(t) &= \beta(n-1)e^{-\beta nt}\int_0^t e^{\beta ns}e^{-\beta s}(1-e^{-\beta s})^{n-2}\,ds \\
&= (n-1)e^{-\beta nt}\int_0^t (e^{\beta s}-1)^{n-2}\beta e^{\beta s}\,ds \\
&= e^{-\beta nt}(e^{\beta t}-1)^{n-1} \\
&= e^{-\beta t}(1-e^{-\beta t})^{n-1}
\end{aligned}$$

Thus, the assertion is true for n whenever it is true for $n-1$, and it follows from the principle of mathematical induction that the assertion is true for all $n \in N$.

3. By Equation 8.9,

$$P'_{n+1}(t) = \beta_n e^{-\beta_{n+1}t}\int_0^t e^{\beta_n s}P_n(s)\,ds = 0$$

since $\beta_n = 0$. Thus, $P'_{n+1}(t) = 0$ and so $P_{n+1}(t) = c$. Using the initial condition $P_{n+1}(0) = 0, c = 0$ and $P(X_t = n+1) = P_{n+1}(t) = 0$ for all t, as would be expected without any calculations.

4. As in Example 8.1,

$$M'(t) = (\beta - \delta)M(t) + \alpha\sum_{n=1}^{\infty} P_{n-1}(t).$$

If there were no deaths in the population, by Theorem 8.3.1 we would have $P(X_t < +\infty) = \sum_{n=1}^{\infty} P_{n-1}(t) = 1$ since the series $\sum_{n=1}^{\infty} 1/(\alpha + \beta n)$ diverges. A fortiori, $P(X_t < +\infty) = \sum_{n=1}^{\infty} P_{n-1}(t) = 1$ in the presence of deaths. Thus,

$$M'(t) = (\beta - \delta)M(t) + \alpha,$$

and so

$$M(t) = \frac{\alpha}{\delta - \beta}(1 - e^{(\beta-\delta)t}) + e^{(\beta-\delta)t}.$$

5. (c) $E[X_t] = n_0 e^{\beta t}$. (d) $P_0(t) \equiv 0, P_n(t) = e^{-\beta t}(1 - e^{-\beta t})^{n-1}$ in the $n_0 = 1$ case.

Exercises 8.4

1. (a) Fix $i \geq 1$. The forward equation for $p_{i,0}(t)$ is

$$p'_{i,0}(t) = \sum_{k\neq 0} p_{i,k}q_{k,0} + q_{0,0}p_{i,0}(t).$$

Since $q_{k,0} = 0$ if $k \geq 1$, the equation reduces to

$$p'_{i,0}(t) = -\lambda p_{i,0}(t),$$

which has the general solution $p_{i,0}(t) = c_i e^{-\lambda t}$ where c_i is a constant. Since the continuity condition requires that $\lim_{t\to 0^+} p_{i,0}(t) = 0, c_i = 0$ and so $p_{i,0} \equiv 0$.
(b) Fix $i \geq 1$. From part (a), $p_{i,0}(t) \equiv 0$. Suppose $p_{i,k}(t) \equiv 0$ for all $k < j \leq i - 1$. The forward equation for $p_{i,j}(t)$ is

$$p'_{i,j}(t) = \sum_{k\neq j} p_{i,k}q_{k,j} + q_{j,j}p_{i,j}(t).$$

Since $q_{j+1,j} = 0$, $q_{j-1,j} = \lambda$, and $q_{k,j} = 0$ for all other cases for which $k \neq j$,

$$p'_{i,j}(t) = \lambda p_{i,j-1}(t) - \lambda p_{i,j}(t).$$

But $p_{i,j-1}(t) \equiv 0$ by the induction hypothesis. Thus,

$$p'_{i,j}(t) = -\lambda p_{i,j}(t)$$

and $p_{i,j}(t) \equiv 0$ as in (a), and therefore $p_{i,j}(t) \equiv 0$ whenever $j < i$.
(c) Suppose $j \geq i$. Then the forward equation for $p_{i,j}(t)$ is

$$p'_{i,j}(t) = \lambda p_{i,j-1}(t) - \lambda p_{i,j}(t)$$

so that

$$\frac{d}{dt}(e^{\lambda t} p_{i,j}(t)) = \lambda e^{\lambda t} p_{i,j-1}(t).$$

Integrating from 0 to t,

$$e^{\lambda t} p_{i,j}(t) - \delta_{i,j} = \lambda \int_0^t e^{\lambda s} p_{i,j-1}(s)\, ds$$

and

$$p_{i,j}(t) = \delta_{i,j} e^{-\lambda t} + \lambda e^{-\lambda t} \int_0^t e^{\lambda s} p_{i,j-1}(s)\, ds.$$

Since $p_{i,i-1}(t) \equiv 0$,

$$p_{i,i}(t) = e^{-\lambda t}.$$

Thus,

$$p_{i,i+1}(t) = \lambda e^{-\lambda t} \int_0^t e^{\lambda s} e^{-\lambda s}\, ds = \lambda t e^{-\lambda t}$$

and

$$\begin{aligned} p_{i,i+2}(t) &= \lambda e^{-\lambda t} \int_0^t e^{\lambda s} \lambda s e^{-\lambda s}\, ds \\ &= \frac{\lambda^2 t^2}{2} e^{-\lambda t}, \end{aligned}$$

and so forth. By induction,

$$p_{i,j}(t) = \frac{\lambda^{j-i} t^{j-i}}{(j-i)!} e^{-\lambda t}, \qquad j \geq i.$$

2.
$$Q = \begin{bmatrix} -\lambda & \lambda & 0 & 0 & \dots & & 0 \\ \mu & -(\lambda+\mu) & \lambda & 0 & \dots & & 0 \\ 0 & 2\mu & -(\lambda+2\mu) & \lambda & \dots & & 0 \\ \vdots & \vdots & \vdots & \vdots & \ddots & & \\ 0 & 0 & 0 & 0 & \dots & m\mu & -m\mu \end{bmatrix}.$$

3.
$$Q = \begin{bmatrix} -\lambda & \lambda & 0 \\ \mu & -(\mu+\lambda) & \lambda \\ 0 & \mu & -\mu \end{bmatrix}.$$

4.
$$Q = \begin{bmatrix} -(\lambda+\lambda_s) & (\lambda+\lambda_s) & 0 \\ \mu & -(\lambda+\mu) & \lambda \\ 0 & \mu & -\mu \end{bmatrix}.$$

Exercises 8.5

1. Since $I^n = I$ for all $n \geq 1$,

$$e^{\alpha t I} = \sum_{n=0}^{\infty} \frac{(\alpha t)^n}{n!} I^n = \sum_{n=0}^{\infty} \frac{(\alpha t)^n}{n!} I = e^{\alpha t} I.$$

2. Since $Q^2 = -(\lambda+\mu)Q, Q^3 = (\lambda+\mu)^2 Q, \dots, Q^n = (-1)^{n-1}(\lambda+\mu)^{n-1}Q, n \geq 1$,

$$P(t) = I + \frac{1}{\lambda+\mu}Q - \frac{e^{-(\lambda+\mu)t}}{\lambda+\mu}Q.$$

If $P(t) = [p_{i,j}(t)]$, then

$$p_{1,1}(t) = \frac{\mu + \lambda e^{-(\lambda+\mu)t}}{\lambda+\mu} \qquad p_{1,2}(t) = \frac{\lambda - \lambda e^{-(\lambda+\mu)t}}{\lambda+\mu}$$

$$p_{2,1}(t) = \frac{\mu - \mu e^{-(\lambda+\mu)t}}{\lambda+\mu} \qquad p_{2,2}(t) = \frac{\lambda + \mu e^{-(\lambda+\mu)t}}{\lambda+\mu}.$$

3. Writing Q as a block matrix,

$$Q = \begin{bmatrix} A & O \\ O & D \end{bmatrix} \qquad \text{where } A = D = \begin{bmatrix} -1 & 1 \\ 1 & -1 \end{bmatrix},$$

$$Q^n = \begin{bmatrix} A^n & O \\ O & D^n \end{bmatrix}, \qquad n \geq 1.$$

By Problem 2 for $n \geq 1, A^n = (-1)^{n-1}2^{n-1}A$ and $D^n = (-1)^{n-1}2^{n-1}D$. Thus,

$$Q^n = \begin{bmatrix} (-1)^{n-1}2^{n-1}A & O \\ O & (-1)^{n-1}2^{n-1}D \end{bmatrix}$$

and

$$P(t) = I + \begin{bmatrix} \left((-1/2)\sum_{n=1}^{\infty}(-2t)^n/n!\right)A & O \\ O & \left((-1/2)\sum_{n=1}^{\infty}(-2t)^n/n!\right)D \end{bmatrix}.$$

Identifying entries in $P(t)$,

$$p_{1,1}(t) = p_{3,3}(t) = \frac{1+e^{-2t}}{2} \qquad p_{1,2}(t) = p_{3,4}(t) = \frac{1-e^{-2t}}{2}$$
$$p_{2,1}(t) = p_{4,3}(t) = \frac{1-e^{-2t}}{2} \qquad p_{2,2}(y) = p_{4,4}(t) = \frac{1+e^{-2t}}{2}.$$

For all other pairs (i, j), $p_{i,j}(t) = 0$.

Exercises 8.6

1. Q_1 is irreducible; Q_2 is not irreducible.
2. $\pi_1 = .995000$, $\pi_2 = .004975$, $\pi_3 = .000025$.
3. $\lim_{t\to\infty} p_{i,1}(t) = 1/6$, $\lim_{t\to\infty} p_{i,2}(t) = 1/24$, $\lim_{t\to\infty} p_{i,3}(t) = 1/8$, $\lim_{t\to\infty} p_{i,4}(t) = 1/3$, $\lim_{t\to\infty} p_{i,5}(t) = 1/3$.
4. $$P(2) = \{p_{i,j}(2)\} = \begin{bmatrix} .285 & .365 & .350 \\ .214 & .419 & .367 \\ .205 & .391 & .404 \end{bmatrix}.$$

   $\pi_1 = 12/53$, $\pi_2 = 21/53$, $\pi_3 = 20/53$.
5. $\pi_1 = .195$, $\pi_2 = .132$, $\pi_3 = .182$, $\pi_4 = .302$, $\pi_5 = .189$.

STANDARD NORMAL DISTRIBUTION FUNCTION

$$\Phi(x) = \frac{1}{\sqrt{2\pi}} \int_{-\infty}^{x} e^{-t^2/2}\, dt, \qquad x \geq 0.$$

For $x < 0$, use the relation $\Phi(x) = 1 - \Phi(-x)$.

	.00	.01	.02	.03	.04	.05	.06	.07	.08	.09
0.0	.5000	.5040	.5080	.5120	.5160	.5199	.5239	.5279	.5319	.5359
0.1	.5398	.5438	.5478	.5517	.5557	.5596	.5636	.5675	.5714	.5753
0.2	.5793	.5832	.5871	.5910	.5948	.5987	.6026	.6064	.6103	.6141
0.3	.6179	.6217	.6255	.6293	.6331	.6368	.6406	.6443	.6480	.6517
0.4	.6554	.6591	.6628	.6664	.6700	.6736	.6772	.6808	.6844	.6879
0.5	.6915	.6950	.6985	.7019	.7054	.7088	.7123	.7157	.7190	.7224
0.6	.7257	.7291	.7324	.7357	.7389	.7422	.7454	.7486	.7517	.7549
0.7	.7580	.7611	.7642	.7673	.7704	.7734	.7764	.7794	.7823	.7852
0.8	.7881	.7910	.7939	.7967	.7995	.8023	.8051	.8078	.8106	.8133
0.9	.8159	.8186	.8212	.8238	.8264	.8289	.8315	.8340	.8365	.8389
1.0	.8413	.8438	.8461	.8485	.8508	.8531	.8554	.8577	.8599	.8621
1.1	.8643	.8665	.8686	.8708	.8729	.8749	.8770	.8790	.8810	.8830
1.2	.8849	.8869	.8888	.8907	.8925	.8944	.8962	.8980	.8997	.9015
1.3	.9032	.9049	.9066	.9082	.9099	.9115	.9131	.9147	.9162	.9177
1.4	.9192	.9207	.9222	.9236	.9251	.9265	.9279	.9292	.9306	.9319
1.5	.9332	.9345	.9357	.9370	.9382	.9394	.9406	.9418	.9429	.9441
1.6	.9452	.9463	.9474	.9484	.9495	.9505	.9515	.9525	.9535	.9545
1.7	.9554	.9564	.9573	.9582	.9591	.9599	.9608	.9616	.9625	.9633
1.8	.9641	.9649	.9656	.9664	.9671	.9678	.9686	.9693	.9699	.9706
1.9	.9713	.9719	.9726	.9732	.9738	.9744	.9750	.9756	.9761	.9767
2.0	.9772	.9778	.9783	.9788	.9793	.9798	.9803	.9808	.9812	.9817
2.1	.9821	.9826	.9830	.9834	.9838	.9842	.9846	.9850	.9854	.9857
2.2	.9861	.9864	.9868	.9871	.9875	.9878	.9881	.9884	.9887	.9890
2.3	.9893	.9896	.9898	.9901	.9904	.9906	.9909	.9911	.9913	.9916
2.4	.9918	.9920	.9922	.9925	.9927	.9929	.9931	.9932	.9934	.9936
2.5	.9938	.9940	.9941	.9943	.9945	.9946	.9948	.9949	.9951	.9952
2.6	.9953	.9955	.9956	.9957	.9959	.9960	.9961	.9962	.9963	.9964
2.7	.9965	.9966	.9967	.9968	.9969	.9970	.9971	.9972	.9973	.9974
2.8	.9974	.9975	.9976	.9977	.9977	.9978	.9979	.9979	.9980	.9981
2.9	.9981	.9982	.9982	.9983	.9984	.9984	.9985	.9985	.9986	.9986
3.0	.9987	.9987	.9987	.9988	.9988	.9989	.9989	.9989	.9990	.9990

SYMBOLS

INDEX